Bioeconomy for Sustainable Development

Chetan Keswani
Editor

Bioeconomy for Sustainable Development

Springer

Editor
Chetan Keswani
Department of Biochemistry
Institute of Science,
Banaras Hindu University
Varanasi, Uttar Pradesh, India

ISBN 978-981-13-9430-0 ISBN 978-981-13-9431-7 (eBook)
https://doi.org/10.1007/978-981-13-9431-7

This Springer imprint is published by the registered company Springer Nature Singapore Pte Ltd.
The registered company address is: 152 Beach Road, #21-01/04 Gateway East, Singapore 189721, Singapore

Kiran Mazumdar - Shaw
Chairperson & Managing Director

Biocon Limited
20th KM Hosur Road
Electronics City
Bangalore 560 100 India
T 91 80 2808 2004
F 91 80 2852 3428

www.biocon.com

Foreword

Global concerns over an ecological meltdown are creating a huge demand for environmentally sustainable solutions. Economies worldwide are seeking innovative ways to meet energy, food and environmental needs without compromising the Earth's future. Biotechnology can provide the much-needed building blocks for a sustainable future by raising agricultural productivity, reducing greenhouse gas emissions, creating innovative biologic therapies and bringing to market biomaterials that are bioresorbable and biodegradable.

India, which has emerged as a notable bioeconomy valued at USD 50 billion by leveraging recombinant DNA technology to deliver genetically engineered agricultural crops, biopharmaceuticals, vaccines and enzymes, is well placed to leverage biotechnology in ensuring a sustainable future.

While our immediate goal is to build a USD 100 billion bioeconomy in India by 2025, our long-term aspiration is to evolve into a global biotechnology hotspot reputed for its high-value, low-cost innovations that incorporate concepts such as smart production with minimal environmental impact.

In the pursuit of our global aspirations, we must invest in our research and innovation efforts to create new scientific and technological knowledge that will enable us to respond to growing worldwide demand.

For long-term success, India will have to achieve a subtle balance between public-sector-supported science push and market and social pull, encouraged by industry and investors as well as policy makers and civil society.

We will need to create an enabling ecosystem with three converging components: financing avenues that will allow biotech entrepreneurs to raise the requisite funds to take 'ideas' to 'market'; a conducive regulatory regime that improves ease of doing business, reduces transaction costs and expedites approval timelines; and a strong market that is accepting of novel products and services. All three components need to come together to create a self-perpetuating virtuous cycle.

The cycle of innovation and business growth will also need greater synchronization of resources, plans, policies and priorities.

This book focuses on the integrated approach for sustained innovation in various areas of bioeconomy. The perspective is largely based on the industrial and socio-legal implications of bioeconomic advances. It takes a comprehensive look not only at the implications of IPR in omics-based research but on what are the ethical and intellectual standards and how these can be developed for sustained innovation.

I congratulate the editor, Dr Chetan Keswani, for integrating the views of global authorities on the upcoming challenges and presenting the most feasible options for translating commercially viable ideas into affordable products and technologies that can ensure a sustainable future.

I believe biotechnology is key to long-term sustainability and we will need to do all that is necessary to position India at the forefront of this global opportunity.

Kiran Mazumdar-Shaw

Date: 1 April, 2019
Place: Bengaluru

Foreword by Prof. M.S. Reddy, Auburn University, USA

Dear Authors and Readers,

It gives me immense pleasure in writing this foreword for the book titled *Bioeconomy for Sustainable Development*. The book is edited by a well-known scientist in the fields of agriculture and plant pathology and in research on plant growth-promoting rhizobacteria (PGPR). Food, chemicals, and industrial sectors are challenged with the growing population, increasing longevity, and quality of life. The increasing demand in these major sectors of economy will increase the consumption of fossil energy sources, agricultural land, and drinking water. This demand could lead to the irreversible changes in climate with unpredictable consequences. A possible direction to address this challenge sustainably is increasing the efficiency of currently used processes and displacement of fossil fuel energy sources by production of useful biomass. Over the past three decades, the landmark discoveries in molecular biology have revolutionized multidimensionally. Contemporary bioeconomic strategies also support an integration of science, technology, economy, environmental issues, rural and industrial developments, regulatory processes, and social sciences initiatives. This is a great opportunity for molecular and engineering sciences to map and engineer the uncharted territories of molecular transformations that are the key to economic growth. Bioeconomy is not just a post-petro economy in which biomass feedstock replaces fossil fuel, but a thorough understanding of the underlying science and engineering is needed. It is also crucial to reinforce the innovation range from fundamental understanding to enabling technologies and applications, such as increasing the effectiveness of biomass production chain or enhancing the value chain from biomass to product targets. Bioeconomy will evolve by incorporating concepts such as smart production with less impact on the environment, as well as by reducing and recycling biowaste into its objectives. Finally, it should be emphasized that biological resources offer unique new functions and properties in comparison with non-biological resources, so that in all contexts, particularly considering the economy, the aspect of innovation must be highly considered. Excellent innovative and mission-oriented research requires long-term support and encouragement from all stakeholders in science, industry, and society to develop smart bioeconomies. Without this, there will be no "fuel" for the bioeconomy to maintain its innovativeness. The foundations and future of the bioeconomy are based on strong science and technology bases, and a dynamic and innovative approach that responds to the constant demands for new scientific and technological knowledge needs to be created. Furthermore, bioeconomy requires transforming basic knowledge into successful industrial production and agriculture, including food, novel bioproducts, and bioenergy. To ensure this, it will be vital to strengthen and stimulate mechanisms for creating an entrepreneurial atmosphere, whether from the public sector or from the private sector. Bioeconomy requires a subtle balance between public sector-supported science push and market and social pull, encouraged by industry and investors as well as policy-makers and civil society.

The proposed book focuses on the integrated approach for sustained innovation in various areas of bioeconomy. The outlook of this book is based to a great extent on industrial and socio-legal implications of bioeconomic advances. The book takes a comprehensive look not only on the implications of IPR in omics-based research but on what are the ethical and intellectual standards and how these can be developed for sustained innovation. I congratulate the editor for synchronizing with global authorities on the subject to underline the upcoming challenges and present most viable options for translating commercially viable ideas into easily affordable products and technologies.

I wish Dr. Keswani every success with the launch of this book and thank him for his dedication to agricultural sustainability around the world.

Sincerely,

Prof. M. S. Reddy
prof.m.s.reddy@gmail.com

Founder Chairman of Asian PGPR Society
Hyderabad, Telangana, India
Professor, Department of Entomology
and Plant Pathology
Auburn University
Auburn, AL, USA
April 10, 2019

Preface

The current era of incredible innovations toward the zeal to chase the heights of development has made science and technology one of the most powerful tools to accomplish the tasks of incremental prosperity for human welfare and sustainable development. The development of biotech industry in any given country is shaped by the characteristics of the technology, particularly its close relation to scientific knowledge, and by country-specific factors—the level and nature of the scientific knowledge base, the institutional setup, and the role assumed by the government—which influence the country's ability to exploit new opportunities and appropriate the respective results.

This volume focusing on the integrated approach for sustained innovation in various areas of biotechnology encourages the use of biotechnology to address global environmental issues by supporting international agreements to create and sustain markets for environmentally sustainable biotechnology products. The outlook of this book is based to a great extent on industrial, socioeconomic, and legal implications of biotechnological advances. The book takes a comprehensive look not only on the implications of IPR in omics-based research but on what are the ethical and intellectual standards and how these can be developed for sustained innovation.

Adopting a unique approach, this book integrates science and business for an inside view on the biotech industry. Peering behind the scenes, it provides a thorough analysis of the foundations of the present-day industry for students and professionals alike.

The proposed book is divided into three sections dealing with agricultural, industrial, and pharmaceutical biotechnology.

Varanasi, Uttar Pradesh, India — Chetan Keswani

Contents

Contributors

Cahyono Agus Faculty of Forestry, Universitas Gadjah Mada, Yogyakarta, Indonesia

Hareram Birla Department of Biochemistry, Institute of Science, Banaras Hindu University, Varanasi, India

Cristine B. Brown National Institute of Molecular Biology and Biotechnology, University of the Philippines, Los Baños, Philippines

Marilyn B. Brown National Institute of Molecular Biology and Biotechnology, University of the Philippines, Los Baños, Philippines

Dhanay M. Cadillo Chandler Faculty of Law, Turku Institute of Advanced Studies, University of Turku, Turku, Finland

Danuta Ciechanska Institute for Sustainable Technologies – National Research Institute, Radom, Poland

Juan Manuel Zaldívar Crúz Colegio de Postgraduados en Ciencias Agricolas, Cardenas, Tabasco, Mexico

Hagera Dilnashin Department of Biochemistry, Institute of Science, Banaras Hindu University, Varanasi, India

Kumari Divyanshu Department of Botany, Institute of Science, Banaras Hindu University, Varanasi, India

Ruhi Dixit Department of General Surgery, Institute of Medical Sciences, Banaras Hindu University, Varanasi, Uttar Pradesh, India

Karim Ennouri Digital Research Centre of Sfax, Technopark of Sfax, Sfax, Tunisia

Laboratory of Amelioration and Protection of Olive Genetic Resources, Olive Tree Institute, University of Sfax, Sfax, Tunisia

W. G. Dilantha Fernando Department of Plant Science, University of Manitoba, Winnipeg, MB, Canada

April S. Gislason Department of Microbiology, University of Manitoba, Winnipeg, MB, Canada

Alexander Golberg Porter School of Environment and Earth Sciences, Tel Aviv University, Tel Aviv, Israel

Ankush Gupta Department of Biochemistry, Institute of Science, Banaras Hindu University, Varanasi, India

Kathleen Hefferon Department of Cell and Systems Biology, University of Toronto, Toronto, Canada

Akira Homma Bio-Manguinhos, Oswaldo Cruz Foundation, Rio de Janeiro, Brazil

Lav Kumar Jaiswal Department of Biochemistry, Institute of Science, Banaras Hindu University, Varanasi, India

Li Jiang Shanghai International College of Intellectual Property, Tongji University, Shanghai, People's Republic of China

Abdelaziz Kallel Digital Research Centre of Sfax, Technopark of Sfax, Sfax, Tunisia

Lazaros Karaoglanoglou National Technical University of Athens, Athens, Greece

Chetan Keswani Department of Biochemistry, Institute of Science, Banaras Hindu University, Varanasi, India

Atif Khan Water & Steam Chemistry Division, BARC Facilities, Kalpakkam, Tamil Nadu, India

Homi Bhabha National Institute, Mumbai, India

Teresa R. de Kievit Department of Microbiology, University of Manitoba, Winnipeg, MB, Canada

Emmanuel Koukios National Technical University of Athens, Athens, Greece

Sunil Kumar Department of Botany, Institute of Science, Banaras Hindu University, Varanasi, India

Gabriel Olivo Locatelli CAV – Academic Centre of Vitória. Rua Alto do Reservatório, UFPE – Federal University of Pernambuco, Vitória de Santo Antão, PE, Brazil

UNIBRA – Centro Universitário Brasileiro, Recife, PE, Brazil

Jorge Lima de Magalhães Farmanguinhos, Oswaldo Cruz Foundation, Rio de Janeiro, Brazil

Nasir Mahmood Department of Biochemistry, University of Health Sciences, Lahore, Pakistan

Department of Cell and Systems Biology, University of Toronto, Toronto, Canada

Adam Mazurkiewicz Institute for Sustainable Technologies – National Research Institute, Radom, Poland

Gloria Isela Hernández Melchor Sociedad Iberoamericana de Bioeconomia y Cambio Climático. Asociación Civíl, Cárdenas, Tabasco, México

Flavia Maria Lins Mendes School of Chemistry, Federal University of Rio de Janeiro, Rio de Janeiro, Brazil

Andreas Meyer Department of Water, Environment, Civil Engineering, and Safety, Working group "Ecological Engineering", University of Applied Sciences Magdeburg-Stendal, Magdeburg, Germany

Sarah Bushra Nasir Abdus Salam School of Sciences, Nusrat Jahan College, Rabwah, Chiniot, Pakistan

Robert A. Nepomuceno National Institute of Molecular Biology and Biotechnology, University of the Philippines, Los Baños, Philippines

Chandra Sekhara Rao Nuthalapati Institute of Economic Growth, University of Delhi Enclave, New Delhi, India

Ruslana Rachel Palatnik Department of Economics and Management, and SEED – the Sustainable Economic and Environmental Development Research Center, The Max Stern Yezreel Valley College, Afula and Nazareth, Israel

NRERC-Natural Resource and Environmental Research Center, University of Haifa, Haifa, Israel

Manoj Pandey Department of Surgical Oncology, Institute of Medical Sciences, Banaras Hindu University, Varanasi, Uttar Pradesh, India

Punam Pandey Department of General Surgery, Institute of Medical Sciences, Banaras Hindu University, Varanasi, Uttar Pradesh, India

Kay Plat SIKT, project "Greenhub", Leipzig University, Leipzig, Germany

Cristina Possas Bio-Manguinhos, Oswaldo Cruz Foundation, Rio de Janeiro, Brazil

Beata Poteralska Institute for Sustainable Technologies – National Research Institute, Radom, Poland

Meghanath Prabhu Porter School of Environment and Earth Sciences, Tel Aviv University, Tel Aviv, Israel

Sachchida Nand Rai Department of Biochemistry, Institute of Science, Banaras Hindu University, Varanasi, India

Mateus Pinheiro Ramos School of Chemistry, Federal University of Rio de Janeiro, Rio de Janeiro, Brazil

Aaina Singh Rathore Department of Biochemistry, Institute of Science, Banaras Hindu University, Varanasi, India

Anna Sacio-Szymańska Institute for Sustainable Technologies – National Research Institute, Radom, Poland

Petra Schneider Department of Water, Environment, Civil Engineering, and Safety, Working group "Ecological Engineering", University of Applied Sciences Magdeburg-Stendal, Magdeburg, Germany

V. K. Shukla Department of General Surgery, Institute of Medical Sciences, Banaras Hindu University, Varanasi, Uttar Pradesh, India

Juliana De Simone Morais School of Chemistry, Federal University of Rio de Janeiro, Rio de Janeiro, Brazil

Saumitra Sen Singh Department of Biochemistry, Institute of Science, Banaras Hindu University, Varanasi, India

Surya Pratap Singh Department of Biochemistry, Institute of Science, Banaras Hindu University, Varanasi, India

Ángel Sol Sánchez Colegio de Postgraduados en Ciencias Agrícolas, Cárdenas, Tabasco, México

Adelaide Maria de Souza Antunes School of Chemistry, Federal University of Rio de Janeiro, Rio de Janeiro, Brazil

National Institute of Industrial Property (INPI), Rio de Janeiro, Brazil

Mohamed Ali Triki Laboratory of Amelioration and Protection of Olive Genetic Resources, Olive Tree Institute, University of Sfax, Sfax, Tunisia

Yashoda Nandan Tripathi Department of Botany, Institute of Science, Banaras Hindu University, Varanasi, India

Tomasz Twardowski Institute of Bioorganic Chemistry, Polish Academy of Sciences, Poznan, Poland

Ram Sanmukh Upadhyay Department of Botany, Institute of Science, Banaras Hindu University, Varanasi, India

Davide Viaggi Department of Agricultural and Food Sciences, University of Bologna, Bologna, Italy

Ewa Woźniak Institute of Bioorganic Chemistry, Polish Academy of Sciences, Poznan, Poland

Walia Zahra Department of Biochemistry, Institute of Science, Banaras Hindu University, Varanasi, India

Tomasz Zimny Institute of Law Studies, Polish Academy of Sciences, Warsaw, Poland

Meiron Zollmann Porter School of Environment and Earth Sciences, Tel Aviv University, Tel Aviv, Israel

About the Editor

Chetan Keswani is a Postdoctoral Fellow in the Department of Biochemistry, Institute of Science, Banaras Hindu University, Varanasi, India. He a has keen interest in intellectual property, regulatory, and commercialization issues of agriculturally important microorganisms. He is an elected Fellow of the Linnean Society of London, UK. He received Best Ph.D. Thesis Award from the Uttar Pradesh Academy of Agricultural Sciences, India, in 2015. He is an Editorial Board Member of several reputed agricultural microbiology journals.

Part I

Food and Agricultural Biotechnology

Understanding Bioeconomy Systems: Integrating Economic, Organisational and Policy Concepts

1

Davide Viaggi

Abstract

The bioeconomy is recognised as a major area of development of current economies. Though defined in different ways and focusing on different sectors depending on countries and areas of the world, a key qualifying feature of the bioeconomy is the sustainable use of biological resources building on a wide range of modern technologies. Biotechnology is the leading type of technology in this context, but attention is much wider. This paper develops a conceptual approach on the development of the bioeconomy based on the integration of three perspectives: investment in biological capital, organisational trends and configuration of policy.

Keywords

Bioeconomy · SETVEWS · Biological capital · Biotechnology

1.1 Introduction

The bioeconomy is recognised as a major area of development of current economies. Though defined in different ways and focusing on different sectors depending on countries and areas of the world, a key qualifying feature of the bioeconomy is the sustainable use of biological resources building on a wide range of modern technologies. Biotechnology is the leading type of technology in this context, but attention is much wider. The economic and business literature is working to identify structuring concepts allowing a better understanding of this idea (Viaggi 2018).

In the EU, a new boost to the bioeconomy has been given by the launch of the revised bioeconomy strategy the EU Commission in October 2018 (European

D. Viaggi (✉)
Department of Agricultural and Food Sciences, University of Bologna, Bologna, Italy
e-mail: davide.viaggi@unibo.it

C. Keswani (ed.), *Bioeconomy for Sustainable Development*,
https://doi.org/10.1007/978-981-13-9431-7_1

Commission 2018). In this case the bioeconomy is defined in a comprehensive way as the aggregate of sectors using biological resources: 'The bioeconomy covers all sectors and systems that rely on biological resources (animals, plants, micro-organisms and derived biomass, including organic waste), their functions and principles. It includes and interlinks: land and marine ecosystems and the services they provide; all primary production sectors that use and produce biological resources (agriculture, forestry, fisheries and aquaculture); and all economic and industrial sectors that use biological resources and processes to produce food, feed, bio-based products, energy and services. To be successful, the European bioeconomy needs to have sustainability and circularity at its heart. This will drive the renewal of our industries, the modernisation of our primary production systems, the protection of the environment and will enhance biodiversity' (European Commission 2018).

While most sectors of the bioeconomy already benefit of a well-established economic literature, the bioeconomy as a whole is struggling to find a clear economic interpretation. One of the main reasons is certainly the complexity of the sector. Economic analysis, on the other hand, is advocated as a major need for the bioeconomy in view of making the sector profitable and competitive with respect to competing sectors in particular those based on fossil resources (Singh et al. 2016a, b, 2019).

This paper develops a conceptual approach to the development of the bioeconomy based on the integration of three perspectives: investment in biological capital, organisational trends and configuration of policy. The paper is organised in four sections. Section 1.2 illustrates the bioeconomy as a trade-off between investment and exploitation of biological capital in a sustainability context and public goods perspectives. This is used to develop a mathematical specification of different economic conditions for the development of the bioeconomy in Sect. 1.3. The bioeconomy is here described as a sort of black box. Section 1.4 gets into the black box illustrating organisational concepts that can support the bioeconomy development and that are shown to converge towards the SETVEWS (socioecological technological value-enhancing web system) idea. Section 1.5 illustrates related policy trends and instruments, especially observing that the bioeconomy is largely based on complex policy mixes that need to be understood in their interplays and on incentive-based policy instruments that need to be embedded in participatory co-constructing processes. Section 1.6 discusses the practical implications and concludes.

1.2 Background Conceptual Representation: Exploitation vs. Investment in Biological Capital

We use the concept of anthropised biological capital (ABC) (Viaggi 2018). Actions can be taken both to exploit this capital for the production of goods and to increase the capital. This section develops a stylised model of the bioeconomy based on a purely neoclassical approach, focusing on two major distinctive factors:

1. The production of economic goods requires the (creation and) maintenance of an ABC.
2. Consumer utility (demand) is directly affected by both the production of market goods and the status of the ABC, in relation to its public goods features and its ability to ensure production in the future.

These can be seen as two different aspects of capital characteristics. The first point is connected to the idea of ABC as an 'artificial' concept of biological resources that is managed and determined by human actions. In principle this would need to be treated in a dynamic or at least multiperiod framework. However, in order to keep things simple and to focus on the main variables, we assume that it can be treated in a partial equilibrium framework, without explicit consideration of time or, better, as a single-period decision. The second point derives from the observation that the state of ABC can affect utility beyond the effect of the consumption of the derived market goods. This may happen in different ways. One is by way of externalities or public goods components. It can also include existence values or option values generated by ABC. This may also include societal risk aversion, acceptability and ethical issues. It is also relevant to note that such effects may, in principle, be both positive and negative in terms of utility. This is not necessarily only a value that is expressed outside the market, but can also affect expectations, willingness to pay by consumers and financial market behaviour.

In order to approach the problem, we describe it as a trade-off between exploitation and investment in natural capital. The term 'exploitation' is used here as comprising harvesting, cultivation and use in processing, considering the fading distinction between these terms.

In principle, having enough financial resources (i.e. lack of financial constraints), the optimal level of effort and stock may be derived by the joint optimisation of both exploitation and investment. However, the relationship and the related trade-offs may be more evident if we assume financial constraints. Based on this assumption, in the short term and given a certain amount of resources, the decision maker (society as a whole assumed to take the form of a single decision maker) has the choice between investment in the stock of ABC (in view of future income and hence envisaging a benefit represented by the present value of money invested) and for exploitation effort. This is depicted in Fig. 1.1, which illustrates the optimal allocation of resources in a given year.

The upper part of the figure illustrates the issue using total revenue curves, while the bottom part illustrates the same idea with the marginal values of revenues. The horizontal bar gives the amount of resources, so that point k illustrates the allocation between investment (growing from left to right) and exploitation effort (growing in the other direction).

In the top part, the total revenue achieved with investment effort is given by the function Ri, while the total revenue achieved by exploitation effort is given by Re. In the upper part of the figure, the optimal allocation of resources (costs) is represented by the point in which the total revenue is higher, i.e. the top of the total revenue function (Rtot), which is the vertical sum of the two revenue functions.

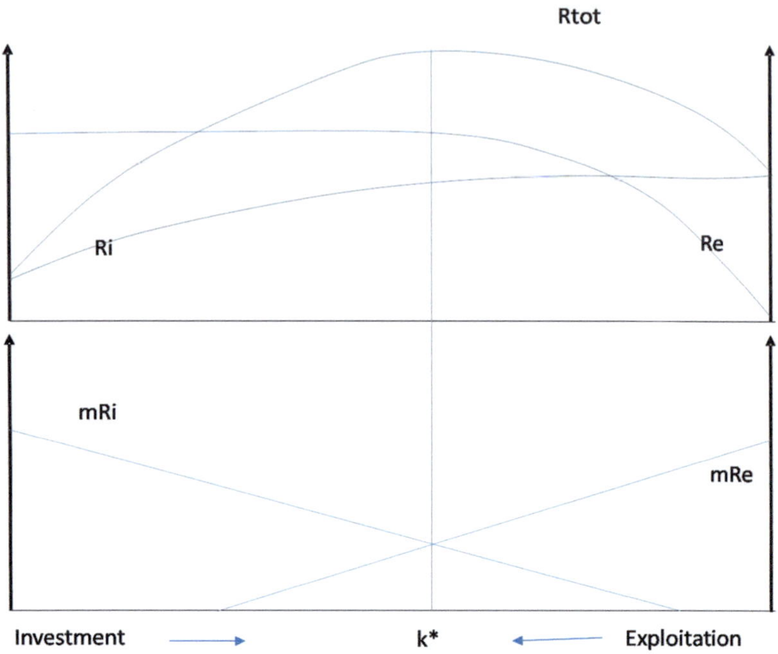

Fig. 1.1 Optimal allocation of resources between exploitation and investment in a given year

Formally, in the bottom part, the optimal level is given by the allocation in which the marginal revenue from costs of harvesting effort meets (is equal to) the marginal revenue from investment costs (resource allocation).

This representation holds assuming fixed resources and variable yield. Let us address the same issue in a different way, assuming now unconstrained but costly resources and a number of potential total revenue levels (Fig. 1.2).

Y1 is an iso-revenue curve in the harvesting-investment effort space, i.e. the combination of harvest and investment yielding the same revenue; each line k is the set of combinations of effort and investment that can be accomplished with the same total cost, i.e. given some amount of economic resources available. If effort and investment are expressed in physical units, k is determined by the amount of investment and effort each multiplied by its price (intended as unitary cost). If investment and effort are expressed in monetary units, the line is a 45° straight line. Given a certain iso-revenue curve, the optimal combination of effort and investment is when the line k is tangent to the iso-revenue curve, i.e. the point in which each specific level of revenue can be achieved at the minimum cost. Assuming resource constraints, e.g. k1 (similar to the previous figure), the highest curve touched would determine the yield achieved, and the combination of effort and investment at the point of tangency the optimal distribution of effort between those two. Assuming that this constraint is removed, different optimal combinations may be envisaged for achieving different yields, and the optimal one would be the one achieving higher profits (difference between revenue and cost). Assuming decreasing marginal

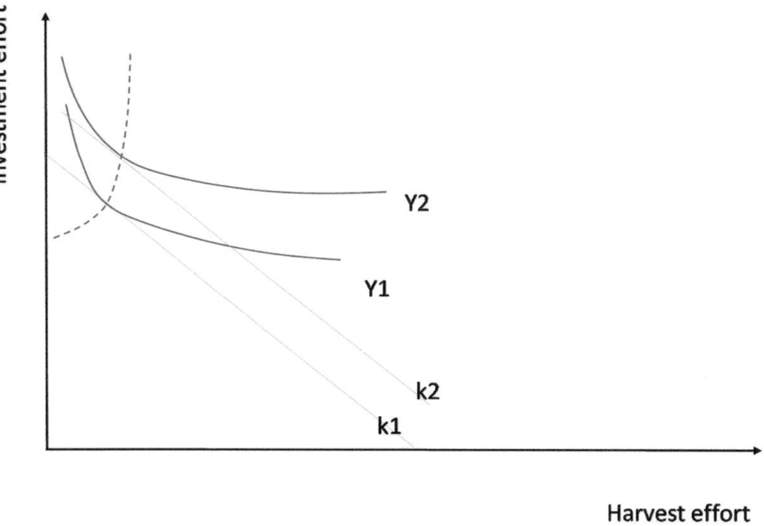

Fig. 1.2 Optimal allocation of investment and harvesting effort with optimal expansion pathway. (Source: Viaggi 2018)

returns in both exploitation and investment, there will be indeed one point on the expansion pathway in which this difference is maximised, which will yield the optimal level of capital-exploitation.

This expansion pathway shows the optimal trade-off between effort and investment as a function of the resource available (maximum yield given the cost) or could be represented the other way round as a traditional cost function when we have minimum costs (as results of the optimisation above) for each level of yield. This can be depicted by a traditional cost function representation. In this way, the optimal equilibrium combination of revenue, investment effort and exploitation effort can be represented and ends up yielding the optimal level of revenue production from biological resources in the short and long run, which may depend on the total opportunity value of biological resources and the (opportunity) cost of resources needed for the exploitation and management of ABC.

The framework as described until now is linked but does not consider explicitly externalities or public goods. There are indeed two types of public goods here: those related to the amount of ABC and those related to the exploitation itself. Considering these public goods would modify the optimal level of capital stock and the optimal effort, by modifying the iso-revenue curve. For example, if some amount of public goods is attached to capital stock, a higher level of such capital stock would become optimal. Even more generally, the iso-revenue is considered as an iso-social utility, representing the flow of utility derived by the combination of exploitation and ABC attributes at a point in time, including private and public goods components.

The framework of Fig. 1.2, besides explaining costs and connecting them with anthropised capital stocks, may be the basis for explaining the effects of innovation.

Innovation would modify the trade-offs between exploitation and investment in the next time period. This will be non-neutral with respect to stock-effort of resources, or better biological capital vs. other production factors. One possibility is to go in the direction of a more flat right arm, which means that harvest effort may contribute poorly to revenue, while investment will contribute more; this may be the current trend, but innovation can also go in the other directions. As a result, the share of allocation to investment or harvesting changes over time.

This may also depend on external drivers, e.g. climate change, rate of return (which affects future benefits from investment) and stock of knowledge (which affects the return on additional investment in innovation). This may also yield a 'zig-zag' trend pushing the balance between effort and investment in one direction or the other depending on different periods and trends in external conditions. This may be used to devise an evolution pathway for the bioeconomy possibly identifying different stages.

1.3 A Mathematical Representation of Investment in Anthropised Biological Capital

Let us now address the same issue in a mathematical way, dealing with the trade-off between exploitation and investment in ABC. Assume a bioeconomy in which we have one market product and one type of anthropised biological capital, and human activities can be exerted in both exploitation and maintenance of the anthropised biological capital, with y = production, e = exploitation effort, m = maintenance effort and c = capital quantity/quality (we will use the term quantity for the sake of simplicity in the following). The amount of total effort is not constrained but has a cost k, assumed linear, but different in e and m, respectively, k_e and k_m. Product y has a price p.

Maintenance effort may imply different types of activities directly linked to modify natural capital, including biotech activities aimed at improving genetic range and capability of plants, biodiversity protection, ecosystem service improvements, etc.

The market product y is generated through a production function f that is a function of exploitation effort and capital quantity $y = f(e, C)$, with $f = (0) = 0, f'(\cdot) > 0$ and $f''(\cdot) \leq 0$. Also, C is a function of yearly capital maintenance and upkeep efforts, with $C = g(m)$, with $g'(\cdot) > 0$ and $g''(\cdot) \leq 0$. Given the above, it also holds that $y = f(e, g(m))$.

Assuming a unique profit-maximising decision maker, i.e. a centralised capitalist led by profit or a benevolent dictator, and linear costs for harvesting and maintenance efforts, the optimal production level is then given by the maximisation of the following total profit function:

$$\max\ \pi = \mathrm{pf}\left(e, g\left(m\right)\right) - k_e e - k_m m \tag{1.1}$$

Taking first derivatives of (1.1) with respect to both effort levels and ensuring first-order conditions, with f_g being the first derivative of f in g, yield:

$$p = \frac{k_e}{f_e\left(e, g(m)\right)} = \frac{k_m}{f_g\left(e, g(m)\right)g'(m)} \tag{1.2}$$

This equation means that, at the optimum, the price of the market product has to be equal to the ratio between cost of exploitation and the derivative of the production function with respect to effort (i.e. marginal revenue must be equal to the marginal cost of effort) and, at the same time, this must also equate the ratio between maintenance costs and the products of the marginal effect of maintenance on capital and its effect on productivity (i.e. the marginal revenue from capital maintenance must equate its cost). In other words each euro spent in effort or maintenance must have the same productivity, and both would be equal to the price of the good. Note that the price of the good is connected to demand and may at the same time be dependent on substitutes such as non-renewable products or renewable non-bio-based products.

Based on Eq. (1.2), four main cases can be identified.

Case (a) has $e > 0$ and $m > 0$; this occurs when $p > \frac{k_e}{f_e\left(0, g(m)\right)}$ and $p > \frac{k_m}{f_g\left(e, g(0)\right)g'(0)}$, and the levels of e and m reflect the trade-offs illustrated in Eq. (1.2). At the optimum, the price equals the cost of producing one unit of the good by spending money on exploitation activities that are also equal to the costs of producing one unit of the good by spending money on maintenance activities. The latter depends on the effect of maintenance expenditure on capital and, in turn, on the effect of capital quantity on production of the private good. A higher relative effort in either exploitation or maintenance is justified for lower respective costs.

Case (b) has $e = 0$ and $m = 0$; this occurs when $p < \frac{k_e}{f_e\left(0, g(m)\right)}$ and $p < \frac{k_m}{f_g\left(e, g(0)\right)g'(0)}$; in this case there is no relationship between human activities and the exploitation of the biological capital, i.e. on the one hand human activities do not modify the biological capital quantity, and, on the other hand, the capital's market products are not exploited. In this case the biological capital involved is not in fact anthropised at all.

Case (c) has $e > 0$ and $m = 0$; this occurs when $p > \frac{k_e}{f_e\left(0, g(m)\right)}$ and $p < \frac{k_m}{f_g\left(e, g(0)\right)g'(0)}$; the profitability of maintenance effort will depend on the assumption about the function g; with $g(0) = 0$ and $f(e, 0) = 0$, some amount of effort in maintenance of the anthropised biological capital will be needed in order to achieve some production of the private good, though this will not be activated when the initial marginal cost is so high that the production cost is higher than the price. If$g(0) > 0$, production can occur without maintenance activities. Maintenance will be activated if $p > \frac{k_m}{f_g\left(e, g(0)\right)g'(0)}$. Hence, this sets the threshold for the antropisation of biological capital; note that in this formulation this only makes sense if k is positive and both $f_g(e, g(0))$ and $g'(0)$ are positive. In this case, biological capital is exploited, but no maintenance activity is performed; this is very much a case analogous to the present treatment of most biological resources, in which management decisions basically concern different levels and modalities of exploitation, but no direct effort is considered for the capital maintenance/modification (except through the consequences of exploitation, e.g. the amount of harvest).

Case (d) has $e = 0$ and $m > 0$; this would theoretically occur when $p < \frac{k_e}{f_e(0, g(m))}$ and $p > \frac{k_m}{f_g(e, g(0))g'(0)}$; however, as the benefits of the anthropised capital only occur through the market good, this case only applies in very special situations in which the market good is produced without need of exploitation effort, i.e. $f_g(0, g(0)) > 0$.

Let us now add the issues addressed in point 2 of the previous section (externalities and public good-type benefits) to the above analysis related to point 1. This means considering explicitly the role attribute to natural capital in terms of utility for society. Instead of a given price p for the private good, let us assume a utility function u of society in which

$$U = u(y, C) \tag{1.3}$$

This function assumes that utility is the result of both the amount of goody and the quantity of anthropised biological capital (of course this could be extended to quality or other issues).

The overall social optimum will be given by maximising the net social benefit ($\prod$), as the difference between the utility generated and the costs entailed by the mix of exploitation and maintenance effort of anthropised natural capital:

$$\max \prod u\left(f\left(e, g(m)\right), g(m)\right) - k_e e - k_m m \tag{1.4}$$

First derivatives and first-order conditions with respect to e yield, after some rearrangement:

$$u_f\left(f\left(e, g(m)\right), g(m)\right) = \frac{k_e}{f_e\left(e, g(m)\right)} = \frac{k_m}{g_m(m) f_g\left(e, g(m)\right)} = \frac{u_g\left(f\left(e, g(m)\right), g(m)\right)}{f_g\left(e, g(m)\right)} \tag{1.5}$$

The understanding of the level of e and m can again follow the four cases identified around the discussion based on Eq. (1.2), but conditions are slightly more complex.

Case (a) has $e > 0$ and $m > 0$; this occurs when both $u_f\left(f\left(0, g(m)\right), g(m)\right) > \frac{k_e}{f_e\left(0, g(m)\right)}$ and $u_f\left(f\left(e, g(0)\right), g(0)\right) > \frac{k_m}{g_m(0) f_g\left(e, g(0)\right)} - \frac{u_g\left(f\left(e, g(0)\right), g(0)\right)}{f_g\left(e, g(0)\right)}$. At the optimum, the relationship depicted by Eq. 1.5 is fully in place. The utility generated by each unit of product is equal to the cost per unit of product generated by the exploitation effort. This also is equal to the cost of producing one unit of the product through maintenance activities, minus the direct utility generated by the production of anthropised capital quantity. The component related to maintenance effort depends on both the contribution to production and the direct utility generated by the status of anthropised natural capital. When utility from capital change caused by higher *m* is negative (negative externalities, negative values associated to antropisation, etc.), the right-hand side becomes a sum, which raises the total cost for the production of the good *y*, other things being equal, and causes an accompanied lower level of production. It is worth noting that higher productivity of effort causes higher effort at the optimum, i.e. increase technology efficiency translates in higher exploitation effort (similar to the well-known rebound effect), other things equal.

Case (b) has $e = 0$ and $m = 0$; it reflects the opposite conditions with respect to (a). While the condition concerning *e* is rather straightforward, the one concerning *m* is more complex. In particular it derives from the fact that $u_f\left(f\left(e,g(0)\right),g(0)\right) < \frac{k_m}{g_m(0)f_g\left(e,g(0)\right)} - \frac{u_g\left(f\left(e,g(0)\right),g(0)\right)}{f_g\left(e,g(0)\right)}$. This may be primarily due to a high cost per unit of production; as positive externalities compensate for high costs, there are different combinations of costs and externalities which make the relationship to occur. For example, high costs may be justified in presence of high positive externalities, while low positive externality cases may be bearable only in case of low costs as well; in case of negative externalities, only low-cost maintenance may be acceptable. In turn, both conditions depend on production of private good when *m* is zero, i.e. the productivity of natural capital without any maintenance.

Case (c) has $e > 0$ and $m = 0$; in this case biological capital is exploited, but no maintenance activity is performed; as with model 1, this is another case analogous to the current treatment of most biological resources, in which management decisions basically concern different levels and modalities of exploitation, but no direct effort is considered for the capital maintenance. However, in this case, this also requires that there be no direct utility from the capital itself, i.e. no positive value or negative value attached to changes in biological capital due to anthropic maintenance.

Case (d) has $e = 0$ and $m > 0$; unlike model 1, this case is relevant here and concerns activities in which biological resources are maintained for the direct utility related to human perception (e.g. externalities from public goods, cultural and existence values, etc.) but do not yield utility through the production of private consumption goods.

1.4 SETVEWS as a Unifying Organisational Concept

The relationships above are not managed by a single decision maker but rather by complex societal interactions. Several organisational concepts have been proposed in order to deal with the relationship between societies and ecosystems, and some of them are now being mutated to describe the bioeconomy. Among them some of the most interesting and tempting are those related to socioecological systems (SES) (Ostrom 2009) and value webs (Scheiterle et al. 2016; Virchow et al. 2016).

Viaggi (2018) proposes an overarching concept partially integrating these approaches and called it socioecological technological value-enhancing web system (SETVEWS). This concept largely draws from the merging of two approaches: the approach of socioecological systems and the approach of value webs. However, it integrates ideas from more general technological views of the bioeconomy and ecosystems and public goods perspectives. The view of bioeconomy systems according to this view can be described as in Fig. 1.3.

According to (Viaggi 2018), a SETVEWS can be described as:

1. "A system of biophysical and social components that interact among each other;
2. The linkages among sub-systems, identified by both material and value flows on multiple product chains and socio-ecological systems;

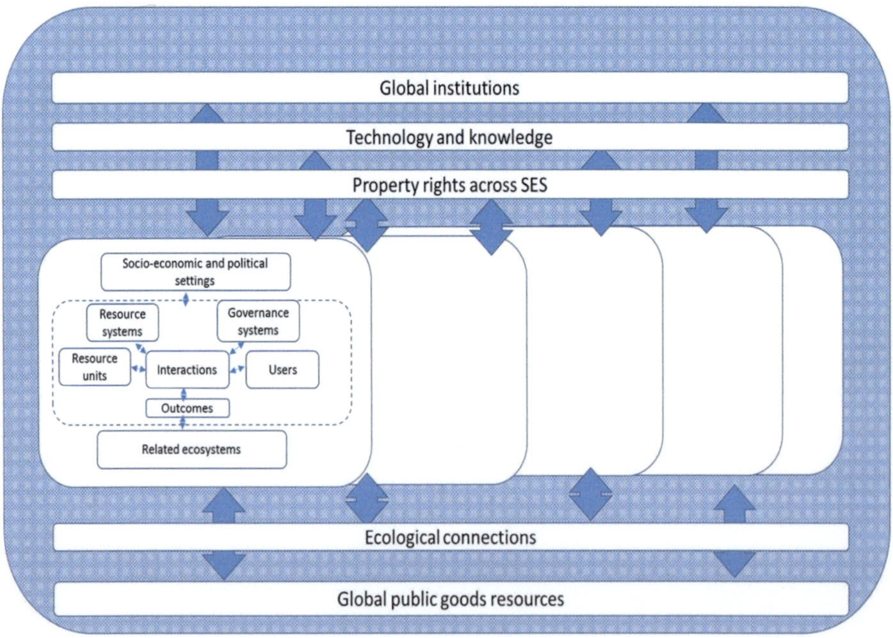

Fig. 1.3 A simplified structure of the SETVEWS

3. A complex system of adaptation solutions from the interaction between its multiple components, including feedback mechanisms;
4. A set of knowledge and technologies".

Several partial (local) SETVEWSs can be identified, linked together in a global bioeconomy SETVEWS. Having in mind at the same time local and global systems makes it possible to better account for interrelationships that have a global dimension and local effects, such as climate change effects on productivity and the reverse.

Similar to SES, the approach emphasises the interactions between social systems and ecological systems and highlights the importance of social and economic interactions in producing value. Compared with SES, however, SETVEWS explicitly integrates the chain and interchain (web) connections, hence providing a more technical structure of the flows of value within an economy. It also expands the concept to the global bioeconomy. In a way, it describes a bundle of relationships between different SES, explicitly accounting for the global geographical complexity of production processes. Another distinctive feature is that SETVEWS is intended to support prescriptive and predictive diagnoses, while SES are largely oriented towards analysing ex post the sustainability of socioecological systems.

In common with value webs (and in contrast to global chain analysis), SETVEWS is not a product-oriented concept. This does not exclude that a final product can be one of the agglomeration criteria. More generally, agglomeration would follow bundles of products or, even more generally, bundles of attributes (characteristics). This links directly SETVEWS with attribute-based theory of public goods (Viaggi 2018). However, value webs are mainly supply-related concepts in their origin, while SETVEWS also integrates demand-side components (households, etc.), the institutional context of production and consumption as well as the relevant environmental components.

The system works integrating action of individual and institutions. Individuals affect their context through direct actions, purchasing actions and contribution to governance decisions. Institutions provide collective rules and behaviour and include also those devoted to the supply of public goods, as well as to ensuring resilience and sustainability. The approach explicitly integrates the role of technology in this system. Technology affects the functional links among different components of the system, including the flows of biomass, the relationship between biomass and non-bio-based components and communication.

By using the term institutions, the approach goes beyond the mere dichotomy between markets and state and rather builds on the role of intermediate types of institutions. The state is confined to one specific type of institution identified by contractual relationships with a geographical reference. The nature and role of the firm are also embedded in this category. While the economics of the firm has already identified several ways of looking at it, here firms become more and more a temporary aggregate of contracts, driven by a value proposition. Hence, the importance taken by business models in identifying economic structures, especially when needed, interprets innovation actions. Altogether, the SETVEWS can be seen as being navigated by 'aboema objects'. They can be identified as institutions and being classified in categories, such as states, firms, networks, clusters, projects, etc. Consumers-citizens configurations and networks are central in this idea. Each one is identified by some stabilising/identifier mission and topological feature (e.g. location) but characterised by a shape and a set of linkages that are by nature dynamic. The issue of how to define/identify fixes, i.e. characters that can allow identifying these entities, becomes a need and a problem at the same time. Using existing economic interpretations, these 'can be identified in bundles of property rights and actions (routines) oriented to ensuring self-sustainability of goals, as well as defined goals. However, fixity is not a value; rather change; asymmetric relationships producing rents and adaptation to change are considered as strong points. More generally, these objects (the nodes of the web) can be identified as mission-oriented bundles of rights and capabilities' (Viaggi 2018). The orientation to value proposition can be added to this and become the driving descriptor, even if it is often difficult to describe beyond the qualitative level.

From the spatial point of view, SETVEWS allow to understand the governance of complex economic systems in their regional/interregional dimension, with a basis of global common goods, flexible boundaries and a number of more or less formalised interconnections. This may accommodate biomass, water and nutrient flows, climate change governance, capital flows, '…migration of labour/households, multinational corporations and international linkages of small firms, sustainability and resilience based on interplay across complementary areas, networks of various kind, flexible specialisation based on cheap production locations, different access to resource and knowledge.' It can also allow addressing the circular economy concept in a multi-scale and multiregional sense, considering different degree of closeness of local ecosystems.

The ambition of SETVEWS is to explain value creation by linking in a flexible way the value generated by consumer preferences and the value generated by increasing biological capacity of production. This value is expressed through transactions, assets and property rights. One important feature of this concept is to go beyond simplistic dichotomies, such as those of promissory vs. real value systems, production vs. consumption and local vs. global.

The link between value and location can be also expressed, for example, using indicators of density of value, capital or density of the web itself, as well as by the degree of connection between this value and a specific territory. However, this information remains functionally embedded in a global view, especially in relation to value creation that depends on global value flows.

1.5 Policies

One of the distinguishing aspects of the bioeconomy is complexity and public goods features. One of the consequences is the pervasive role of policies. In addition, given its differences across countries and the variety of maturity of technologies involved, the bioeconomy tends to benefit of a number of different types of policy actions and instruments. Table 1.1 provides an overview for OECD countries.

Table 1.1 Overview of policies in OECD countries

Key points	Policy measures	Canada	EU	France	Germany	UK	Japan	USA
(a) Promoting innovation	Basic research and applied research	x		x	x	x	x	
	Pilot and demonstration plants	x		x	x	x		
	R&D		x					
	Key enabling technologies		x					
	Clusters		x		x			
	Public-private partnerships		x					x
	International collaborations				x			
	Bioeconomy research							x
	Cross-cutting technology research							x
	Interdisciplinary research							x
	Promoting innovative research							x
(b) Infrastructure	Cross-cutting technology			x				
	Centres of competence				x			
	Research networks and training specialists				x			
	Key enabling technology infrastructure					x		
	Rural development					x	x	x
	Education					x		x
	Biomass suppliers						x	
	Vocational training and further education							x
	Cyberinfrastructure							x

(continued)

Table 1.1 (continued)

Key points	Policy measures	Canada	EU	France	Germany	UK	Japan	USA
(c) Commercialisation	Marketing	x						
	Support for biomass producers	x						
	Financing and venture capital		x		x			
	Private innovation capital			x				
	Market development				x		x	
	Startup funding				x			
	Feasibility studies					x		
	Advice to business					x		
	Innovation capital						x	x
	Lab-to-market plans							x
	Market readiness and penetration							x
	Globalisation of businesses							x
(d) Demand-side instruments	Public procurement	x	x	x			x	x
	Standards and labels		x					
	Support for biomass producers				x		x	
	Support for biofuel producers						x	
	Tax relief and sustainable investments			x				
	Labels			x	x		x	x
	Information and social dialogue				x			
	Support to producers					x		
	Use of bioenergy							x

(e) Policy framework conditions	Policy coherence		x		x			
	Plans for industrial renaissance			x				
	Participation and political representation			x				
	Green taxes			x		x	x	
	Laws and regulation			x				
	Access to renewable resources				x			
	Recycling						x	
	Legislation for and approval of new technologies							x

Source: German Bioeconomy Council (2015)

As a starting point, it should be noted that many countries have now comprehensive strategies, but none has really comprehensive bioeconomy policies. The bioeconomy action is rather embedded in subsector policies (e.g. agriculture or biofuel). A very important policy area is that of framework conditions, especially in relation of background property right, regulation and incentives. Innovation is a typical area at the core of bioeconomy intervention, especially in developing and promoting new technologies. This may depend a lot on infrastructure. Infrastructure may actually relate to innovation itself or other components, including supply and commercialisation. Many markets for novel bio-based products are being developed and hence benefit of intervention to stimulate market creation. Demand-side intervention is also possible, ranging from public procurement to support to the creation of demand through information, awareness and dialogue.

The most suitable policy mix is dependent on a number of local conditions and path dependency. The fact that the bioeconomy is linked to a so complex mix of policies makes policy consistency and participated co-construction key elements to ensure success. Two classical examples of problems in this sense have been biotech and bioenergy. In the first case, research and public perception have not gone at the same pace in every area of the world, and (lack of) acceptance has been and remains a key issue. In the second case, strong intervention affecting energy markets has had a number of indirect effects questioning agriculture and food systems.

1.6 Discussion and Conclusions

The joint development of adequate economic, organisational and policy concepts is at the core of the ability to develop a sustainable bioeconomy. Sustainable means here also economically competitive with fossil-based solutions. In this chapter, we provide a review of current trends and further develop the related concept of bioeconomy as a trade-off between exploitation and investment in biological capital, discuss the SETVEWS approach to describe the organisation of this system and link this to the current policy approach. The further integration of these lines of thinking is key for the future development of the bioeconomy and needs to rely on an inclusive view of policy into the decision-making system and the system view as a supporting logic of social interaction concerning biological resources.

References

European Commission (2018) A sustainable bioeconomy for Europe: strengthening the connection between economy, society and the environment. Updated Bioeconomy Strategy. Bruxelles. Available at: https://ec.europa.eu/research/bioeconomy/pdf/ec_bioeconomy_strategy_2018.pdf#view=fit&pagemode=none

German Bioeconomy Council (2015) Bioeconomy policy (Part I) Synopsis and analysis of Strategies in the G7

Ostrom E (2009) A general framework for analyzing sustainability of social-ecological systems. Science 325(5939):419. https://doi.org/10.1126/science.1172133

Scheiterle L, Ulmer A, Birner R, Pyka A (2016) From commodity-based value chains to biomass-based value webs: the case of sugarcane in Brazil's bioeconomy. J Clean Prod. https://doi.org/10.1016/j.jclepro.2017.05.150

Singh HB, Jha A, Keswani C (2016a) Intellectual property issues in biotechnology. CABI, Wallingford. 304 pages, ISBN-13: 9781780646534

Singh HB, Jha A, Keswani C (2016b) Biotechnology in agriculture, medicine and industry: an overview. In: Singh HB, Jha A, Keswani C (eds) Intellectual property issues in biotechnology. CABI, Wallingford, pp 1–4

Singh HB, Keswani C, Singh SP (2019) Intellectual property issues in microbiology. Springer-Nature, Singapore. 425 pages, ISBN- 9789811374654

Viaggi D (2018) The bioeconomy delivering sustainable green growth. CABI Publishing, Wallingford

Virchow, D., Beuchelt, T. D., Kuhn, A., Denich, M. (2016) Biomass-based value webs: a novel perspective for emerging bioeconomies in Sub-Saharan Africa. In: Technological and institutional innovations for marginalized smallholders in agricultural development. https://doi.org/10.1007/978-3-319-25718-1_14

Agri-biotechnology: Legal and Economic Aspects of Using GMOs in EU

2

Ewa Woźniak, Tomasz Zimny, and Tomasz Twardowski

Abstract

The development of the bioeconomy, including biotechnology, has significant meaning for many regions in the European Union. Nineteen member states already have a bioeconomy strategy (or a similar strategic document) in place or are in the process of developing a strategy (Haarich S, Bioeconomy development in EU regions: mapping of EU Member States/regions' research and innovation plans and strategies for smart specialisation (RIS3) on bioeconomy. Final Report, Framework Contract: 2014.CE.16.BAT Lot 2. Specific Contract: RTD/F1/PP-03681-2015. European Commission, Brussels, 2017). Currently, the development of science and technology creates new opportunities that make bioeconomy one of the most dynamic sectors in the European economy. Biotechnology has the greatest potential impact in the sectors of agriculture, health and industry. The global human population is estimated to reach nine billion people by 2050, which creates a serious challenge for achieving global food security and adequate responses to the effects of climate change. Progress may be achieved by applying knowledge of molecular and genetic mechanisms to create or improve agricultural processes and food production. This chapter aims to review current knowledge on agri-biotechnology, within the context of the legal and economic aspects of using GMOs (genetically modified organisms). The area of GM crops in EU countries in 2017 was marginal. GM plants were cultivated in a few EU countries: Spain (0.1 million hectares), Portugal (less than 0.1 million hectares)

Authors Ewa Woźniak and Tomasz Zimny have equally contributed to this chapter.

E. Woźniak · T. Twardowski (✉)
Institute of Bioorganic Chemistry, Polish Academy of Sciences, Poznan, Poland
e-mail: twardows@ibch.poznan.pl

T. Zimny
Institute of Law Studies, Polish Academy of Sciences, Warsaw, Poland

C. Keswani (ed.), *Bioeconomy for Sustainable Development*,
https://doi.org/10.1007/978-981-13-9431-7_2

and marginally in Slovakia and Czech Republic. As a consequence, EU countries import raw materials and GM products from other countries. Finally, we highlight the legal aspects of using GM plants in agriculture, including GM products that are registered as food and feed. In the last chapter, we present perspectives on the development of the bioeconomy and biotechnology in the European Union.

Keywords
Bioeconomy · GMOs · Agri-biotechnology · European economy · Sustainable development

2.1 Bioeconomy in the European Union and in the World

The importance of the bioeconomy has gradually increased in recent years. Countries that are developed in agriculture and biotechnologies have paid attention to the development of *the knowledge-based bioeconomy* (KBBE). According to Patermann and Slusarczyk (2009), this concept combines knowledge from natural sciences and biotechnology that coincide with other technologies, such as nanotechnology, chemistry and information technology, which are used in the bioeconomy, including sectors that produce, manage and use biological resources (such as biowaste). There were two European conferences that gave the basis for the development of the bioeconomy concept. In 2005, at the conference *New Perspectives on the Knowledge-Based Bio-Economy*, the importance of progress in the field of natural sciences and biotechnology for applications in the development of knowledge-based bioeconomy was stressed. However, the term *bioeconomy* was not specified at the conference. Instead, it was formulated that the bioeconomy means a sustainable, ecologically effective transformation of renewable biological resources into food, energy and other industrial products. Medical and industrial biotechnology were recognized as activities belonging to strengths of Europe, whereas agricultural biotechnology was defined as a weak point in the development of Europe, mainly due to controversies that have been caused by genetically modified organisms (GMOs) (McCormick and Kautto 2013). The aim of the second conference in 2007, called *En Route to the Knowledge-Based Bio-Economy*, was to determine the prospects for the development of the bioeconomy over the next 20 years. The result of the work was a report that summarizes the most important conclusions from the conference (European Union 2007). It was stated that biotechnology will be a critical element of the European economy until 2030. The development of medical biotechnology and the production of bioenergy in the industry were suggested. According to Hilgartner (2007), the concept of bioeconomy was relatively new and poorly disseminated in 2007. The rapid dissemination of this term took place during the period 2010–2013.

2.1.1 Development of the Bioeconomy and Its Definitions

Despite the large interest, one common, generally accepted, definition of the bioeconomy is missing. The way to define this concept has a significant meaning in the formulation of policies and programmes or strategies for the development of the bioeconomy. The first attempts at defining bioeconomy took place in 1997–1998. Enriquez and Martinez, at a genomics seminar, presented a definition that initiated interest in the bioeconomy of the EU (Maciejczak and Hofreiter 2013). According to them, the bioeconomy encompasses all forms of economic activity based on scientific activity, a focus on understanding the mechanism and processes at the genetic (molecular) level and their application in industrial processes (Martinez 1998). Such definition of bioeconomy resulted, among others, from the dynamic development of biology and biotechnology. In later years, the links between the bioeconomy and the natural environment were noticed, and the concept of eco-development and sustainable development was developed. In 2005, the Directorate-General for Research and Innovation in European Commission formulated the definition of bioeconomy, which is an environmentally friendly, ecologically efficient transformation of renewable biological resources for food, energy and other industrial products (European Commission 2005; Schmidt et al. 2012).

Currently, bioeconomy can be defined as a form of economy where the main elements of production, such as the production of materials, chemical products and energy, are based on renewable biological resources (McCormick and Kautto 2013). Bioeconomy uses biomass in the production processes. It should be noted that often in the terminology explaining the concept of bioeconomy, biomass appears as a renewable resource, by which new useful products are obtained.

Attempts to determine the meaning of "bioeconomy" were also made by organizations (such as the OECD) that adapted different working definitions. For instance, in the publication published by the OECD *Bioeconomy to 2030: Designing a Policy Agenda* (2009), the term bioeconomy refers to an innovative approach to production processes. It means:

> the transformation of life science knowledge into new, sustainable, eco-efficient and competitive products. (OECD 2009)

Moreover the role of biotechnology was highlighted by the OECD. It is said that the bioeconomy is a world wherein biotechnology has a significant share of economic output. Bioeconomy offers technological solutions that can respond to challenges facing the world. Biotechnology can be applied in agriculture (agri-biotechnology), health (medical biotechnology) and industry (industrial biotechnology).

In 2012, the European Commission (EC) presented the first European bioeconomy strategy (European Commission 2012). According to this document, bioeconomy encompasses

> the production of renewable biological resources and their conversion into food, feed, bio-based products and bioenergy. (European Commission 2012)

The bioeconomy covers many industries, including agriculture, forestry, fisheries, food and pulp and paper production, as well as parts of chemical, biotechnological and energy sectors. These sectors have a strong innovation potential considering their use of a wide range of sciences (such as life sciences, ecology and food science), enabling and industrial technologies (biotechnology and nanotechnology) and local and tacit knowledge (Haarich 2017). The definition presented by the EC is very general and covers a wide spectrum of bioeconomic activities. In recently updated European Commission (2018d):

> the bioeconomy covers all sectors and systems that rely on biological resources (animals, plants, micro-organisms and derived biomass, including organic waste), their functions and principles.

Bioeconomy involves far more than using biomass for bioenergy. Von Braun (2018) highlighted that bioeconomy not only concentrates on resource flows and resource management but aims to transform society and the economy towards its "biologization". As a result, new products and solutions that simplify a sustainable humanity are created (von Braun 2018).

The bioeconomy can be defined as a sector that covers traditional branches of the national economy, such as agriculture, forestry, fishing and industry (e.g. food, furniture, wood products), and modern activities, which include agricultural, medical, industrial and environmental biotechnology. The traditional bioeconomy uses natural resources, whereas modern activities integrate biological and engineering sciences.

There are many definitions of the bioeconomy, and its range of mutual interactions does not have clearly defined boundaries, due to the diversity of entities. However, all cited definitions show similarities, such as using innovations, biological resources and inter-sectoral activities.

2.1.2 The Background on Bioeconomy

As noted by McCormick and Kautto (2013), the discussion and interest in bioeconomy appeared at the beginning of 2000, when the Lisbon Agenda was established (McCormick and Kautto 2013). In the conference report *New Perspectives on the Knowledge-Based Bio-Economy*, it was noted that:

> the bioeconomy is one of the oldest economic sectors known to humanity, and the life sciences and biotechnology are transforming it into one of the newest. (European Commission 2005)

The development of the bioeconomy is based on knowledge and research. Innovations are the effect of research and provide the possibility of creating new products and methods of production.

The progress of knowledge in the area of sustainable management, production and use of biological resources is the basis for creating safe, ecologically effective

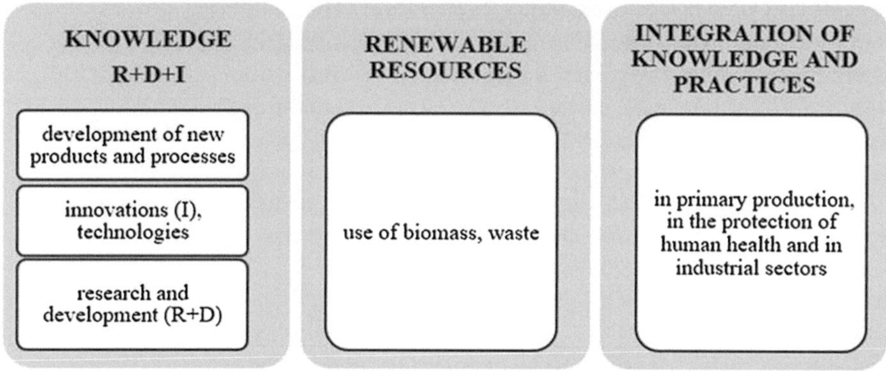

Fig. 2.1 The elements of a knowledge-based bioeconomy. (Source: European Commission 2010)

and competitive products and services for agriculture, medicine, food industry, etc. Renewable resources, mainly biomass, are an element of the knowledge-based economy, which is used for the production of bio-energy or biofuels. An important factor in the development of the bioeconomy is the integration of knowledge and practice (Fig. 2.1). It is worth mentioning that the expenditures on R&D&I activity are substantial.

Minimizing the environmental impact of manufactured products is a challenge for Europe. It is important to choose the ingredients that will allow product reuse. In this context, the concept of the *circular economy* should be noted. It is a popular concept promoted by the EU, by several national governments and by many businesses around the world (Korhonen et al. 2018). A circular economy is recommended as an approach to economic growth that is in line with sustainable environmental and economic development (Ellen MacArthur Foundation 2012). In 2015, the EC adapted an ambitious Circular Economy Package, which includes measures that will help stimulate Europe's transition towards a circular economy. Moreover, it will help to boost global competitiveness, foster sustainable economic growth and generate new jobs (European Commission 2015). Circular economy is:

> an economic model based inter alia on sharing, leasing, reuse, repair, refurbishment and recycling, in an (almost) closed loop, which aims to retain the highest utility and value of products, components and materials at all times. (European Commission 2016a)

The concept of circular economy is related to the environmental challenges and promotion of sustainable development. As the development of the bioeconomy concept, the document *A European Strategy for Plastics in a Circular Economy* was published by the EC in 2018 (which is based on the 2015 *Circular Economy Action Plan*). In the *strategy*, the importance of a negative impact of plastics on the environment, society and economy was highlighted. It is said that the global production of plastics has increased 20-fold since the 1960s, reaching 322 million tonnes in 2015 (European Commission 2018d). In 2015, in the EU, the plastics sector employed 1.5 million people; however, reuse and recycling of end-of-life plastics are still very low. It is estimated that 25.8 million tonnes of plastic waste are generated in Europe every year (Plastic Europe 2016). Less than 30% of waste is collected for recycling. The majority of plastic waste comes from packaging (59%). A significant share of plastic waste from the EU is sent to third-world countries, where there are different environmental standards. It is estimated that in the EU, 150 000–500 000 tonnes of plastic waste enter the oceans every year (Sherrington et al. 2016).

The EC has pinpointed a range of measures to implement the plastics strategy. Several are relevant for research and innovation. For the period 2018–2020, the remaining years of Horizon 2020, the EU's research and innovation framework programme, approximately €100 million has been allocated to projects that are directly related to the plastics strategy. This funding comes on top of the €250 million that was already spent on plastic-related projects through Horizon 2020 (European Commission 2018a).

2.1.3 Bioeconomy as a Political Strategy

In 2012, the EC presented the first European bioeconomy strategy, named *Innovating for Sustainable Growth: A Bioeconomy for Europe* (European Commission 2012). The importance of bioeconomy for smart and ecological growth was highlighted. The bioeconomy strategy and its action plan aim to create a more innovative, resource-efficient and competitive society that reconciles food security with the sustainable use of renewable resources for industrial purposes while ensuring environmental protection (European Commission 2012). This policy influences the development of research and innovation in the bioeconomy sectors and their effect on the creation of a cohesive political environment. The strategy defines social challenges and actions that let us use research and innovation in the bioeconomy sector in a more effective approach (Fig. 2.2).

In 2018, the EC published an updated bioeconomy strategy, named *A sustainable Bioeconomy for Europe: Strengthening the connection between economy, society and the environment* (European Commission 2018d). This strategy is part of the EC's efforts to boost employment, growth and investment in the EU. The aim is to increase and improve sustainable use of renewable resources to address global and local challenges, such as climate change and sustainable development. Bioeconomy has the potential to generate one million new green jobs by 2030 (European Commission Database 2018). The EC proposed three key objectives related to strengthening and scaling up the bio-based sectors, which would rapidly deploy bioeconomies across Europe as well as protect and understand the ecological limitations of bioeconomy (European Commission 2018d).

According to Aguilar et al. (2018) at the beginning of this decade, there was not a single bioeconomy strategy available in the world. Now, dozens have been developed across all continents, and their number is constantly growing. A report from the German Bioeconomy Council (2015) presented an overview of the strategies of selected countries from around the world that relate to the bioeconomy. This report identified 45 countries in total that have developed national policy strategies with a

Fig. 2.2 Bioeconomy strategy 2012 and action plan in the development of bioeconomy. (Source: European Commission 2012)

significant impact on bioeconomic development. The report highlighted the state, organizations and regions whose strategies in a comprehensive way were referred to as the bioeconomy. These included the EU, Finland, Germany, Japan, Malaysia, South Africa, the USA and the west Nordic countries (German Bioeconomy Council 2015). As reported by Aguilar et al. (2018), there will be no single bioeconomy but, rather, as many bioeconomies as there are ecosystems and socioeconomic models. According to the report of Haarich (2017), 19 member states already have a bioeconomy strategy or a similar strategic document in place or are in the process of developing a strategy (Table 1.1). Moreover, 49 of the analysed regions have developed a regional bioeconomy strategy or a similar comprehensive document (Haarich 2017).

Table 2.1 shows documents related to the different aspects of bioeconomy in EU countries, such as innovation, industry, climate, marine and biotechnology. The newest bioeconomy strategy was published by the Ministry of Agriculture, Republic of Latvia, in 2017 (European Union 2017).

Many bioeconomy-related policy actions have emerged in the world. For instance, in sub-Saharan Africa, the German government supports the Biomass Web project in Ethiopia, Ghana, Kenya and Nigeria (Biomass Web 2018). The aim of this project is to improve food security in Africa through increased system productivity of biomass-based value webs. Agri-food sector is under pressure to increase substantially food security. Biomass Web (2018) highlighted three pillars of food security: (1) food availability, through increased biomass productivity; (2) access to food, generation of income from non-food biomass production; and (3) the use of food to increase nutritional value. Projects covered the years 2013–2018. Moreover, in Eastern African countries such as Burundi, Ethiopia, Kenya, Rwanda, Tanzania and Uganda, research is led by the Swedish development agency SIDA, which supports the development of

Table 2.1 Strategies related to the bioeconomy in EU countries

No.	Country	Document	Year of establishment
1.	Austria	Policy paper on bioeconomy	2013
2.	Belgium	Bioeconomy in Flanders	2014
3.	Czech Republic	Strategic framework Czech Republic 2030	2017
4.	Denmark	Denmark as growth hub for a sustainable bioeconomy	2014
5.	Estonia	The regional development strategy of Estonia 2014–2020	2014
6.	Finland	Finnish bioeconomy strategy	2014
7.	France	The French bioeconomy strategy	2017
8.	Germany	Bioeconomy 2030 research strategy	2010
9.	Ireland	Developing the green economy in Ireland	2009
		Delivering our Green Potential	2012
10.	Italy	Bioeconomy in Italy	2017
11.	Latvia	Latvian Bioeconomy Strategy 2030	2017
12.	Lithuania	National Industrial Biotechnology Development Programme 2007–2010	2007
13.	Netherlands	The Bio-based Economy in the Netherlands	2013
14.	Norway	Research Programme on Sustainable Innovation in Food- and Bio-based Industries 2012–2022	2012
		National Strategy for Biotechnology	2011
		Marine bioprospecting – a source of new and sustainable wealth growth	2009
15.	Poland	National Programme for Low-Emission Economic Development	2015
16.	Portugal	Green growth commitment	2015
17.	Slovenia	Slovenian industrial policy 2014–2020	2013
18.	Spain	The Spanish bioeconomy strategy – 2030 Horizon	2015
19.	Sweden	Swedish Research and Innovation Strategy for a Bio-based Economy	2012
20.	UK	Building a high value bioeconomy: opportunities from waste	2015

Source: Own elaboration

bio-innovation in regions. Western African countries, such as Nigeria, Mali and Senegal, focus on promoting bioenergy policies. In regard to dedicated bioeconomy strategies in African countries, only South Africa stands out. The bioeconomy strategy was introduced there in 2013. Bioeconomy-related strategies were presented in Kenya, Tanzania, Uganda (concerning high-tech), Mali, Mozambique, Nigeria, Senegal (concerning bioenergy) and Namibia (research and innovation) (Biomass Web 2018).

The South American countries promote bioeconomy as an alternative model for sustainable development of a green economy. The continent has a huge potential for biomass production as well as utilization. There are no dedicated national bioeconomy policy strategies. Most of the countries have prepared documents that are related to the bioeconomy. Brazil and Argentina are leaders in bioenergy production and, furthermore, are countries in the top five users of genetically modified crops (with the USA, India and Canada). There are few politics in each country that relate to the bioeconomy. While in South American countries policies in terms of bioeconomy are related to the bioenergy and high-tech, in North America, the perspective of bioeconomy is more bioresources-driven (e.g. *A forest bioeconomy framework for* Canada or *the farm bill* in the USA) (German Bioeconomy Council 2015).

In Asia and the Pacific region, bioeconomic development is strongly oriented towards high-tech and industrial innovations (German Bioeconomy Council 2015). Bioeconomy-related policies were adapted in many countries, such as Russia (Osmakova et al. 2018), China, India, Indonesia, Japan, Malaysia, South Korea, Sri Lanka, Thailand, New Zealand (Wreford et al. 2018) and Australia. According to the Biooekonomierat (2018), Japan, Thailand and Malaysia have dedicated bioeconomy strategies.

2.2 Agri-biotechnology

Agri-biotechnology, called green biotechnology or agricultural biotechnology, means the use of technology based on molecular biology in the agriculture, forestry and food industries. Tyczewska et al. (2018) highlighted a few milestones in agriculture that significantly increased agricultural productivity. In the beginning, farmers selected plants that were more resistant to droughts and pest; they chose good-quality plants with high yields (Zilberman et al. 2018). The introduction of chemical fertilizers (as early as the nineteenth century) and then pesticides and herbicides (after the Second World War) increased crop yields (Paarlberg and Paarlberg 2008). Another meaningful boost in crop yields, called *the Green Revolution*, was developed in the 1940s in Mexico by Dr. N. Borlaug. The new varieties obtained by traditional breeding were not only disease-resistant but also better at adapting to various growth conditions, thus having an increased yield potential. Another milestone of agricultural production was related to the use of a wide range of biotechnology tools (Singh et al. 2016a, b, 2019; Tyczewska et al. 2018).

2.2.1 Biotech-Improved Plants

The most common and controversial form of biotechnology in agriculture is the cultivation of new varieties of plants, called genetically modified (GM) plants (Tyczewska et al. 2018). The global area of GM crops increased from 1.7 million hectares in 1996 to 189.8 million hectares in 2017 (Fig. 2.3) (ISAAA 2018).

In 2017, of the 24 countries cultivating GM plants, 19 were developing countries (53%), and 5 were industrialized countries (47%). Approximately 50% of GM plants were grown in North and South America, 33% in Asia, 8% in Europe and 8% in Africa. The largest area of GM crops were in countries such as the USA (75 million hectares), Brazil (50.2 million hectares), Argentina (23.6 million hectares),

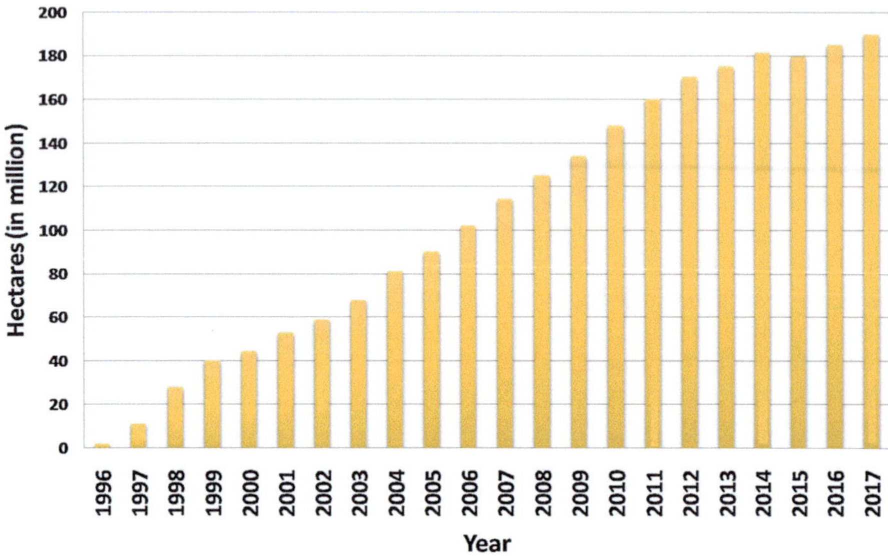

Fig. 2.3 Global area of biotech plants in 1996–2017. (Source: Data based on ISAAA 2018)

Canada (13.1 million hectares) and India (11.4 million hectares). GM crops were also grown in European Union (EU) countries, including Spain (0.1 million hectares), Portugal (<0.1 million hectares), Slovakia and the Czech Republic (small areas, several hundred ha). Compared to GM crops from other parts of the world, this is a low share. The GM crop area in the USA was 40% of the global GM crop area. In this country, GM soybean and GM maize accounted for the largest area, 34.05 million hectares and 33.84 million hectares, respectively (ISAAA 2018).

According to the ISAAA (2018), the most planted biotech crops in 2017 were soybean, maize, cotton, and canola, which accounted for 99% of global GM planting (Paull 2018). Although only a 3% increase was observed in the planting of biotech soybean, it maintained its high adaptation rate of 50% of the global biotech crops (94.1 million hectares). This area is nearly 80% of the total soybean production worldwide. Biotech maize occupied 59.7 million hectares globally, which was 32% of the global maize production in 2017. A negligible decline (1%) in the biotech maize area from 2016 can be explained by unfavourable weather conditions in Latin America, the low market price, limited pest incidence, high year-end stocks and counterfeit seeds in the Philippines (ISAAA 2018).

Biotech cotton was planted across 24.1 million hectares in 2017, a decrease of 8% from 2016. This was a result of the improvement of the global market value and the high adaptation rate of insect-resistant/herbicide-tolerant cotton in 2017. In turn, biotech canola increased by 19% (from 8.6 million hectares in 2016 to 10.2 million hectares in 2017). This increase was connected with an increasing area of biotech canola in the USA, Canada and Australia, thereby addressing the demand for edible oil (ISAAA 2018).

2.3 Importance of Food Security

One of the societal, economical and scientific challenges of the modern world is the constantly growing human population. It has been estimated that by 2050, the global population will exceed nine billion people and food demand will increase by 60% (Fabbri 2017). Unfortunately, current food production levels are not sufficient to meet the demand of such a large population. Achieving global food security by 2050 is very challenging, and unless society, politicians and leaders take vital steps to prevent devastating ripple effects, the agricultural sector will be affected by changes in climate and soil deterioration that will further challenge the productivity of the world's agricultural resources.

According to the FAO (2003), food security exists when all people, at all times, have access to sufficient safe and nutritious food to meet their dietary needs and food preferences for an active and healthy life. The broader definition of food security includes access to non-food inputs, such as clean water, sanitation and health care (FAO 2009).

Food and nutrition security is a priority for EU policy, as presented in *Food 2030.* In this document, the EU research and innovation policy is presented, which includes sustainable development goals (SDGs) and aims by the Conference of the Parties (COP21) related to climate change. This policy is a tool to help address global hunger as well. According to food security in the EU, there are four priorities, such as nutrition for sustainable and healthy diets, climate smart and environmentally sustainable food system, circularity and resource efficiency of food systems and innovation and empowerment of communities (European Commission 2018b). Genetic engineering technology has great potential to contribute to food and nutrition security; the need for both is urgent, as was stressed by Potrykus (2017).

According to GMO and non-GMO standards, significant concern is a consumer *right-to-know* issue. Consumers need full information about products in the marketplace, including the processes used to make those products, not only for food safety or scientific reasons but also for making choices in line with their personal ethics (Castellari et al. 2018). Biosafety regulatory system and the development of education networks are necessary to ensure the information risks and benefits of GM technology (Potrykus 2017).

The regulations on GMOs in the EU and in the USA start from the same presumption: GMOs that receive approval are safe for human consumption and environment. However, the regulations are not the same in regard to the definition of GMO or the labelling standards (Smart et al. 2017). In July 2016, President Obama signed bill S.764, which requires announcement of GMO food in the USA to avoid a patchwork system in which each state has its own labelling law. As reported by EU legislation, a producer must label a product as GMO if it contains more than 0.9% of EU-approved GMO material (Castellari et al. 2018). Products that fall below this threshold do not need to be labelled if the presence of GMOs was adventitious or technically unavoidable. Deliberate use of GMOs in production always results in an obligation to label a product as containing GMOs.

In the EU and in the USA, non-GMO standards were introduced by retailers and food processors in the early 2000s. In the USA, since 2005, a non-profit organization, *Non-GMO Project*, sets voluntary private standards for labelling products as non-GMO. The project allows a threshold of 0.9% of GMO sources in food ingredients and 5% for GMO traces in the feed (the Non-GMO Project 2018). In Europe, legislation related to non-GMO or *GMO free* does not exist (Castellari et al. 2018). In 2016, non-GMO standards were created by a non-profit organization, *Donau Soja*, which associates 15 countries along the Danube river (Krön and Bittner 2015).

Only three countries, Benin, Serbia and Zambia, implemented an official ban on GE food, both imported and cultivated. Most of the countries in the EU, Australia and New Zealand implemented mandatory labelling, with up to a 1% labelling threshold of GMO content. Countries such as Brazil, South Africa, Ukraine, China, Japan, Sri Lanka, South Korea, Malaysia and Indonesia established mandatory food labelling on many GE foods; however, the threshold for the GMO content is higher (more than 1%) (Macahilo 2018).

The US Department of Agriculture (USDA) announced the National Bioengineered Food Disclosure Standard on December 20, 2018. The National Bioengineered Food Disclosure Law, passed by Congress in July 2016, directed the USDA to establish this national mandatory standard for disclosing foods that are or may be bioengineered (BE). Bioengineered foods contain genetic material that has been modified through certain lab techniques and cannot be created through conventional breeding or found in nature (USDA 2018).

Recently, the Ministry of Agriculture and Rural Affairs of China approved five GM crops for importation, including RF3 canola (originally developed by Bayer, now owned by BASF), Monsanto's glyphosate-tolerant MON 88302 canola, DuPont Pioneer DP4114 corn, Syngenta's SYHT0H2 soybean and Dow AgroSciences' DAS-44406-6 soybean. It is worth mentioning that China had not released approvals for GM crops since July 2017 (ISAAA 2019a).

Researchers at the University of Washington developed GM houseplants, such as pothos ivy, to help clean the air inside homes (UW NEWS 2018). GM pothos ivy removes chloroform and benzene from the air. Both benzene and chloroform exposure have been linked to cancer (Zhang et al. 2019). Pothos ivy does not flower in temperate climates, so the GM plants will not spread via pollen.

In December 2018, Calyxt (a consumer-centric, food- and agriculture-focused company from the USA) received a patent from the European Patent Office (EPO) for the use of CRISPR/CAS9 for genome editing in plants (UW News 2018).

2.4 Legal Aspects of Using GM Plants in Agriculture

Development of bioeconomy requires the development of biotechnology. The latter is being fostered by a multitude of factors, including research financing from both the public and private sectors and a friendly legal environment that provides for biosafety on the one hand and provides legal certainty and development conditions on the other. An efficient system of intellectual property protection allows for return on investments and provides conditions for the transfer of technology and affordable access to said technology for the end user. The opportunities for public financing of biotechnology programmes have been provided in the united Europe since at least the 1980s (Patermann and Aguilar 2018). This decade was also marked with rapid changes in patent laws and attempts to draft European legislation aimed at securing biosafety and regulating conditions for using newly developed products of genetic engineering, which resulted in the adaptation of legislation at the beginning of the 1990s. Thirty years later, the prospects for development of some innovative technologies do not look too promising, especially in the field of agri-biotechnology.

Since their first commercial introduction in 1994 (Ricroch et al. 2014), GMOs are being widely adapted in agriculture worldwide. Both the global use of GM products and the overall acreage grow. This phenomenon does not seem to occur in the EU, in regard to the cultivation of GMOs, where the area seems to be on a decline and the number of countries where GMOs are grown has dropped (ISAAA 2018). This rather low adaptation rate for cultivation in the EU (as mentioned earlier, grown on approximately 135,000 ha, in two countries – Spain and Portugal only in 2017) is contrasted with a rather broad usage of GM products in other areas, in particular, for feed production. Estimates show that 30 million tonnes of GM soybean is imported into the EU on a yearly basis (approximately 85% of total soybean imported into the EU) (European Commission 2016b). While one of the reasons for a relatively low adaptation rate of GMOs in the EU may be a general scepticism of consumers towards such technology (Eurobarometer 2010), another might be connected with a rather strict policy towards the marketing of GMOs in the EU.

The use of GMOs is regulated by several acts of EU law, the most important being the ELAW (2001) and Regulation-EC (2003). The legislation on food and feed generally requires authorization of all GM products before they can enter the market. Said authorization requires performance of risk assessment of any product (be it for cultivation, food or feed purposes), a subsequent positive opinion of the European Food Safety Authority (EFSA) and, later, a decision of the European Commission. The legislation is based on a precautionary approach that mandates applying risk-limiting measures before taking a certain decision, if the scientific data as to the outcomes of such decision is incomplete or nonconclusive, while prima facie of such outcomes could be harmful. Hence, any GM product that is to enter the EU market has to be proven to be safe for consumption and the environment. This assessment is done on a case-by-case basis.

All food or feed products containing or produced from GMOs placed on the EU market have to be labelled as such. This requirement does not apply to products where GM ingredients are present proportion no higher than 0.9% of the food or feed ingredients that are considered individually or food or feed consisting of a single ingredient, provided that this presence is adventitious or technically unavoidable. A deliberate addition of a GM ingredient creates an obligation to label a given product, regardless of the content (Regulation-EC 2003, Art. 12 and 24).

Despite formally being based on scientific criteria for biosafety assessment, the authorization process is costly and lengthy due to delays in the process itself. It is estimated that an authorization for food or feed purposes takes approximately 5 years on average (Smart et al. 2017). An estimate for approval times of GMOs for cultivation purposes cannot be reliably performed, as there was only one transformation event approved in the EU so far, namely, the MON 810 maize, in 1998 (European Commission 1998). Nevertheless, approval of a GM variety for cultivation would also require field studies on the interactions between the planted GMOs and the environment, further raising the cost (EFSA Panel on Genetically Modified Organisms 2010). Regardless of the outcomes of such study, after the adaptation of Directive 2015/412/EU, EU member states are at liberty to introduce bans on the cultivation of GMOs, regardless of the safety assessment outcomes (Directive 2015). The aforementioned directive added Art. 26b to Directive 2001/18/EC, according to which member states can limit the territorial scope of an authorization or limit an already existing one as to their territory. The abovementioned factors consider the development of the GM variety for the EU market a rather risky one. Nevertheless, there are numerous GM products registered in the EU for purposes other than cultivation.

2.4.1 GM Products Registered as Food or Feed in the EU

According to the database of the International Service for the Acquisition of Agri-Biotech Applications (ISAAA), there are currently 507 transformation events registered worldwide for various purposes, including food, animal feed or cultivation. The species registered span from alfalfa or apples to wheat (ISAAA 2019b). As already mentioned above, so far, only MON 810 maize has been approved for cultivation in the EU. Briefly, the potato Amflora was also authorized for cultivation (in 2010); however, the decision on authorization was later annulled by the General Court (in 2013, case T-240/10), and the producer (BASF) decided to withdraw the product from the market. There are, however, numerous other GM products registered for the EU market.

Currently, over 60 transformation events are registered as food or feed for the European market. They belong to five plant species: cotton, maize, oilseed rape, soybean and sugar beet (EU GMO Registry 2019). Many of those events are also registered as uses other than food or feed, excluding cultivation. Although most of the aforementioned products are registered as food, GM food products are rather rare and essentially unavailable to consumers. Most of the products registered are used as animal feed, predominantly soybean and maize.

The labelling regulations do not require products such as meat or milk of animals fed with GMOs to be labelled as containing GMOs. However, since GM products are treated with reluctance by the consumers, some food producers attempt to take advantage of this reluctance by labelling products as "GMO free". Such labels are not regulated by EU law; however, in some cases, domestic law regulates issues connected with usage of such labels (e.g. Austria, Germany or France). In these cases, products such as meat or dairy are labelled as "GMO free", if GM feed was not used in production (European Commission 2015). This can, however, mean different things from country to country (e.g. GM feed was not used at all vs. it was not used for a certain time before slaughtering). Some countries (e.g. Poland) attempt to outright ban the use of GM feed on their territories; so far, such bans have not been emanated (Art. 15.1.4 and 65 of Polish *Act on Animal Feed*, OJ of RP 2017.453 with further amendments).

GM products or ingredients cannot be used in organic farming or production in the EU. This is expressly excluded by Art. 9 of the 834/2007/EC *Regulation on organic production and labelling of organic products*. When adhering to this prohibition, producers may rely on the labels accompanying a product or any other accompanying document, affixed or provided pursuant to Directive 2001/18/EC or Regulation 1829/2003/EC, including the labelling thresholds.

Increased regulatory pressure on a certain industry may effectively decline in its development in spite of this industry being capable of providing potentially profitable solutions in the field of bioeconomy (Hall et al. 2018). Strict regulations impose costs not only connected with authorization but also delays (Smyth et al. 2014), costs of separation of GM material from non-GM material or actions aimed at GMO detection. In case of some essential crops, such as soybean, the cost of non-modified alternative can be as much as € 200/t higher than that of the modified counterpart. This phenomenon depends on the species; however, the price premium is quite volatile (FEFAC 2015). Hence, the economic viability of products of modern agri-biotechnology may differ from case to case; yet, the additional costs of their marketing are often considerably higher than of those developed with traditional methods.

As follows from the information presented above, despite being extensively used as animal feed, GMOs are used only in a limited scope within the EU. Those limitations are connected not only with apparent unpopularity of such products but also, to a large part, with rather strict authorization requirements and inefficiency of the authorization process. Since it will often be economically unjustified to develop GM plants for the EU market, to question the definition of GMO or the scope of the GMO legislation becomes crucial (plants not considered to be GMOs are not subjected to most of the aforementioned limitations).

According to the definition contained in Art. 2 pt. 2 of the *Directive 2001/18/EC*, a GMO is an organism, with the exception of human beings, in which the genetic material has been altered in a way that does not occur naturally by mating and/or natural recombination (Art. 2 pt 2). Annex 1 A part 2 to the directive mentions techniques that do not lead to the creation of GMOs, while annex 1 B mentions techniques of genetic modification, whose products are not covered by the directive. One such methods is mutagenesis – a method that, in its random form (either radiation or chemical to induce random mutations in genetic materials), has been used successfully in plant breeding since the 1950s. The scope of this exclusion has become subject of dispute with the advent of some new plant breeding techniques (NBTs), in particular, genome editing techniques that are based on inducing precise, point mutations in the plant genome (e.g. utilizing site-directed nucleases – SDNs – and clustered regularly interspaced short palindromic repeats, CRISPR) (Custers 2017). If the mutagenesis exception applies to products of certain NBTs, such products would not be subjected to GMO legislation and hence enjoy easier access to the EU market. In a judgement from July 2018, the Court of Justice of the European Union ruled that annex 1 B to the directive "must be interpreted as meaning that only organisms obtained by means of techniques/methods of mutagenesis which have conventionally been used in a number of applications and have a long safety record are excluded from the scope of that directive", thus adding another layer of uncertainty to the question about whether certain new breeding techniques should be subjected to the restrictions that are applied to GMOs or not (Court of Justice 2018).

2.4.2 Intellectual Property Protection

Innovation in the field of agri-biotechnology can be protected in several ways but chiefly through plant variety protection and patent law. Currently, plant variety protection is regulated by 2100/94/EC *Regulation on community plant variety rights*, which creates a unified EU system of legal plant variety protection, and also by domestic laws of many member states. Protection of inventions is regulated at the EU level by 98/44/EC *Directive on the protection of biotechnological inventions*, which harmonized member states' laws and provides for patent protection of biological material and methods for their modification. The provisions of that directive have been implemented not only into the legal orders of the member states but also into the *Convention on the Grant of European Patents* (EPC). The provisions of European patent law expressly provide for patentability of biological material (from DNA sequences to whole living organisms) and methods for its modification, including microbiological processes, while expressly excluding the patentability of plant varieties and, essentially, biological processes. The scope of protection of biological material extends to all material in which the product is incorporated and in which the genetic information is contained and performs its function (Art. 9 of the Directive 98/44/EC).

A recent amendment to the implementing regulations to the EPC provided some clarifications as to the scope of the latter exclusion by stating that plants that also developed through essential biological processes are not patentable nor are the processes. Currently, after the US Supreme Court judgement in the *Myriad Genetics* case (JUSTIA 2013), where the court expressly excluded products of nature (e.g. DNA sequences) as patentable subject matter, there is a disparity between the US and EU patent law, where the patentability of such products is, under certain conditions, guaranteed by law. The practical effects of such disparity are yet to be known (Aboy et al. 2017). Both the EU patent law and the plant variety protection law allow for a farmer's privilege that allows farmers to save seed of the most important crop species for reuse in further years. Large-scale farmers are obliged to pay for such reuse; however, the fee is supposed to be lower than the original royalty (Art. 11 of the 98/44/EC Directive and Art. 14 of the 2100/94/EC Regulation).

The intellectual property protection provisions in the EU are, thus, rather inventor friendly and cannot be viewed as a major hindrance to the development of bioeconomy. It is rather easy to protect an innovation but, often, prohibitively difficult to market it later.

2.5 Perspectives on the Bioeconomy and Biotechnology in the European Union

According to the Bioeconomy Knowledge Centre (2016), the European bioeconomy generated an estimated turnover of approximately €2.2 trillion and employed 18.2 million people in 2015. The growing demand due to global population and income growth requires significant increases in food and feed production. Bioeconomy investments, especially in developing countries, can provide agricultural growth, energy security, economic development and employment (von Braun 2015).

The EC set a range of initiatives targeted to develop the bioeconomy. One of them is BIOEAST – The Central-Eastern European Initiative for Knowledge-based Agriculture, Aquaculture and Forestry in the Bioeconomy (BIOEAST 2018). Activity for this initiative was started by Visegrad Group countries (Czech Republic, Hungary, Poland and Slovakia) and joined by Bulgaria, Croatia, Latvia, Lithuania, Republic of Estonia, Romania and Slovenia. The aim of this initiative is to work towards sustainable bioeconomies in Central and Eastern European countries. The development of the bioeconomy at the macro-regional level is essential to achieve sustainability. It is worth mentioning that this initiative is agreeable with Action 3.2.3., *Set up an EU Bioeconomy policy support facility and a European Bioeconomy Forum for Member States* (European Commission 2018c).

Another initiative is EBCL (2016). It promotes the concept of developments of bioeconomy in local bio-communities (bio-villages, bio-cities, bio-regions). A declaration was provided by Central and Eastern European regions and stakeholders from companies, academia, non-government organizations and farmers at the European Bioeconomy Congress in Lodz in 2016. The declaration is a strategic document for developments in the Central and Eastern European Bioregions Forum.

The International Bioeconomy Forum (IBF) was announced by Commissioner Moedas in 2016; however, it was officially launched at the end of 2017 in Brussels with a first plenary meeting. The meeting was co-chaired by the EC and Agri-Food Canada. Nine countries from five continents participated. Canada, the USA, Argentina, South Africa, India, China and New Zealand joined the IBF as members. Australia and South Korea joined as observers (European Commission 2018c).

As noted by Aguilar et al. (2018), for the development of bioeconomy and innovation, research needs to be supported long term and stimulated by all stakeholders in science, industry and government. Authors have indicated that the foundation and future of bioeconomy rely on strong science and technology, especially for new advancements.

2.6 Conclusion

Development of the bioeconomy requires innovativeness in the field of biotechnology. Living organisms, in particular plants, are used as more than food; therefore, new technologies that would allow for harnessing the full potential of crops need to be developed. The use of many of such technologies in the EU is strongly limited, although some sort of favourability towards their development is being declared by the authorities. The strongest limitations, in particular, occur in the field of agri-biotechnology and are connected with legal obstacles for marketing and further sale of such technologies as effects of genetic engineering. Legal uncertainty, delays and often large costs connected with compliance with existing regulations create a high-entry threshold that is particularly difficult to overcome for small entrepreneurs. Recent questions as to the status of new breeding techniques further worsen the situation, despite there being a favourable climate for the protection of innovations in the described field. This leads to a confusing situation, whereby the protection of certain innovations is available, yet the conditions for commercialization of such innovations are difficult to meet, if at all possible.

Plant production is critically important for food (and feed) production. To guarantee food security (as well as security within production of medicines, materials and many other goods), we need development and broad use of innovative technologies, particularly agri-biotechnology. Unjustified regulation and people lay opinions cannot block the progress.

As reported by winners of the Nobel Prize for Chemistry, Professor Frances Arnold and Sir Gregory Winter, concerns about GM foods are preventing society from benefiting from technology (Falzone 2018). Genetic engineering has a potential to create economically important crop varieties. Taking into account the concerns related to GM crops, decisions must be based on reliable science-based information. What is more, people have to be educated, they have the right and privilege of full information about products they buy.

Acknowledgement This work was supported by grants from the National Science Centre, Poland (nos. 2012/06/A/NZ9/00125 and 2014/15/B/NZ9/02312), and by funding from the ERANET-CORNET #22/87/2017 research project named "Innovative processing technology of rapeseed products for poultry nutrition" and was co-funded by the National Centre for Research and Development and the Ministry of Science and Higher Education of the Republic of Poland by the KNOW programme.

References

Aboy M, Liddicoat J, Liddell K, Jordan M, Crespo C (2017) After Myriad, what makes a gene patent claim "markedly different" from nature? Nat Biotechnol 35:820–825

Aguilar A, Wohlgemuth R, Twardowski T (2018) Perspectives on bioeconomy. New Biotechnol 40:181–184

BIOEAST (2018) Central and Eastern European initiative for knowledge-based agriculture, aquaculture and forestry in the bioeconomy. http://www.bioeast.com. Accessed 5 Jan 2019

Bioeconomy Knowledge Centre (2016) Infographic on jobs and growth of the bioeconomy 2015. https://ec.europa.eu/knowledge4policy/sites/know4pol/files/2018_09_01_bioeconomi_infographic_update_en.pdf. Accessed 20 Dec 2018

BiomassWeb (2018) Improving food security in Africa through increased system productivity of biomass-based value webs. https://biomassweb.org/. Accessed 1 Jan 2019

Biooekonomierat (2018) International bioeconomy strategies. http://biooekonomierat.de/en/international/. Accessed 1 Feb 2019

Castellari E, Soregaroli C, Venus TJ, Wesseler J (2018) Food processor and retailer non-GMO standards in the US and EU and the driving role of regulations. Food Policy 78:26–37

Court of Justice of the European Union (2018) Judgment in case C-528/16 confédération paysanne and Others v Premier minister and Ministre de l'Agriculture, de l'Agroalimentaire et de la Forêt. Court of Justice of the European Union

Custers R (2017) The regulatory status of gene-edited agricultural products in the EU and beyond. Emerg Top Life Sci 1:221–229

Directive 2015/412 of the European Parliament and of the Council of 11 March 2015 amending Directive 2001/18/EC as regards the possibility for the Member States to restrict or prohibit the cultivation of genetically modified organisms (GMOs) in their territory

EBCL (2016) Lodz declaration of bioregions. http://scanbalt.org/wp-content/uploads/2016/11/LODZ-BIOCOMMUNITIES-DECLARATIONfinalEN6.pdf. Accessed 01 Jan 2019

EFSA Panel on Genetically Modified Organisms (2010) Guidance on the environmental risk assessment of genetically modified plants. EFSA J 8:1879

ELAW (2001) Directive 2001/18/EC of the European Parliament and of the Council of 12 March 2001 on the deliberate release into the environment of genetically modified organisms and repealing Council Directive 90/220/EEC. ELAW, Oregon

Ellen MacArthur Foundation (2012) Towards the circular economy. EMAF, Cowes

EU GMO Registry (2019) Genetically modified organisms. https://webgate.ec.europa.eu/dyna/gm_register/index_en.cfm. Accessed 14 Jan 2019

Eurobarometer (2010) Biotechnology: special eurobarometer 341. TNS Opinion & Social, Brussells

European Commission (1998) MON-00810-6 EU register of genetically modified food and feed. http://ec.europa.eu/food/dyna/gm_register/gm_register_auth.cfm?pr_id=11. Accessed 16 Sept 2018

European Commission (2005) New perspectives on the knowledge-based bio-economy. European Commission, Brussels

European Commission (2010) The knowledge based bio-economy (KBBE) in Europe: achievements and challenges. European Commission, Brussels

European Commission (2012) Innovating for sustainable growth: a bioeconomy for Europe. European Commission, Brussels

European Commission (2015) Closing the loop: an EU action plan for the Circular Economy. European Commission, Brussels

European Commission (2016a) Closing the loop: new circular economy package. European Parliamentary Research Service, Brussels

European Commission (2016b) Genetically modified commodities in the EU. European Commission, Brussels

European Commission (2018a) The commission's plan for plastics. Plastic in a circular economy. https://ec.europa.eu/info/research-and-innovation/research-area/environment/plastics-circular-economy_en. Accessed 1 Feb 2019

European Commission (2018b) Directorate-general for research and innovation. Food 2030. European Commission, Brussels

European Commission (2018c) The international bioeconomy forum. https://ec.europa.eu/research/bioeconomy/index.cfm?pg=policy&lib=ibf. Accessed 19 Dec 2018

European Commission (2018d) A sustainable bioeconomy for Europe: strengthening the connection between economy, society and the environment. European Commission, Brussels

European Commission Database (2018) A new bioeconomy strategy for a sustainable Europe. http://europa.eu/rapid/press-release_IP-18-6067_en.htm. Accessed 12 Dec 2018
European Union (2007) En route to the knowledge-based bio-economy KBBE. https://dechema.de/dechema_media/Cologne_Paper-p-20000945.pdf. Accessed 13 Dec 2018
European Union (2017) Latvian bioeconomy strategy 2030. http://www.llu.lv/sites/default/files/2018-07/Latvian-Bioeconomy-Strategy-Summary-WEB_0.pdf. Accessed 19 Dec 2018
Fabbri K (2017) Food 2030. EU research and innovation for tomorrow's nutrition and food systems. https://ec.europa.eu/health/sites/health/files/nutrition_physical_activity/docs/ev_20171129_co04_en.pdf. Accessed 2 Jan 2019
Falzone J (2018) Nobel prize-winning chemists: misguided GMO fears could hinder technology's societal benefits. http://www.biotech-now.org/food-and-agriculture/2018/12/nobel-prize-winning-chemists-misguided-gmo-fears-could-hinder-technologys-societal-benefits. Accessed 1 Jan 2019
FAO (2003) Trade reforms and food security, conceptualizing the linkages. http://www.fao.org/3/a-y4671e.pdf. Accessed 1 Jan 2019
FAO (2009) The state of food insecurity in the world 2009. FAO, Rome
FEFAC (2015) Economic impact assessment on the European GM authorisation "opt-out" proposal. FEFAC, Brussels
German Bioeconomy Council (2015) Bioeconomy policy (Part II): synopsis of national strategies around the world. Office of the German Bioeconomy Council, Berlin
Haarich S (2017) Bioeconomy development in EU regions: mapping of EU Member States/regions' research and innovation plans and strategies for smart specialisation (RIS3) on bioeconomy, Final Report, Framework Contract: 2014.CE.16.BAT Lot 2. Specific Contract: RTD/F1/PP-03681-2015. European Commission, Brussels
Hall J, Matos S, Gold S, Severino LS (2018) The paradox of sustainable innovation: the 'Eroom' effect (Moore's law backwards). J Clean Prod 172:3487–3497
Hilgartner S (2007) Making the bioeconomy measurable: politics of an emerging anticipatory machinery. BioSocieties 2:382–386
ISAAA (2018) Global status of commercialized biotech/GM crops in 2017: biotech crop adoption surges as economic benefits accumulate in 22 years, ISAAA Brief No. 53. ISAAA, Ithaca
ISAAA (2019a) Chinese agri ministry approves import of 5 GM crops. http://www.isaaa.org/kc/cropbiotechupdate/article/default.asp?ID=17146. Accessed 10 Jan 2019
ISAAA (2019b) GM approval database. http://www.isaaa.org/gmapprovaldatabase/advsearch/default.asp?CropID=Any&TraitTypeID=Any&DeveloperID=Any&CountryID=Any&ApprovalTypeID=Any. Accessed 14 Jan 2019
JUSTIA (2013) Association for molecular pathology v. Myriad Genetics, Inc., 569 U.S. 576. https://supreme.justia.com/cases/federal/us/569/576/. Accessed 15 Jan 2019
Korhonen J, Honkasalo A, Seppälä J (2018) Circular economy: the concept and its limitations. Ecol Econ 143:37–46
Krön M, Bittner U (2015) Danube soya – improving European GM-free soya supply for food and feed. OCL 22:D509
Macahilo MM (2018) GMO labeling laws per country. https://globalfoodsafetyresource.com/gmo-labeling-laws/. Accessed 1 Jan 2019
Maciejczak M, Hofreiter K (2013) How to define bioeconomy? Ann Pol Assoc Agric Agribus Econ 15:243–248
Martinez J (1998) Genomics and the world's economy. Science 281:925–926
McCormick K, Kautto N (2013) The bioeconomy in Europe: an overview. Sustainability 5:2589–2608
OECD (2009) The bioeconomy to 2030: designing a policy agenda. https://www.oecd-ilibrary.org/economics/the-bioeconomy-to-2030_9789264056886-en. Accessed 20 Dec 2018
Osmakova A, Kirpichnikov M, Popov V (2018) Recent biotechnology developments and trends in the Russian Federation. New Biotechnol 40:76–81
Paarlberg D, Paarlberg P (2008) The agricultural revolution of the 20th century. Iowa State University Press, Ames

Patermann C, Aguilar A (2018) The origins of the bioeconomy in the European Union. New Biotechnol 40:20–24
Patermann C, Slusarczyk H (2009) The knowledge-based bio-economy – from concept to practice: experiences in Germany and Europe. http://www.owwz.de/fileadmin/Biotechnologie/BioVeranst/Novosibirsk_2009/Patermann.pdf. Accessed 20 Dec 2018
Paull J (2018) Genetically modified organisms (GMOs) as invasive species. J Environ Prot Sustain Dev 4:31–37
Plastic Europe (2016) Association of plastics manufacturers. www.plasticseurope.org.eu. Accessed 27 Dec 2018
Potrykus I (2017) The GMO-crop potential for more, and more nutritious food is blocked by unjustified regulation. J Innov Knowl 2:90–96
Regulation-EC (2003) No 1829/2003 of the European Parliament and of the Council of 22 september 2003 on genetically modified food and feed (Text with EEA relevance)
Ricroch A, Chopra S, Fleischer SJ (2014) Plant biotechnology. Springer, Cham
Schmidt O, Padel S, Levidow L (2012) The bio-economy concept and knowledge base in a public goods and farmer perspective. Bio-based Appl Econ 1:47–63
Sherrington C, Darrah C, Hann S, Cole G, Corbin M (2016) Study to support the development of measures to combat a range of marine litter sources. Report for European Commission DG Environment. http://ec.europa.eu/environment/marine/good-environmental-status/descriptor-10/pdf/MSFD%20Measures%20to%20Combat%20Marine%20Litter.pdf. Accessed 5 Dec 2018
Singh HB, Jha A, Keswani C (eds) (2016a) Intellectual property issues in biotechnology. CABI, Wallingford. 304 pages. ISBN:9781780646534
Singh HB, Jha A, Keswani C (2016b) Biotechnology in agriculture, medicine and industry: an overview. In: Singh HB, Jha A, Keswani C (eds) Intellectual property issues in biotechnology. CABI, Wallingford, pp 1–4
Singh HB, Keswani C, Singh SP (2019) Intellectual property issues in microbiology. Springer, Singapore. 425 pages. ISBN:9789811374654
Smart RD, Blum M, Wesseler J (2017) Trends in approval times for genetically engineered crops in the United States and the European Union. J Agric Econ 68:182–198
Smyth SJ, McDonald J, Falck-Zepeda J (2014) Investment, regulation, and uncertainty: managing new plant breeding techniques. GM Crops Food 5:44–57
The Non-GMO Project (2018) Non-GMO project standard. https://www.nongmoproject.org/product-verification/the-standard-2-2/. Accessed 3 Jan 2019
Tyczewska A, Woźniak E, Gracz J, Kuczyński J, Twardowski T (2018) Towards food security: current state and future prospects of agrobiotechnology. Trends Biotechnol 36:1219–1229
USDA (2018) BE disclosure. https://www.ams.usda.gov/rules-regulations/be. Accessed 10 Jan 2019
UW News (2018) Researchers develop a new houseplant that can clean your home's air. https://www.washington.edu/news/2018/12/19/new-houseplant-can-clean-air/?utm_source=UW%20News&utm_medium=tile&utm_campaign=UW%20NEWS. Accessed 27 Dec 2018
von Braun J (2015) Bioeconomy – science and technology policy to harmonize biologization of economies with food security. In: Sahn D (ed) The fight against hunger and malnutrition. Oxford University Press, Oxford, pp 240–262
von Braun J (2018) Bioeconomy – the global trend and its implications for sustainability and food security. Glob Food Secur 19:81–83
Wreford A, Bayne K, Edwards P, Renwick A (2018) Enabling a transformation to a bioeconomy in New Zealand. Environ Innov Soc Transit. https://doi.org/10.1016/j.eist.2018.11.005
Zhang L, Routsong R, Strand SE (2019) Greatly enhanced removal of volatile organic carcinogens by a genetically modified houseplant, pothos ivy (*Epipremnum aureum*) expressing the mammalian cytochrome P450 2e1 gene. Environ Sci Technol 53:325–331
Zilberman D, Holland GT, Trilnick I (2018) Agricultural GMOs—what we know and where scientists disagree. Sustainability 10:1514

Agricultural Biotechnology in the Philippines: Prospects and Challenges

3

Marilyn B. Brown, Cristine B. Brown, and Robert A. Nepomuceno

Abstract

As early as 1979, the Philippine Government envisioned that biotechnology would play a pivotal role in solving the long-standing problems on agricultural sustainability, incurable diseases, and climate change. With the burgeoning human population on an exponential rise, there is a dire need for a new paradigm to encourage agricultural sustainability. Biotechnology can be adopted and has been identified as one of the solutions to address these growing concerns. As an important tool, it has wide applications in agri-industry as it offers opportunities and promising economic potential for enhancing productivity and developing competitive products. There are groups in the country which abhor biotechnology, particularly the use of GMOs. They are very visible and articulate in various forums intended to discuss issues on biotechnology outputs. Support to research and development on biotechnology would suffer if their effort succeeded in urging the government to make a misinformed decision on policies concerning biotechnology.

Keywords

Biotechnology · Sustainable development · Bioeconomy · GMOs

3.1 Introduction

Biotechnology, in essence, is the use of living organisms or parts thereof in technologies, particularly with an industrial, agricultural, or medical application. The very first recorded use of biotechnology is in cheese production (Verma et al. 2011).

M. B. Brown (✉) · C. B. Brown · R. A. Nepomuceno
National Institute of Molecular Biology and Biotechnology, University of the Philippines, Los Baños, Philippines

C. Keswani (ed.), *Bioeconomy for Sustainable Development*,
https://doi.org/10.1007/978-981-13-9431-7_3

Back then, the stomach of ruminants is used for transporting liquids such as milk. As rennet is present in the stomach sack, milk curding will inevitably occur. Evidence of milk production is evident on what resembles to be cheese strainer of the Urnfield culture situated near Lake Neuchâtel dated 6000 BCE (Toussaint-Samat 2008). Cheese is considered the first product of biotechnology as its production made use of rennet which is found in ruminant's stomach. In the field of medicine, the ancient Chinese in 500 BCE made use of moldy soybean curds as a primitive antibiotic as a treatment for boils (Hughesman 2004). It took more than two millennia for humanity to take advantage of antibiotics where it was the first mass-produced and had great impact in saving the lives of the injured during the Second World War (Quinn 2013). The widespread use of antibiotic contributed greatly to the longevity and quality of life.

Contemporary use of biotechnology makes use of the examination of the DNA. The use of which has revolutionized several fields of study such as in the field of forensics. Seemingly minute biological evidence would play a pivotal role in crime investigation. Not only is the methodology sensitive enough to make use of the smallest of biological samples, but the results are with almost absolute certainty. The advent of DNA technology in forensics has led to solving cold cases which remained unsolved for more than 20 years (Pillay et al. 2007; Singh et al. 2016a, b).

Cutting-edge biotechnology now makes use of manipulation of DNA to tailor specific characteristics of an organism that suit one's needs. In agriculture, one can simply edit a gene to increase its shelf life such as in the case of CRISPR-edited mushroom (Gill 2016). The good thing about this methodology is that initial ruling does not consider CRISPR-edited organisms as genetically modified which means that it is not subject to the stringent policies surrounding the use of GMOs (Waltz 2016).

As early as 1979, the Philippine Government has envisioned that biotechnology would play a pivotal role in solving our long-standing problems on agricultural sustainability, incurable diseases, and climate change. With the burgeoning human population on an exponential rise, there is a dire need for a new paradigm to solve agricultural sustainability. Biotechnology can be adopted and has been identified as one of the solutions to address these growing concerns. As an important tool, it has wide applications in agri-industry as it offers opportunities and promising economic potential for enhancing productivity and developing competitive products. Hence, in 1979, the National Institute of Biotechnology and Applied Microbiology at UP Los Baños (BIOTECH-UPLB) the first center for biotechnology in the ASEAN region was established. Currently, it is known as the National Institute of Molecular Biology and Biotechnology at UP Los Baños (BIOTECH-UPLB) (Fig. 3.1).

The institute is regarded as a premier research and development institution for basic and applied researches on molecular biology and biotechnology mission and is mandated "to develop cost-effective and environment-friendly technologies for the production of goods and services that are comparable or better alternatives to conventional products for their use in the following sectors: agriculture, forestry, environment, energy and industry sectors." To date, BIOTECH has developed 38 cost-effective products such as biofertilizers, microbial-based enzymes, animal

Fig. 3.1 The National Institute of Molecular Biology and Biotechnology – University of the Philippines, Los Baños (BIOTECH-UPLB)

probiotics, animal vaccines, microbial pesticides, and various diagnostic kits and provided services that contribute significantly to (a) instruction, (b) training/extension, (c) research, (d) resource generation, and public service programs for the university and national development.

Three other branches of biotechnology were established within the University of the Philippines system such as the National Institute of Molecular Biology and Biotechnology, Diliman (NIMBB-Diliman) which focuses on industrial biotechnology; the National Institute of Molecular Biology and Biotechnology, UP Visayas (BIOTECH-UP Visayas) which focuses on marine biotechnology; and the National Institute of Health-Institute of Molecular Biology and Biotechnology (BIOTECH-UP Manila) which focuses on human health biotechnology.

There are also other institutions in UP Los Baños undertaking research using biotechnology such as the Institute of Plant Breeding where it spearheads the research on the breeding of important agricultural crops, preservation of crop germplasm, and extending the technological breakthroughs of the institution through extension activities. Also, the Institute of Biological Sciences (IBS), Institute of Animal Sciences (IAS), and College of Forestry and Natural Resources (CFNR) also conduct research on biotechnology. Moreover, outside of the UP system are the Philippine Rice Research Institute (PhilRice) and the International Rice Research Institute (IRRI). Newly established centers include the Philippine Genome Center which is the product of a collaborative project between the University of the Philippines (UP) and Department of Science and Technology (DOST). It is comprised of several facilities and research laboratories including DNA Sequencing Core Facility, Core Facility for Bioinformatics, Biobank Facility (BF), and Agriculture/Livestock/Fisheries aiming to alleviate technological status of the agricultural system, forensics, and disease prevention and management (Luminoan 2014).

3.2 Prospects of Biotechnology in the Philippines

3.2.1 Crop Biotechnology

Biotechnology is still widely regarded in the world as the flagship of research and development in agriculture, natural resources, health, and other fields. We all know the miracles brought by the green revolution in the 1970s. However, the productivity improvement brought by the green revolution, characterized by improved varieties heavily dependent on high fertilizer and pesticide use, has reached a plateau. There is an ever-decreasing gain from traditional breeding in terms of yield. In fact, an overall decline in productivity has already been noted. It is therefore widely regarded that the second wave of green revolution may only be possible only through biotechnology. IRRI, for instance, has spearheaded a global collaborative effort to improve the photosynthetic efficiency of rice through the C_4 Rice Project. The project aims to convert the photosynthetic machinery of rice from C_3 to C_4. The effort will be conducted through engineering and expressing the genes responsible for C_4 photosynthesis into rice.

3.2.2 Bioprospecting

The Philippines is considered one of the mega biodiversity hotspots in the world, harboring more than the usual number of species and endemic species relative to other areas; this translates to a high number of novel species and thus represents a rich genetic resource. Novel genes, enzymes, and drugs can be harnessed from microorganisms, plants, and animals. This is both a prospect and a challenge; it is imperative to conserve and protect this unique resource for posterity's sake, and at the same time we are compelled to utilize the resource for economic benefit or from the sheer necessity to discover a more potent alternative to existing products. However, the need to preserve shouldn't prevent the utilization of the resource to its full potential. Convoluted and overly protective ordinances are counterproductive; a balance should be achieved. Moreover, a collaboration between other countries should be promoted, and bureaucratic red tapes should be dismantled. For instance, in joint bioprospecting endeavors, it is required that the prospecting of biological and genetic resources shall be only allowed when the national institution from a foreign country, on recommendation of the Inter-Agency Committee on Biological and Genetic Resources, has entered into a Research Agreement with the Philippine Government, represented by the Department of Environment and Natural Resources (DENR), Department of Health (DOH), Department of Agriculture (DA), or Department of Science and Technology (DOST). Given the bureaucratic inefficiencies, foreign researchers would just give up trying to go through the process; it is simply not worth the time.

3.2.3 Rapid Diagnostics

Rapid diagnostic test is imperative in almost all fields, from medical applications to veterinary and agricultural. There is a need to know what the disease or the affliction is at the soonest so that doctors or even farmers can take immediate action and prevent further damage. Sending samples for laboratory analysis sometimes is not an option; time is of the essence. Moreover, the archipelagic nature of the Philippines makes inaccessibility to laboratory services a significant constraint in extending services to those in need. The development of rapid detection kits using biotechnology is a major prospect in extending the benefits of biotechnology to the general public. NIMBB-Diliman has developed a highly sensitive dengue detection kit facilitating a faster response to patients that require immediate treatment for the disease. BIOTECH-UPLB, on the other hand, is on the process of developing a rapid diagnostic kit for mango anthracnose. Since the disease is latent, development of black lesions in mango fruits occur late, and the black blemish severely affects the marketability of the fruits especially if the mangoes are considered for export.

3.2.4 Vaccines

Traditional vaccinology relies on biochemical, microbiological, and immunological methods to identify an antigen of interest; but in most cases these methods fail to deliver an antigen which could be used for vaccines. The use of biotechnology specifically through the screening of potential antigens extracted from the genomic data offers a comprehensive means to identify the most effective antigen for vaccine development. The method even offers an understanding of the intricacies of the pathogenesis of bacteria and has proved successful in identifying the antigen for the vaccine against *Neisseria meningitis* serogroup B (Serruto et al. 2004).

3.2.5 Microbial Inoculants

BIOTECH-UPLB is the leading institution in terms of applied microbiology in agriculture forestry, industrial, and environmental biotechnology. It is mainly focused on the development of microbial inoculants for the increased absorption of macronutrients such as vesicular-arbuscular mycorrhizal root inoculant (Brown et al. 2016), micronutrients such as zinc solubilizing inoculant for rice and corn (Nepomuceno et al. 2018), and minerals in the soil (Fig. 3.2). The microbial inoculants aim to reduce or even prevent the wide usage of inorganic fertilizers since the improper application of it tends to degrade soil health and fertility rendering other nutrients inaccessible to plant absorption (Singh et al. 2016c, 2017, 2019).

Most of the inoculants are mycorrhiza-based (Fig. 3.2e, h), plant growth-promoting microorganisms (Fig. 3.2a–c), and bioorganic fertilizers (Fig. 3.2d)

Fig. 3.2 Some microbial inoculant products of the National Institute of Molecular Biology and Biotechnology – University of the Philippines Los Baños (BIOTECH)

3.2.6 ASEAN Economic Integration

The opening of the trade boundaries offers attractive opportunities for the commercialization of BIOTECH products and processes, but it may also bring threats and increase the risk of doing commercialization engagements, specifically on the entry of competing products in the country.

3.2.7 Funding

There is perhaps no other research discipline that could match biotechnology in terms of private sector interest. Local and multinational corporations in agriculture are into biotechnology. Relative to other disciplines, it is much easier to tap the resources of the private sector to further the interest biotechnology. We do not see any conflict with government-private sector collaboration in BIOTECH R&D, provided we safeguard the interest of the general public in process.

3.3 Challenges

3.3.1 Misinformation Campaign on Biotechnology

There are groups in the country which abhor biotechnology, particularly the use of GMOs. They are very visible and articulate in various fora intended to discuss issues on biotechnology outputs. There are well-intent oppositions, but there are also "mercenary" advocacy groups whose only purpose is to distort the facts. We have to be on close guard against these groups lest they would be successful in putting biotechnology in bad light. Support to research and development on biotechnology would suffer if their effort succeeded in urging the government to make a misinformed decision on policies concerning biotechnology.

3.3.2 Brain Drain

There is an increasing number of well-trained staff seeking for a greener pasture; a high percentage of staff that were allowed for graduate studies and training in a prestigious university can be easily recruited to institutions abroad. There are a number of reasons for this: (a) Loss of plantilla items of retired staff resulting in the lack of items for young staff for mentoring or for staff who have been mentored for a succession plan; (b) yet there is an increasing number of retiring staff and lack of items for young staff for mentoring; (c) manpower concerns such as slow/stagnant promotion process for research personnel and underutilization of expertise; and (d) increasing number of returning scholars for Ph.D. and master's degrees without assistants and aides.

3.3.3 Stringent Regulatory Policies Surrounding the Use of GMOs

Mendoza et al. (2008) characterized the regulatory body for biotechnological products in the Philippines as overly strict and inefficient in the processing of the applications. Also, the process is too cost-prohibitive and often delayed directly affecting the adoption of technology. It is imperative that a revision is made to the guidelines taking into consideration the results of the studies supporting the safety of the products accrued for more than 30 years of field testing.

3.3.4 Red Tapes in Procurement Process in the Institutions Under the Government

One of the most important resources in virtually any endeavor is time. This is true, especially in research. Discoveries are made by the day, and the credit goes to who published it first. Streamlined acquisition of supplies is imperative to maintain global competitiveness. Due to the convolutedness of the bureaucracy surrounding the procurement process in institutions under the Philippine Government, a simple purchase of reagents, chemicals, and even equipment can take up to 2 years. The Government and the private sector can invest hundreds of millions of dollars in infrastructure, human resource development, and equipment, but if getting a supply takes years to do, then every bit of it is a waste; a chain is no stronger than its weakest link.

3.4 Threats and Opportunities

The agricultural and industrial sectors are now confronted by challenges arising from low productivity, climate change, natural resource constraints, the need to provide safe and high-quality food for the growing population, the need to look for an alternative energy source due to fossil fuel depletion, biodiversity conservation, and ASEAN economic integration and sustainable development.

3.4.1 Climate Change

Agriculture is particularly vulnerable to the adverse effect of climate change. The agricultural sector has inevitably been affected specifically in the production of crops, livestock, and fisheries. Breeding of climate-resilient crops has been conducted such as the case in submarine rice developed by IRRI. The submarine rice can tolerate prolonged submerged conditions without affecting its viability but only if the flooding occurs before flowering.

3.4.2 Food Security

There is thus a need to utilize biotechnology, for a more efficient, productive, and sustainable food production. Also, the use of biotechnology in prolonging the shelf life of agricultural products is imperative.

3.4.3 Food and Feed Safety

Chemical residues resulting from the inappropriate application of chemical pesticides, chemical fertilizers, animal drugs, or food additives, as well as industrial wastes, may also affect food/feed safety.

3.4.4 Sustainable Agricultural Productivity

The yield ceiling for most of our staple crops has been reached; the gains from traditional plant breeding is in a plateau. The use of crops developed through biotechnology and application of synergistic microorganisms are not mutually exclusive and thus should be explored.

3.4.5 Low Rate of Adoption of Technologies

Overly strict guidelines in testing and registering products developed through biotechnology should be revised. It is human nature to fear the unknown; thus, campaign to educate the people on the matter should be taken seriously.

3.4.6 Energy Shortage

The world's supply of fossil is significantly depleting over time, and it is clear that we have to find alternative sources to avoid energy shortages in the future.

References

Brown MB, Brown CMB, Nepomuceno R (2016) Regulatory requirements and registration of biopesticides in the Philippines. In: Singh H, Sarma B, Keswani C (eds) Agriculturally important microorganisms: commercialization and regulatory requirements in Asia. Springer, Singapore

Gill C (2016) A CRISPR mushroom. Penn State Ag Sci Mag. https://agsci.psu.edu/magazine/articles/2016/fall-winter/a-crispr-mushroom. Accessed Jan 2019

Hughesman C (2004) The business of biotechnology. Sci Creat Q. https://www.scq.ubc.ca/the-business-of-biotechnology/. Accessed Jan 2019

Luminoan ML (2014) DOST-supported genomics research hub to rise in up Diliman. Retrieved from: http://dost.gov.ph/knowledge-resources/news/34-2014-news/68-dost-supported-genomics-research-hub-to-rise-in-up-diliman.html

Mendoza EMT, Laurena AC, Botella JR (2008) Recent advances in the development of transgenic papaya technology. Biotechnol Annu Rev 14:423–462

Nepomuceno RA, Gargarino AMP, Elca LJG, Brown CMB, Pedro MS, Brown MB (2018) Isolation and characterization of zinc solubilizing fungi derived from the rhizospheric soil of rice, corn, and cassava grown in zinc-deficient areas of the Philippines. Mycological Society of the Philippines, Inc. "20th Annual Scientific Meeting and Symposium". Institute of Biological Sciences, University of the Philippines Los Baños

Pillay G, Menezes RG, Krishnaprasad R, Pillay M (2007) Biotechnology in forensic science: the revolution continues. Nepal Med Coll J: NMCJ 9:57–62

Quinn R (2013 March) Rethinking antibiotic research and development: world war II and the penicillin collaborative. Am J Public Health 103(3): 426–434

Serruto D, Adu-Bobie J, Capecchi B, Rappuoli R, Pizza M, Masignani V (2004) Biotechnology and vaccines: application of functional genomics to Neisseria meningitidis and other bacterial pathogens. J Biotechnol 113(1–3):15–32

Singh HB, Jha A, Keswani C (eds) (2016a) Intellectual property issues in biotechnology. CABI, Wallingford. 304 pages, ISBN-13:9781780646534

Singh HB, Jha A, Keswani C (2016b) Biotechnology in agriculture, medicine and industry: an overview. In: Singh HB, Jha A, Keswani C (eds) Intellectual property issues in biotechnology. CABI, Wallingford, pp 1–4

Singh HB, Sarma BK, Keswani C (eds) (2016c) Agriculturally important microorganisms: commercialization and regulatory requirements in Asia. Springer, Singapore. 336 pages, ISBN-13:978-9811025754

Singh HB, Sarma BK, Keswani C (2017) Advances in PGPR research. CABI, Wallingford. 408 pages, ISBN:9781786390325

Singh HB, Keswani C, Singh SP (eds) (2019) Intellectual property issues in microbiology. Springer-Nature, Singapore. 425 pages, ISBN:9789811374654

Toussaint-Samat (2008) A history of food, 2nd, New and Expanded Edition. ISBN:978-1-405-18119-8

Verma AS, Agrahari S, Rastogi S, Singh A (2011) Biotechnology in the realm of history. J Pharm Bioallied Sci 3(3):321–323. https://doi.org/10.4103/0975-7406.84430

Waltz E (2016) Gene-edited CRISPR mushroom escapes US regulation. Nat News 532:293. https://www.nature.com/news/gene-edited-crispr-mushroom-escapes-us-regulation-1.19754. Accessed Jan 2019

Biological Control as a Tool for Sustainable Development: For Increase the Distribution and Income Generation

4

Gabriel Olivo Locatelli

Abstract

With a world population of more than 7.5 billion people, today we are able to produce enough food to support the entire population of the planet. However, in the next few years, the population will continue to grow and the challenges are even greater, limitation in expanding new agricultural frontiers, besides the demand of conscious consumers who seek healthy sustainable produced food. In this scenario, the biological control seems to be one solution and one opportunity to reduce the dependence of the oligopoly on agrochemicals. Based on the strategic innovation model of the triple helix, the decentralization of investments in research and innovation, with the approach of universities and research institutes, of the rural producers, fomented by the government, boosts the search of solutions of biological control, supporting the regional development, with the increase the income generation and socioeconomic contributions in diverse countries. This way, we will be able to break the circle of poison.

Keywords

Agricultural production · Bioeconomics · Biological control · Sustainability · Triple helix

G. O. Locatelli (✉)
CAV – Academic Centre of Vitória. Rua Alto do Reservatório, UFPE – Federal University of Pernambuco, Vitória de Santo Antão, PE, Brazil

UNIBRA – Centro Universitário Brasileiro, Recife, PE, Brazil
e-mail: gabriel.locatelli@ufpe.br

C. Keswani (ed.), *Bioeconomy for Sustainable Development*,
https://doi.org/10.1007/978-981-13-9431-7_4

4.1 Introduction

The agriculture as it exists today would not exist without the use of pesticides and chemical fertilizers. With the green revolution that happened after World War II, new technologies and scientific advances allowed today the production of enough food to support the growing population of the planet (FAO 2017a). However, the population will continue to grow, reaching 9.8 billion by 2050, being able to pass the 11 billion in 2100, before it begins to decline (ONU 2013; FAO 2017b). Furthermore, one-third of the food produced in the world is still lost due to the attack of pests and diseases. Allied with the waste and bad distribution of income, they are responsible for chronic hunger and malnutrition that reaches about 10% of the population (one in nine), the vast majority of whom live in developing countries (FAO et al. 2015; OECD/FAO/UNCDF 2016).

Throughout history, several authors have warned over the inability of the Earth's capacity to sustain its growing population. Each time, the crisis has been averted due to technological advances in plant breeding, fertilization, crop protection, and agronomic management (Parnell et al. 2016). Nevertheless, this exponential growth in the world's agricultural production has neglected the consequences of environmental impacts, with excessive deforestation, the irrational use of soil and natural resources.

The indiscriminate use of chemical fertilizers has increased the salinity of soils, mostly in arid and semiarid regions, where it seriously threatens agricultural sustainability and food security (Keswani et al. 2016; Ram et al. 2018; Singh et al. 2016a, 2019). The most limiting soil nutrients for plant growth are nitrogen and phosphorus. Although many soils contain ample quantities of these nutrients, most are not readily accessible for plant growth (Schachtman et al. 1998; Rai 2006). This reinforces the need to find alternative strategies to improve acquisition and use efficiently these nutrients (Laplase et al. 2018). Consequently, microbial products can be developed to increase the availability of nitrogen or phosphorus to crops, thereby maximizing the efficient, sustainable use of nutrients (Parnell et al. 2016).

In addition, the intensification of agriculture, the low diversity of cultivated species, and the excessive agrochemicals used have been increasing in selective pressure, favoring mutations and the propagation of resistant phytopathogenic organisms. On the other hand, billions in investments and decades of research have ensured the discovery and development of new chemical molecules that are capable of controlling the most diverse and resistant unwanted organisms in crops, avoiding billions of dollar losses worldwide (FAO 2017a). However, chemical molecules have shortened their useful lives, with the emergence of the new varieties of pests and diseases, resistant at the newer agrochemical, making unfeasible the billions of investments necessary for its development (Lucas et al. 2015). To break this cycle of resistant organisms and new and more potent chemical products, the biological control arises as an alternative way to control pests and diseases without the emergence of resistance in phytopathogens (Singh et al. 2016b, c, 2017).

On these perspectives, agriculture and food production should increase by 70% to be able to feed all world population until 2100 (FAO 2017b), which will require profound changes and adaptations. In addition, with global development and an increasingly enlightened population that seeks for a healthy diet, it is expected that this increase in agriculture production must be sustained by ecological agriculture linked to environmental preservation. As we face our next challenge, it is critical that we continue to discover new sustainable cropping system solutions to produce more with fewer resources (Parnell et al. 2016).

Concerned about the need to increase the food supply, but without neglecting the risks to the environment and population health, several countries have established stricter legislation over the use of chemical pesticides. The EU has banned the use of various chemical molecules in the territory and reducing the maximum residue limit (MLR) of pesticides in food, water, and soil. They have the strictest regulatory system in the world about pesticides, although the largest pesticide companies are based in European countries (FCEC 2012; European Commission 2017).

On the other hand, the main world food producers (USA, China, and Brazil) have less restrictive legislation regarding the use of chemical pesticides (more flexible and tolerant). In Brazil, the world's largest consumer of agrochemicals (19% of the world market), despite being only the third largest food producer, 30% of chemical pesticides used are banned in the EU. In addition, the MRLs are 400 times higher for food and 5000 times higher for glyphosate in drinking water (Bombardini 2017).

The permissibility and flexibility of these large global food producers have effects around the world, characterizing the circle of poison. Many European countries that are the largest pesticide producers are also among the larger importers of agricultural commodities (Galt 2008; Bombardini 2017). However, despite the fact that many pesticide residues return to these countries through food, we cannot forget that the greatest damage is caused by residues that remain in the soil and water, affecting the biodiversity and health of the population, in countries such as Brazil, which pride themselves on being the world's breadbasket. Thus, we become simple producers of commodities, putting our land, our people, our natural wealth, and our biodiversity at the service of those who can pay for what they do not want to produce. We swap tons of commodities for grams of poison.

Today, with the current knowledge, we cannot maintain the same developmental line of the last century. We need to look for means of agricultural production integrated with the environment and sustainability, respecting nature to serve our production. Nevertheless, we understand that agriculture will not change dramatically in a short time, nor definitively leave aside the use of chemical pesticides. Agriculture is gradually evolving, incorporating new technologies such as integrated pest management, assessment of the level of economic damage for rational and conscious use of agrochemicals, even as the use of biological control agents, which together will be the key to sustainable food production.

4.2 Biological Control and the Global Market for Agrochemicals

Biological control can be defined as the use of an organism to reduce the population density of another organism, and this includes the control of animals, weeds, and diseases (De Bach 1964; Bale et al. 2008). Thus, biopesticides are products developed with biocontrol organisms, which can be microorganisms (fungi, bacteria, and viruses), microscopic animals (nematodes), and macroorganisms (predatory parasites, insects, and mites) or natural products derived from living organisms, used for the protection of plants or animals (Bettiol 2011).

Compared to most of the current chemical pesticides, the biopesticides have a more complex mode of action, carved by billions of years of development. They are able to control pests and diseases in crops without the emergence of resistance (Hubbard et al. 2014; Senthil-Nathan 2015). This happens because the biological control reduces rather than eradicates pests, making the pest and natural enemy remain in the agroecosystem at low densities. A number of important pests can be kept at a low population density by biological control over long periods of time. In other cases, populations of pests are significantly reduced by natural enemies, but repeated releases or additional methods are needed to achieve an adequate level of control. These methods include, for example, resistant plants, cultural techniques, physical barriers, semiochemicals, and, as a last resource, the use of selective chemicals; this is the fundamental philosophy of integrated pest management (Bale et al. 2008).

In general, biological control can be classified as the form of occurrence, but, firstly, it needs to be distinguished from natural control. The biological control is the use by men of an organism to reduce the population density of another organism, whereas natural control is the reduction in numbers of the population of a species by a naturally occurring natural enemy with no human intervention. Thus, there are three main techniques of biological control:

(1) *Classical* – when an exogenous species is introduced into a new ecosystem, imported from another region or country. It is used mainly against "exotic" pests that have established in new countries or regions of the world.
(2) *Augmentative* – refers to all forms of biological control in which natural enemies are periodically introduced and usually requires the commercial production of the released agents.
(3) *Conservative* – refers to the use of indigenous biological control agents, usually against native pests. Various measures are implemented to enhance the abundance or activity of the natural enemies (Bale et al. 2008; Polanczyk and Pratissoli 2009; Roderick et al. 2012).

The techniques of biological control *classical* and *augmentative* requires the development of products that guarantee the delivery of the agent in the field. For their development, several steps are required: collect, isolation, selection, identification, characterization of the microorganism, production optimization, production scale-up, product formulation and application studies, shelf life, registration, and commercialization of the product. However, despite this long process to get to the market, the cost of developing biopesticides is much less expensive than the development of new chemical molecules, which can cost hundreds of millions of dollars and take decades of research (Cumagun 2014).

Biological control, despite being known and studied for more than 100 years, still presents low diversity and availability of products in the market. However, the interest in biopesticides has increased, because the use of biological control agents as part of the IPM program represents a change in the model of agriculture being practiced in the country. If these technologies were considered a myth in the universe of large plantations, today MIP is a rational method for pest and disease control, perfectly aligned with the use of biological control agents (Bueno et al. 2012; Faria 2017).

The global biopesticide market in 2013 was estimated at approximately USD 3 billion, or 5% of total crop protection market, and it is expected to grow more than USD 4.5 billion by 2023 (Olson 2015; Kumar et al. 2018), whereas the global pesticides market, in 2017, estimated at USD 61 billion. This was achieved with a growth above 15% a year from 2011, and today, in Brazil, from the total sales of agricultural defenses, it is estimated that only 1–2% are represented by biopesticides. In the USA biological control products already represent 6% of the market and in Europe 14–16%, and it is being used in more than 50% of the planted area in the system of agroecological of production of Cuba (Altieri and Funes-Monzotefr 2012; Abcbio 2016; Locatelli et al. 2018).

With an eye on this growing market, the largest agrochemical companies have invested in the development of new biopesticides, which can be evidenced by patent registrations in this area. The major companies follow the *classical* biological control, searching for one species, through selection or genetic manipulation, capable of controlling several pests and diseases, in different cultures in different regions of the world. Unlike the agrochemical, the biological control agents present peculiarities when used in the field. Their efficiency can be influenced by seasonal and intrinsic characteristics of the soil. Biological control agents that have high efficiency in pest control in a given region may not be as effective in another country or region.

In addition, many questions were not answered about biological control agents. Can this biological agent introduced into a new ecosystem generate environmental imbalances? Can it destroy native species of nontarget organisms? Can they cross with nearby species? Can this introduction of a new species in another region cause a reduction of genetic biodiversity (Roderick et al. 2012)? Thus, how to ensure the safety in the use of biological control and increase your participation in the world agrochemicals market?

Usually, beneficial and pathogenic organisms are naturally present in the soil, living in balanced conditions that do not cause economic damage at crops, but some factors may favor the onset of a pest or disease. From the environmental perspective, it is more sustainable and safe the development of personalized or regionalized bio-defensives, based on the concept of *augmentative* and *conservative* biological control. Thus, seeking to isolate the environment where the disease occurs, the natural enemy present in that ecosystem is able to control the disease, without the need of introducing a new species from another region to control pests. Therefore, we benefit from millions of years of coevolution between biological control agent and phytopathogenic organism. It is smarter, more economical, and safer to search for the biological control agent at the very site of disease manifestation, developing a personalized product for the disease that is occurring in that region or country. This model can be better understood with the example of Cuba agroecological agriculture.

4.2.1 Cuba as Reference in Agroecological Agriculture

Cuba is today a reference in agroecological agriculture. With economic embargoes imposed on Cuba and especially after the collapse of trading relations with the former Soviet Bloc in 1989, Cuba restricted its access to chemical pesticides and agricultural inputs. Then they had to look for another way for agricultural development and food production to support its population, converting its modern conventional agriculture to semi-organic farming on a large scale. Thus, a country with a highly industrialized agricultural system technologically similar to that in California (dominated by monocultures) has had to dramatically increase food production without significantly affecting earnings from export agriculture, all virtually without the chemical inputs and machinery on which it had become dependent (Rosset 1997; Altieri and Funes-Monzotefr 2012).

The strategy adopted by the government was to mobilize Cuba's substantial scientific infrastructure – both physical and human resources. Since 1982 Cuba has focused on integrating pest management (IPM), with an emphasis on biological control. This meant biopesticides and natural enemies to combat insect pests, resistant varieties, crop rotations, and microbial antagonists to combat phytopathogens. The chemical fertilizers were replaced by biofertilizers (microbial products) and other forms of compounds and other organic fertilizers (Rosset 1997; Nicholls et al. 2002).

The Cuban program involves decentralized regional laboratories, territorial plant protection centers scattered throughout the country, more than 200 units of the Center for Reproduction of Entomophages and Entomopathogens, guaranteeing the use of biological control in more than 50% of the country's agricultural area (Shishkoff 1993; Rosset 1997; Almeida et al. 2001) which, for example, was able to increase its production by 145% to general vegetables and 351% to beans, same with a decrease in agrochemicals of 72% and 55% for both crops, respectively. Cuba has sufficient land to produce enough food with agroecological methods to satisfy the nutritional needs of its 11 million inhabitants (Nicholls et al. 2002; Altieri and Funes-Monzotefr 2012).

A rich knowledge of agroecology science and practice exists in Cuba, the result of accumulated experiences promoted by researchers, professors, technicians, and farmers. This legacy is based on the experiences within rural communities that contain successful "agroecological lighthouses" from which principles have radiated out to help build the basis of an agricultural strategy that promotes efficiency, diversity, synergy, and resiliency (Altieri and Funes-Monzotefr 2012). We leave aside the political and economic discussions, to verify the success of this agroecological system. This model can serve as a basis for the construction of strategies to increase the participation of biological control agents in the world market of agrochemicals.

4.3 Strategies for Increasing the Use of Biopesticides in a Global Market Dominated by Agrochemicals

Before we delineate a strategy that allows greater participation of biological control in the market of agrochemicals, we must know some concepts:

Strategy – can be defined as the determination of the long-term goals and objectives of an enterprise and the adoption of courses of action and the allocation of resources necessary for carrying out these goals (Chandler 1962).

Innovation – the concept of innovation from the economic perspective was defined by Schumpeter (1943), as "an innovation was something that changed the marketplace in a profound way. The innovating organization was, thus, likely to become the new market leader and to gain an immense advantage over its competitors." Already Drucker (1985) defined innovation as "the act that endows resources with a new capacity to create wealth". This way the "innovation transforms insight and technology into novel products, processes and services that create new value for stakeholders drive economic growth and improve standards of living" (González 2004).

Based on these concepts, we can understand the concept of innovation strategy, which can be defined as the strategy that generates innovation within a company or organization because, although we can talk about accidental technical inventions, we can hardly use the same qualification for innovations, which requires a continuous and complex effort to develop a new product, process, or service. Such an effort requires a long-term vision, commitment of resources and time, and an integrated structure of decisions. That is, innovation needs a strategy (Porto 2013).

Thus, the innovation strategy applied by Cuba to reach its goals of productivity and food self-sufficiency based on agroecological crops were diversification and decentralization. They made available the physical and intellectual structure of the research institutes and universities of the rural producers. This generated disruptive innovations that allowed the breakdown of the dependence of inputs (agrochemicals and fertilizers) of the monopoly of large companies that dominate the sector, besides increasing the productivity of several crops and guaranteed self-sufficiency in food for its population.

This innovation model seeks to convert academic knowledge, generated in universities and research institutes, in applications that bring social and economic development, which translates the concept of the triple helix. Government and industry, the classic elements of public-private partnerships, have been recognized as important spheres of society since the eighteenth century. The triple helix model is that the university is no longer having a secondary social role, though important, of providing higher education and research, but is assuming a primordial role equivalent to that of industry and government as the generator of new industries and enterprises. Thus, the University has an almost governmental performance, as an organizer of local or regional technological innovation. This triad model of innovation and development allows the presence of a conciliator in the individual's interests, giving greater sustainability to the cooperation between the parties (Dzisah and Etzkowitz 2008; Porto 2013; Etzkowitz and Zhou 2017).

In the model exemplified by the success of agroecological agriculture in Cuba, we have research institutes and universities as protagonists of innovation, rural producers in the role of industry, transforming knowledge into product generation, and government in its role of mediator and foster of innovations. Leaving aside the economic policy adopted by the country, this model can perfectly foster the economical niche of biological control products, increasing participation in the agrochemical market. That would be capable of creating jobs, generating and distributing income mainly in underdeveloped countries, such as Brazil. The disruptive innovation generated by the niche of biological control will allow decrease of the dependence of products and inputs imported from developed countries, which produce the agrochemicals, but prohibit the application, of the great majority, in their soils. This way, we will be able to break the circle of poison.

4.4 Final Considerations

In this way, it seems clear that the decentralization of investments, to foment different strategies of the use of agents of biological control, bringing together innovation agents, favors the emergence of agribusiness' start-ups. This will allow the increase in the participation of biological control agents in the agricultural defensive market. Start-ups have more flexibility to maintain close relations with the farmers and cooperatives, in the search of innovative solutions, and directed to the solutions of local problems.

However, this requires that each helix of the triple model do its part, the government as a mediator of interests and research financer, guaranteeing a greater plurality of investments in research and development, with the aim of knowing the biodiversity, the problems, and challenges of regional agriculture. The university as a research center and teaching, to generate and disseminate knowledge, which along with producers or companies acted as agents of innovation, transforming knowledge generated in products that solve problems and lead to social development.

Agricultural cooperatives or large independent producers, in partnerships with universities and research institutions, can identify and isolate biological control agents present mainly where the disease is occurring. Many biological control agents, whether microorganisms or insects, can be multiplied and produced by simple methods with a small structure; as an advantage in Brazil and other countries, the multiplication of a biological control agent for own use does not require registration with the official control entities.

The correct triple helix circulation to the proposed niche increases the offer of biological control products, diversity of companies and solutions, reducing the dependence of agribusiness of the dominance of few agrochemicals companies. Thus, the bioeconomy – the set of economic activities relating to the invention, development, production, and use of biological products and processes – became an important tool for major socioeconomic contributions in diverse countries, which contributes to the plan by policy agenda from OECD to Bioeconomy to 2030 (OECD 2009).

References

ABCBIO (2016) Incentivo aos Produtos Biológicos: Panorama Atual e Perspectivas. Agroanalysis 30–35

Altieri MA, Funes-Monzotefr (2012) The Paradox of Cuban Agriculture. Mon Rev: Indep Socialist Mag 63:23–33

Bale JS, Van Lenteren JC, Bigler F (2008) Biological control and sustainable food production. Phil Trans R Soc B363:761–776. https://doi.org/10.1098/rstb.2007.2182

Bettiol W (2011) Biopesticide use and research in Brazil. Outlooks Pest Manag 22(6):280–284. https://doi.org/10.1564/22dec10

Bombardini LM (2017) Geografia do Uso de Agrotóxicos no Brasil e Conexões com a União Europeia. FFLCH – USP, São Paulo, 296p

Bueno AF, Sosa-Gómez DR, Corrêa-Ferreira BS, Moscardi F, Bueno RCOF (2012) Capítulo 8 – Inimigos naturais das pragas de soja. In: Hoffmann-Campo CB, Corrêa-Ferreira BS, Moscardi F (eds) Soja: manejo integrado de insetos e outros artrópodes-praga. Embrapa, Brasília, pp 493–630

Chandler AD (1962) Strategy and structure: chapters in the history of the industrial enterprise. MIT Press, Boston

Cumagun CJR (2014) Advances in formulation of trichoderma for biocontrol (Chapter 39). Biotechnol Biol Trichoderma 527–531. https://doi.org/10.1016/B978-0-444-59576-8.00039-4

de Bach P (1964) Biological control of insect pests and weeds. Reinhold, New York. 844p

Drucker P (1985) Innovation and entrepreneurship. Harper, New York

Dzisah J, Etzkowitz H (2008) Triple helix circulation: the heart of innovation and development. Int J Technol Manag Sustain Dev 7:101–115. https://doi.org/10.1386/ijtm.7.2.101_1

Etzkowitz H, Zhou C (2017) Hélice Tríplice: inovação e empreendedorismo universidade-indústria-governo. EstudosAvançados 31:23–48. https://doi.org/10.1590/s0103-40142017.3190003

European Commission (2017) Pesticides in the European Union authorization and use. Available in: http://europa.eu/rapid/attachment/IP-17-5191/en/Pesticides_factsheet.pdf. Accessed 26 Mar 2019

FAO (2017a) Concerted action needed to stop diseases and pests from ravaging the food chain. Available in: http://www.fao.org/news/story/en/item/469269/icode/. Accessed 26 Mar 2019

FAO (2017b) Estudo revela que Brasil é um dos países mais eficientes no uso da terra e insumos agrícolas em função de sua alta produção. FAO no Brasil. Available in: http://www.fao.org/brasil/noticias/detail-events/en/c/1070557/. Accessed 26 Mar 2019

FAO, IFAD, WFP (2015) The state of food insecurity in the world 2015, in meeting the 2015 international hunger target: taking stock of uneven progress. FAO, Rome. Available in: http://www.fao.org/3/a-i4646e.pdf. Accessed 26 Mar 2019

Faria Jr P (2017) Biocontrol in Brazil: opportunities and challenges. Agropages – Biopesticide Suppl 12–13. Available in: http://img.agropages.com/userfiles/pdf/2017bssl2.pdf. Accessed 26 Mar 2019

FCEC Food Chain Evaluation Consortium (2012) Study on existing monitoring and surveillance activities, communication of the results of these activities to the public and awareness raising programmes put in place by MS on the impacts of use of plant protection products on human health and the environment. European Commission Directorate General for Health and Consumers 145p

Galt RE (2008) Beyond the circle of poison: significant shifts in the global pesticide complex, 1976–2008. Glob Environ Chang 18:786–799
Gomes De Almeida S, Petersen P, Cordeiro A (2001) Crise Socioambiental e Conversão Ecológica da Agricultura Brasileira: subsídios à formulação de diretrizes ambientais para o desenvolvimento agrícola. AS-PTA, Rio de Janeiro
González F (2004) Innovation for the 21st century banking industry. In: Innovation perspectives for the 21st century. BBVA
Hubbard M, Hynes RK, Erlandson M, Bailey KL (2014) The biochemistry behind biopesticide efficacy. Sustain Chem Process 2:2–8. https://doi.org/10.1186/s40508-014-0018-x
Keswani C, Sarma BK, Singh HB (2016) Synthesis of policy support, quality control, and regulatory management of biopesticides in sustainable agriculture. In: Singh HB, Sarma BK, Keswani C (eds) Agriculturally important microorganisms: commercial and regulatory requirement in Asia. Springer, Singapore, pp 167–181
Kumar KK, Sridhar J, Murali-Baskaran RK, Senthil-Nathan S, Kaushal P, Dara SK, Arthurs S (2018) Microbial biopesticides for insect pest management in India: current status and future prospects. J Invertebr Pathol. https://doi.org/10.1016/j.jip.2018.10.008
Laplase L, Sparvoli F, Masmoudi K, Hash CT (2018) Harvesting plant and microbial biodiversity for sustainably enhanced food security. Front Plant Sci Front Microbiol. https://doi.org/10.3389/978-2-88945-444-0
Locatelli GO, Santos GF, Botelho PS, Finkler CLL, Bueno LA (2018) Development of *Trichoderma* sp. formulations in encapsulated granules (CG) and evaluation of conidia shelf-life. Biol Control 117:21–29. https://doi.org/10.1016/j.biocontrol.2017.08.020
Lucas JA, Hawkins NJ, Fraaije BA (2015) The evolution of fungicide resistance. Adv Appl Microbiol 90:29–92
Nicholls CI, Perez N, Vasquez L, Altieri MA (2002) The development and status of biologically based integrated pest management in Cuba. IPM Rev 7:1–16
OECD (2009) The bioeconomy to 2030: designing a policy agenda. 326p. https://doi.org/10.1787/9789264056886-en
OECD/FAO/UNCDF (2016) Adopting a territorial approach to food security and nutrition policy. OECD Publishing, Paris. https://doi.org/10.1787/9789264257108-en
Olson S (2015) An analysis of the biopesticide market now and where it is going. Outlooks Pest Manag 26:203–206. https://doi.org/10.1564/v26_oct_04
ONU (2013) População mundial deve atingir 9,6 bilhões em 2050, diz novo relatório da ONU. Available in: https://nacoesunidas.org/populacao-mundial-deve-atingir-96-bilhoes-em-2050-diz-novo-relatorio-da-onu/. Accessed 26 Mar 2019
Parnell JJ, Berka R, Young HA, Sturino JM, Kang Y, Barnhart DM, Dileo MV (2016) From the lab to the farm: an industrial perspective of plant beneficial microorganisms. Front Plant Sci 7. https://doi.org/10.3389/fpls.2016.01110
Polanczyk RA, Pratissoli D (2009) Biological control of agricultural pests: principles and field applications. Rev Ceres 56:410–419
Porto GS (2013) Gestão da Inovação e Empreendedorismo. Elsevier, Rio de Janeiro
Rai MK (2006) Handbook of microbial biofertilizers. Food Products Press, New York. 42p
Ram RM, Keswani C, Bisen K, Tripathi R, Singh SP, Singh HB (2018) Biocontrol technology: eco-friendly approaches for sustainable agriculture. In: Brah D, Azevedo V (eds) Omics technologies and bio-engineering: towards improving quality of life volume II microbial, plant, environmental and industrial technologies. Academic, London, pp 177–190
Roderick GK, Hufbauer R, Navajas M (2012) Evolution and biological control. Evol Appl 5:419–423. https://doi.org/10.1111/j.1752-4571.2012.00281.x
Rosset PM (1997) Cuba: ethics, biological control, and crisis. Agric Hum Values 14:291–302
Schachtman DP, Reid RJ, Ayling SM (1998) Phosphorus uptake by plants: from soil to cell. Plant Physiol 116:447–453. https://doi.org/10.1104/pp.116.2.447
Schumpeter JA (1943) Capitalism, socialism and democracy. ALLAN and UNWIN, London

Senthil-Nathan S (2015) A review of biopesticides and their mode of action against insect pests. In: Thangavel P, Sridevi G (eds) Environmental sustainability. Springer, New Delhi. https://doi.org/10.1007/978-81-322-2056-5_3

Shishkoff N (1993) Plant diseases and their control by biological means in Cuba. Agric Hum Values 10:24–30

Singh HB, Jha A, Keswani C (eds) (2016a) Intellectual property issues in biotechnology. CABI, Wallingford. 304 pages, ISBN-13:9781780646534

Singh HB, Jha A, Keswani C (2016b) Biotechnology in agriculture, medicine and industry: an overview. In: Singh HB, Jha A, Keswani C (eds) Intellectual property issues in biotechnology. CABI, Wallingford, pp 1–4

Singh HB, Sarma BK, Keswani C (eds) (2016c) Agriculturally important microorganisms: commercialization and regulatory requirements in Asia. Springer, Singapore. 336 pages, ISBN-13:978-9811025754

Singh HB, Sarma BK, Keswani C (eds) (2017) Advances in PGPR research. CABI, Wallingford. 408 pages, ISBN:9781786390325

Singh HB, Keswani C, Singh SP (eds) (2019) Intellectual property issues in microbiology. Springer, Singapore. 425 pages, ISBN:9789811374654

Applications of Remote Sensing in Pest Monitoring and Crop Management

5

Karim Ennouri, Mohamed Ali Triki, and Abdelaziz Kallel

Abstract

Precision agricultural skill has constructed and will still construct the road we are moving into this novel theory of precision agriculture. By increasing the inspection and appliance of inputs on the land, farmers are changing from a usual, standardized treatment of every agricultural land to a perfect treatment for as little as possible districts. Remote sensing processes offer a basis for which vegetal stress and growth reaction can be estimated. Remote sensing research based on terrestrial and spatial domains has demonstrated that numerous kinds of plant illness, through pre-visual infection signs for pathogens, hostile species and also plant health indicators, can be identified through aerial hyperspectral imaging. Inspecting foliage using remote sensing data necessitates understanding of the organization and role of foliage and its reflectance characteristics. Sensors have been ameliorated to calculate the reflectance of incident bright at numerous wavebands and have been associated to plant evolution and plant cover. Remote sensing technology has the major advantage to obtaining data about a given entity or region without having physical exchange and frequently employs surface-based instruments or spatial pictures. Remote sensing would be considered as an economic and relevant instrument for land-scale pest controlling and study.

K. Ennouri (✉)
Digital Research Centre of Sfax, Technopark of Sfax, Sfax, Tunisia

Laboratory of Amelioration and Protection of Olive Genetic Resources, Olive Tree Institute, University of Sfax, Sfax, Tunisia

M. A. Triki
Laboratory of Amelioration and Protection of Olive Genetic Resources, Olive Tree Institute, University of Sfax, Sfax, Tunisia

A. Kallel
Digital Research Centre of Sfax, Technopark of Sfax, Sfax, Tunisia

C. Keswani (ed.), *Bioeconomy for Sustainable Development*,
https://doi.org/10.1007/978-981-13-9431-7_5

Keywords
Remote sensing · Plant · Illness · Reflectance · Precision agriculture

5.1 Introduction

Since the start of agriculture, the huge concept changes are induced through recently developed methods and instruments to cultivate arable land with. To cite a few, plough and the agricultural truck have transformed the approach how humans cultivate the land (Zhang and Kovacs 2012). Nowadays, farmers are continuously developing and revolutionizing their modes of cultivation according to novel investigations and expertise. With the fast enhancement due to the rising examination of applied knowledge and the team-based research, the growth of novel devices in cultivation is happening quickly (Singh et al. 2016a, b). Precision agricultural skill has and will still construct the road farmers and engineers are moving into this novel theory of precision agriculture. By increasing the inspection and appliance of inputs on the land, farmers are changing from a usual, standardized treatment of every agricultural land to a perfect treatment for as little as possible districts (Singh et al. 2017).

Precision agriculture had its beginning in the middle of 1980s with sensors for soil organic matter and is currently developing exponentially. It has quickly evolved to include satellite, aerial and tractor-mounted or handheld sensors. It was not before the 1990s that precision agriculture became commercially available.

Precision agriculture had a number of objectives through its development to where it is now. It commenced with cultivating by means of soil and has improved to site-specific crop organization on the basis of supervision areas and grid sampling. Afterwards, there has been rising importance on synchronized active monitoring with soil-based captors. The correctness of the imagery has turned into better which permits assessment of soil and harvest characteristics at a very well-spatial presentation at the charge of improved data storage and handling requests. Precision agriculture has been inclined to provide improved farm productivity by increasing the harvest production (Gebbers and De Bruin 2010) and throughout increased organization of farm inputs guiding to less ecological contamination (Lake et al. 2008).

The big quantity of data that is being accumulated gives more precise and accurate appliance of the inputs on a terrain. This leads nevertheless again to increased crop yield and ecological value (Schellberg et al. 2008). There has in addition been a change from spatial data investigations and supervision alone to a spatial-temporal data investigations and supervision (Chen et al. 2015). The precision agriculture engages numerous steps of information organization, processing and examining of the information that is accumulated, scientific progresses in processor calculating, production monitoring, land location as well as captor design and remote sensing (Winstead et al. 2010). Precision agriculture implies numerous diverse recent skills and extends over a lot of supervision inputs for the cultivators. Captors and cameras can be climbed to any platform that can take them such as trucks, tools and satellites, which are generally employed.

With the quick expansion of technology, cameras and captors are becoming slighter and lighter, among best imagery quality. Along with these developments, the use and expansion of unmanned aerial motor vehicles offer the potentials of merging these to an aerial structure (Koleshko et al. 2012).

Infection disorders, pest invasion, nutrient lacks and climate changes continually menace to significantly reduce annual productions and consequently financial incomes. Ecological stress can impose damages in modern agronomic production structures (Khosla 2010). Identification of an operational and concrete device to detect and estimate regions of vegetal stress is required. In fact, confirmed growers can recognize various syndromes and pathogen insects; they could not have the possibility to methodically examine total plantations. Knowing the spatial dissemination of ecological stress in areas and the subsequent development, penalties can permit for modifications in organization exercises (Fenghua and Shujuan 2008). What is required is a lucrative approach for farmers to detect harvest difficulties in their soils without physically scouting the complete area. Remote sensing affords an occasion to assess plant communities via bright reflectance (Adam et al. 2010; Campbell and Wynne 2011). The assessment of recent processes and methods employed to understand plant evolution over a diversity of restrictions will imply the application of remote sensing practices for managing region harvests. In this review, we define the remote sensing characteristics and advantages in modern agriculture.

5.2 Remote Sensing: Novel Approach of Agriculture

5.2.1 Generalities

Remote sensing can be described as the discipline and ability of achieving information about an object, region or event via sensors that are not in touch among the segment below examination (Rees and Pellika 2010). Remote sensing actions include a huge number of activities, comprising the operation of satellite coordination, figure data achievement, the succeeding data step, analysis, distribution of the registered data and picture results (Sabins 2007).

On other hand, remote sensing affords an occasion to assess plant communities via bright reflectance (Ozdogan et al. 2010). The assessment of recent processes and methods employed to understand plant evolution over a diversity of restrictions will imply the application of remote sensing practices for managing region harvests. Besides, remote sensing processes offer a basis for which vegetal stress and growth reaction can be estimated. The question with technological yield examination via digital and numerical descriptions only is that recognizing crop difficulties can be stimulating, as technological methods are not very forceful (Ulaby et al. 2014). Multispectral description permits useful data to be matched about diverse bright bands, permitting for much easier recognition of crop obstacles (Schowengerdt 2006).

It is well known that solar spectrum has a waveband ranged from 400 to 3000 nm (Sabins 2007; Thenkabail and Lyon 2016). Within this range, the 4×10^2 to

25×10^2 nm segment is often calculated via a compilation of optical captors varying from multispectral (eg. Landsat) to hyperspectral (ex: AVIRIS) (Frohn and Lopez 2017; Gupta 2017). The Landsat platform offers the extensive constant space-based report of existing planet's area. Landsat satellites furnish major details to serve region managers to make conclusions concerning natural reserves and environmental conditions. Moreover, the airborne visible/infrared imaging spectrometer (called also AVIRIS) sensor compiles data that can be used for characterization of the globe's regions and pressure from geometrically logical spectroradiometric analysis. This information may be employed to works in the domains of environmental disciplines and agronomy. The connection concerning plant development and spectral answer in the observable and infrared wavebands has been well constructed via the quotient of infrared and also red reflectance or else further indices related on this quotient (Landgrebe 2005; Kuenzer et al. 2011). Sensors have been ameliorated to calculate the reflectance of incident bright at numerous wavebands and have been associated to plant evolution and plant cover.

Foliage reflectance characteristics are employed to determine vegetation indices. AAAThese deducted data are built since reflectance values in at least two wavebands through the optical range to study particular specificities of plant coverage, such as chlorophyll and water compositions (Bioucas-Dias et al. 2013; Lillesand et al. 2014). For diverse characteristics and also soil conditions, some indices in a class deliver results with greater validity compared with others.

In addition, the employ of particular indices has permitted operators to detect modifications in reflectance to transformations in canopy specificities (Schowengerdt 2012). Moreover, there are several index, all developed from quotients based on the reflectance of incident bright at particular wavebands. The normalized difference vegetative index (NDVI) has acquired widespread approval on the basis of its interface simplicity, only needing two wavebands, and the vegetal particularities has been combined also. NDVI has been employed to estimate nitrogen level of plant, chlorophyll amount, green foliage volume and grain harvest (Twomey 2013). Spectral data has been employed to calculate micronutrient lack, recognition of insect invasion and disease contamination of vegetal (Sabins 2007; Thilakarathna and Raizada 2018) (Fig. 5.1).

5.2.2 Vegetation Indices

Vegetation indices are known as empirical designs used within agriculture to extract biophysical characteristics such as leaf area index and biomass from remotely sensed images. The leaf area index can be defined as the ratio of green leaf area per area of ground, and biomass is estimated from optical sensors which in turn can be related to yield. The indices are based on the deduction that red light is actively griped by photosynthetic colours (eg. chlorophyll) found within living plantations, whereas near-infrared light either elapses through or else is redirected. As such, on a satellite capture image, zones covered with green plants will be very brilliant in the near-infrared, due to the fact that improved reflectance and also very obscure in

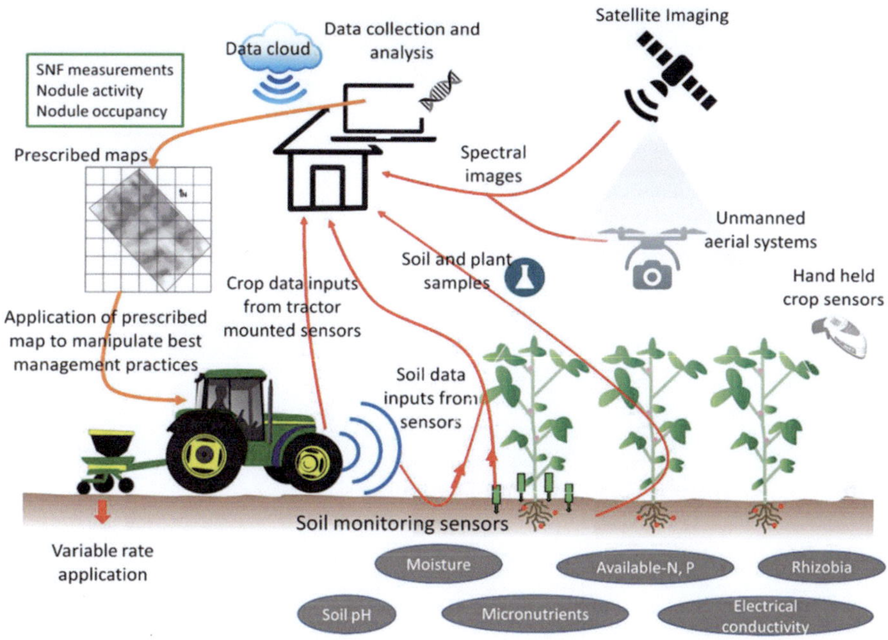

Fig. 5.1 A conceptual model showing how precision agricultural tools can be integrated together at the field level, using on-the-go variable management. (Adapted from Thilakarathna and Raizada 2018)

the red portion of the spectrum due to enhanced absorption. Vegetation indices employ a quotient of the reflected NIR and reflected RED wavelengths in numerous ways to achieve a value which is typical to the quantity of foliage present (Schowengerdt 2012; Lillesand et al. 2014). The most common index is NDVI, which computes the variance in reflectance subdivided by the addition of the reflectance in both wavelengths (Fig. 5.2).

The NDVI can vary from −1 to 1. A surface with a small contrast among the NIR and R channels will obtain a NDVI value nearby to zero, whereas surfaces of great dissimilarity, principally green foliage, will have NDVI rates much closer to one (Thenkabail and Lyon 2016). By investigating the association found among NDVI and the feature of interest, an easy empirical design can be generated.

5.3 Remote Sensing for Plant Disease Detection

Optical remote sensing makes employ of visible, near-infrared and shortwave infrared area of the range to form pictures of the globe's surface by perceiving the solar radiation returned from surface trait. Numerous matters return and absorb in a different ways at diverse wavebands. The reflectance range of a matter is a design of the part of radiation returned as a relation of the incident waveband and considers as

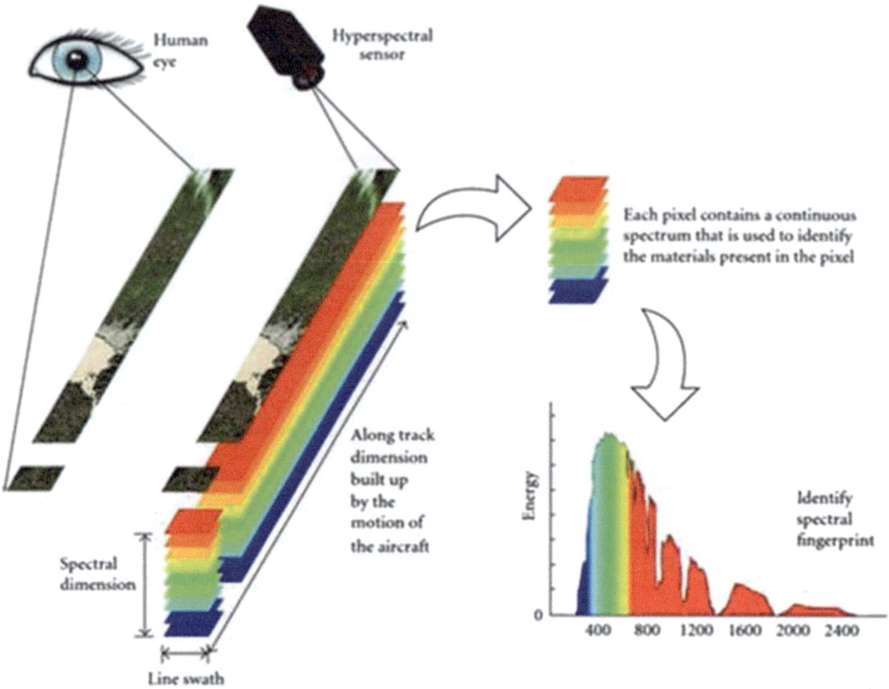

Fig. 5.2 Hyperspectral remote sensing of the earth. (Aiazzi et al. 2012)

a distinctive mark for the matter. Theoretically, a matter can be recognized since its spectral reflectance mark if the sensing structure has satisfactory spectral resolution to differentiate its range from those of other matters. This offers the foundation for multiple spectral remote sensing functions. As a consequence, the objectives can be distinguished via the spectral reflectance marks in the remotely sensed pictures. Crop stress as pest/disease stress or water stress are indicators usually detected as failure of green pigments and transformation in pigment constitution (called also chlorosis) or hurt of tissue as injuries (known as narcosis), or by lack of moisture of tissue or increase in tissue temperature.

Figure 5.3 represents remote sensing sensors which record indicators of forest health in this case, in two consecutive steps: Step 1 records spectral traits (ST) and spectral trait variations (STV); Step 2 distinguishes species, populations, communities and biomes of forest ecosystems depending on the spatial, spectral, radiometric, angular and temporal resolution of the remote sensing techniques, the distribution of ST/STV in space and their temporal changes, the choice of the modelling method (classification: biophysical/chemical parameter estimation), the geographic data representation [pixel-based or geographic object-based] and the appropriateness of the remote sensing algorithm and its assumptions for the given spectral traits.

Suitable detection as well as estimation of crop stress indicators is especially essential. Conventionally, insect and disease evaluation of plants is being done via

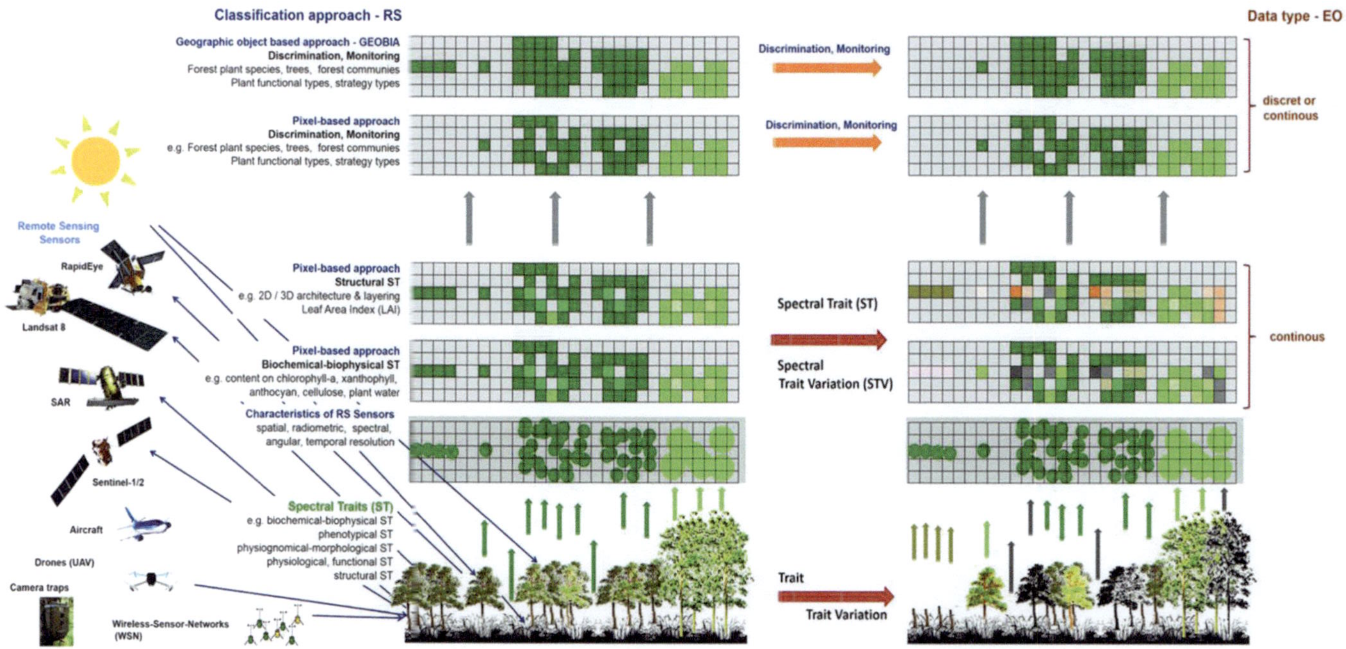

Fig. 5.3 Remote sensing—a physically-based system, which records indicators of vegetation health (ex: forest). (Adapted from Lausch et al. 2016)

an ocular approach, for example, relying up on eyes and mental capacity in order to calculate their occurrence. Nevertheless, the difficulty among the conventional approach is that they are frequently time demanding along with effort intensive. Modern progresses in the domain of remote sensing and radiometry skills attempt sufficient capacity for using these tools towards expanding a switch means that can improve or complete the habitual approaches.

Detection of vegetation stress through remote sensing is principally reliant with the statement that stress features that meddle with photosynthesis mechanism or else the physical organization of the plant influence the assimilation of light power and consequently modify the reflectance range of the plant. Leaf reflectance is influenced by numerous features as well as leaf internal compositions, surface characteristics, the amount and circulation of biochemical constituents, for instance, chlorophyll a and also water amount. Red and green wavelengths are appropriate to detect sign associated to transformation in coloured pigment; near-infrared wavelength is appropriate for tissue loss recognition. Shortwave infrared is responsive to humidity content/drying recognition, and thermal band is appropriate for modification in cover temperature that can be directly linked to photosynthesis activity. As a result, in relation on the signs of disease visible on the vegetation, the spectral mark in particular spectral wavelengths will modify from that of vigorous vegetation. Employ of remote sensing methods for revealing of yield pests and diseases is found on the statement that stress provoked by them meddles by means of photosynthesis and also physical constitution of the studied vegetation and influences the assimilation of light power and consequently changes the reflectance range of the studied vegetation (Anderson et al. 2007; Erener 2011).

Figure 5.4 shows the different processes of vegetation health determination; step (a) represents the impacts of diverse processes and factors such as fragmentation or infestations in forest ecosystems. After the impact of these diverse effects, vegetation reactions in vegetation ecosystems go after; step (b) consists in leading to changes in characteristics and attribute variations. Step (c) reveals that the spectral response can be estimated through close-range as well as airborne and spaceborne remote sensing (RS) data. The diagram (d) represents a case of the hyperspectral band answer, on the basis of fire impact. This procedure of transformation from forest traits to forest trait variances is an elementary theory that runs from close-range remote sensing processes, extending to spaceborne and airborne remote sensing records.

Quick and precise quantization of primary symptoms is significant from a pest supervision approach in addition to attempts at remotely identifying plant stress because disease or insect action employs rules of biophysical remote sensing (Lawley et al. 2016). Vegetation stress typically results in an amplification in observable reflectance caused by a decline in chlorophyll with a resulting decline in assimilation of observable light and a decline within NIR reflectance since transformations in the internal leaf composition (Zargar et al. 2011).

In addition, remote sensing has the immense capacity to be employed as an efficient and economical method to recognize diseased vegetation in a crop, principally contaminated plants comprise several spectral reactions when compared to vigorous

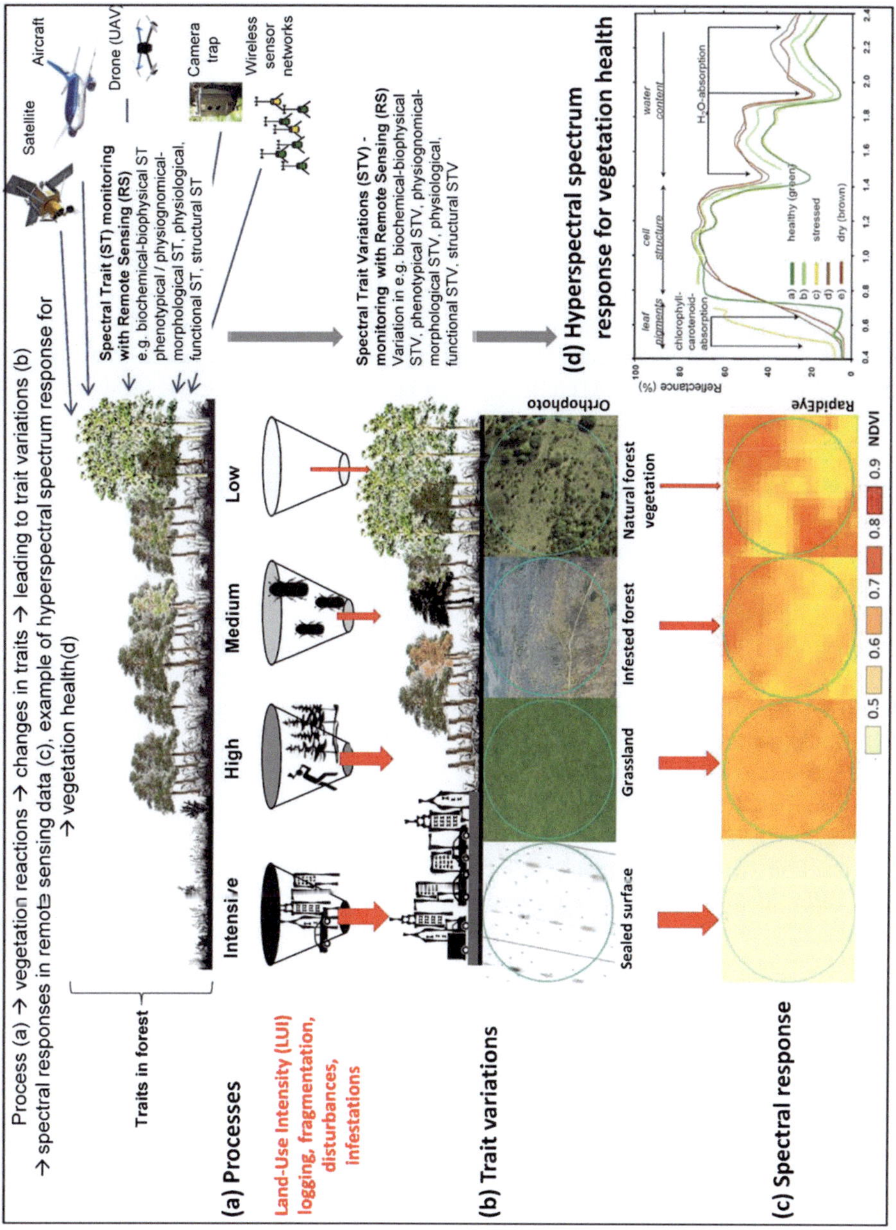

Fig. 5.4 Schematic diagram showing a chain of effects on vegetation health (eg. forest). (Adapted from Lausch et al. 2018)

plants (Agam et al. 2007; Dandois and Ellis 2010). Physiological responses of vegetation caused by infection result in a transformation of spectral reflectance imputable to the reducing chlorophyll quantity and modifying internal organization. Since the chlorophyll quantity is inclined to diminish under disease stress, the occurring solar energy assimilation of the green vegetation usually results in a decline in the visible area. Subsequently, the spectral reflectance normally is superior in the observable green spectrum relying on the disease severity. The potent spectral reflectance of green vegetation in the near-infrared range is essentially imputable to its internal foliar composition.

Vegetation under malady stress furthermore shows different degrees of internal structural modifications, which conduct to a decline of spectral reflectance in the near-infrared range. These spectral characteristics of vegetation are the foundation for remote sensing of malady-stressed vegetations (Wang et al. 2010). Pests and malady can provoke physiological tensions and physical modifications in vegetation, such as yellowing (decrease in vegetative pigment) or else chlorosis, necrosis expressed by damage on cells, etc. (Peijun et al. 2010). Accidentally, these variations can modify the reflectance properties of vegetation in the observable fraction of the electromagnetic range (from 400 to 700 nm); the reflectance of green vigorous vegetation is moderately small in perceptible fraction because of high assimilation via pigments (chlorophyll) in vegetative leaves. Whether there is a diminution in pigments caused by pests otherwise diseases, the reflectance in this spectral area will enhance. In stressed plants, leaf chlorophyll quantity reduces, thus changing the amount of light absorbing pigments, leading to a diminution in the total assimilation of luminosity. These modifications influence the spectral reflectance marks of plants throughout a diminution in green reflection and an enhancement in blue and red reflections, resulting in modifications in the usual spectral reflectance samples of vegetations (Barton 2012).

Recently, considerable advancement is made in remote sensing methods for monitoring diseases at subsequent four echelons: single leaf scale (ground-based), canopy scale (ground-based), field crop scale (aerial) and finally countries/regional scale (satellite-based). Remote sensing data at one leaf, canopy and field crop scale stages offer local and incomplete investigational information, whereas satellite-based remote sensing can offer an adequate and low-cost data base. It furthermore provides the benefit of constantly collected data and accessibility of instantaneous or archived data sets. A number of examples of satellite and other remote sensing methods employed for detecting crop diseases are presented in Table 5.1.

Table 5.1 Satellite and remote sensing methods employed for detecting crop diseases

Hosts	Diseases	Sensors	References
Wheat	Yellow rust	SPOT5 image	Zhang et al. (2009)
Rice	Sheath blight	ADAR system 5500	Zhihao et al. (2003)
Soybean	Cyst nematode	Landsat 7	Nutter et al. (2002)
Oak	Wilt	Hyperspectral satellite imagery	Blake et al. (2005)
Olive	*Xylella fastidiosa*	Hyperspectral/thermal aerial imagery	Zarco-Tejada et al. (2018)

5.4 Advantages and Benefits of Remote Sensing

Remote sensing techniques have been employed in diverse domains such as agroforestry, hydric resources management, land mapping and harvest estimation. The advantages of using remote sensing skill include the following:

- Spatial treatment over a large geographic region and accessibility through all periods.
- Comparatively low cost. In fact, remote sensing is an economically efficient method when replicated filed task is not required, and similarly a huge number of operators can exchange and employ the identical data.
- Tools are able to describe specified spatial distributions of regions under cultivation.
- Technology can incorporate any type of geospatial data into a database to do complete study and explanation; and then new and old data can be obtained without any directorial limits.
- Remote sensing platforms offer simpler techniques to revise and collect, therefore being user friendly in exhibiting the conclusions.

The main benefits of remote sensing are numerous. Firstly, remote sensing permits the achievement of graphical pictures over big geographical range on an appropriate foundation. It facilitates the capture data in various ranges. Moreover, it recovers the similar zones constantly and can be employed to identify transformations. Secondly, data can be saved in diverse wavelengths of the electromagnetic spectrum, which offer precise knowledge about the ground circumstances.

5.5 Conclusion

Precision agriculture is a relatively novel scientific domain that could possibly increase farmer incomes by decreasing pesticide usage, enhancing harvests, and decreasing work and operating expenses. On the basis of this deduction, the general targets were to construct an airborne/spaceborne imaging programme merging economic sensors with economic machines and consequently estimate the efficacy of this model at identifying infection and abiotic stress of plantation harvests. This will serve to the improvement of sophisticated devices that permit farmers to recognize crop complications through observation of their harvests. Remote sensing would be considered as an economic and relevant instrument for land-scale pest controlling and study.

It is clear that, based on remote sensing, it is possible to monitor vegetation health; nevertheless, due to sensor imprecision and image processing uncertainty, it is possible to produce some erroneous results. Therefore, it will be helpful to couple such approaches with artificial intelligence techniques allowing learning from past observations the right relationship linking the anomaly to the image.

References

Adam E, Mutanga O, Rugege D (2010) Multispectral and hyperspectral remote sensing for identification and mapping of wetland vegetation: a review. Wetl Ecol Manag 18(3):281–296
Agam N, Kustas WP, Anderson MC, Li F, Neale CM (2007) A vegetation index based technique for spatial sharpening of thermal imagery. Remote Sens Environ 107(4):545–558
Aiazzi B, Alparone L, Baronti S, Lastri C, Selva M (2012) Spectral distortion in lossy compression of hyperspectral data. J Electrical Comput Eng 2012:3
Anderson MC, Norman JM, Mecikalski JR, Otkin JA, Kustas WP (2007) A climatological study of evapotranspiration and moisture stress across the continental United States based on thermal remote sensing: 2. Surface moisture climatology. J Geophys Res Atmos 112(D11)
Barton CV (2012) Advances in remote sensing of plant stress. Plant Soil 354(1–2):41–44
Bioucas-Dias JM, Plaza A, Camps-Valls G, Scheunders P, Nasrabadi N, Chanussot J (2013) Hyperspectral remote sensing data analysis and future challenges. IEEE Geosci Remote Sens Mag 1(2):6–36
Blake W, Hongjie X, Paul J (2005) Early detection of oak wilt disease in *quercus ssp.*: a hyperspectral approach. Pecora 16 "Global Priorities in Land Remote Sensing" October 23–27, 2005 * Sioux Falls, South Dakota, USA
Campbell JB, Wynne RH (2011) Introduction to remote sensing. Guilford Press, New York
Chen N, Zhang X, Wang C (2015) Integrated open geospatial web service enabled cyber-physical information infrastructure for precision agriculture monitoring. Comput Electron Agric 111:78–91
Dandois JP, Ellis EC (2010) Remote sensing of vegetation structure using computer vision. Remote Sens 2(4):1157–1176
Erener A (2011) Remote sensing of vegetation health for reclaimed areas of Seyitömer open cast coal mine. Int J Coal Geol 86(1):20–26
Fenghua W, Shujuan Z (2008) Research progress of the farming information collections key technologies on precision agriculture. Trans Chin Soc Agric Mach 39(5):112–121
Frohn RC, Lopez RD (2017) Remote sensing for landscape ecology: new metric indicators: monitoring, modeling, and assessment of ecosystems. CRC Press
Gebbers R, De Bruin S (2010) Application of geostatistical simulation in precision agriculture. In: Geostatistical applications for precision agriculture. Springer, Dordrecht, pp 269–303
Gupta RP (2017) Remote sensing geology. Springer
Khosla R (2010) Precision agriculture: challenges and opportunities in a flat world. In: 19th World Congress of Soil Science, soil solutions for a changing world. Brisbane, Australia
Koleshko VM, Gulay AV, Polynkova EV, Gulay VA, Varabei YA (2012) Intelligent systems in technology of precision agriculture and biosafety. In: Intelligent systems. InTech
Kuenzer C, Bluemel A, Gebhardt S, Quoc TV, Dech S (2011) Remote sensing of mangrove ecosystems: a review. Remote Sens 3(5):878–928
Lake JV, Bock GR, Goode JA (2008) Precision agriculture: spatial and temporal variability of environmental quality (vol. 210). Wiley
Landgrebe DA (2005) Signal theory methods in multispectral remote sensing (vol. 29). Wiley
Lausch A, Erasmi S, King DJ, Magdon P, Heurich M (2016) Understanding forest health with remote sensing-part I—a review of spectral traits, processes and remote-sensing characteristics. Remote Sens 8(12):1029
Lausch A, Borg E, Bumberger J, Dietrich P, Heurich M, Huth A et al (2018) Understanding forest health with remote sensing, part III: requirements for a scalable multi-source forest health monitoring network based on data science approaches. Remote Sens 10(7):1120
Lawley V, Lewis M, Clarke K, Ostendorf B (2016) Site-based and remote sensing methods for monitoring indicators of vegetation condition: an Australian review. Ecol Indic 60:1273–1283
Lillesand T et al (2014) Remote sensing and image interpretation. John Wiley & Sons, Hoboken
Nutter FW, Tylka GL Jr, Guan J, Moreira AJD, Marett CC, Rosburg TR, Basart JP, Chong CS (2002) Use of remote sensing to detect soybean cyst nematode-induced plant stress. J Nematol 34(3):222–231

Ozdogan M, Yang Y, Allez G, Cervantes C (2010) Remote sensing of irrigated agriculture: opportunities and challenges. Remote Sens 2(9):2274–2304

Peijun DU, Xingli LI, Wen CAO, Yan LUO, Zhang H (2010) Monitoring urban land cover and vegetation change by multi-temporal remote sensing information. Min Sci Technol (China) 20(6):922–932

Rees WG, Pellika P (2010) Principles of remote sensing. Remote Sensing of Glaciers. London

Sabins FF (2007) Remote sensing: principles and applications. Waveland Press

Schellberg J, Hill MJ, Gerhards R, Rothmund M, Braun M (2008) Precision agriculture on grassland: applications, perspectives and constraints. Eur J Agron 29(2–3):59–71

Schowengerdt RA (2006) Remote sensing: models and methods for image processing. Academic Press, Orlando

Schowengerdt RA (2012) Techniques for image processing and classification in remote sensing. Academic Press, San Diego

Singh HB, Jha A, Keswani C (eds) (2016a) Intellectual property issues in biotechnology. CABI, Wallingford. 304 pages, ISBN-13:9781780646534

Singh HB, Jha A, Keswani C (2016b) Biotechnology in agriculture, medicine and industry: an overview. In: Singh HB, Jha A, Keswani C (eds) Intellectual property issues in biotechnology. CABI, Wallingford, pp 1–4

Singh HB, Sarma BK, Keswani C (eds) (2017) Advances in PGPR research. CABI, Wallingford. 408 pages, ISBN-9781786390325

Thenkabail PS, Lyon JG (2016) Hyperspectral remote sensing of vegetation. CRC Press

Thilakarathna M, Raizada M (2018) Challenges in using precision agriculture to optimize symbiotic nitrogen fixation in legumes: progress, limitations, and future improvements needed in diagnostic testing. Agronomy 8(5):78

Twomey S (2013) Introduction to the mathematics of inversion in remote sensing and indirect measurements (vol. 3). Elsevier

Ulaby, F. T., Long, D. G., Blackwell, W. J., Elachi, C., Fung, A. K., Ruf, C., et al. (2014). Microwave radar and radiometric remote sensing (4, 5, 6). Ann Arbor: University of Michigan Press

Wang J, Sammis TW, Gutschick VP, Gebremichael M, Dennis SO, Harrison RE (2010) Review of satellite remote sensing use in forest health studies. Open Geogr J 3(1):28–42

Winstead AT, Norwood SH, Griffin TW, Runge M, Adrian AM, Fulton J, Kelton J (2010) Adoption and use of precision agriculture technologies by practitioners. In: Proc. the 10th International Conference on Precision Agriculture. pp 18–21

Zarco-Tejada PJ, Camino C, Beck PSA, Calderon R, Hornero A, Hernández-Clemente R, Gonzalez-Dugo V (2018) Previsual symptoms of Xylella fastidiosa infection revealed in spectral plant-trait alterations. Nat Plants 4(7):432

Zargar A, Sadiq R, Naser B, Khan FI (2011) A review of drought indices. Environ Rev 19.(NA.333–349

Zhang C, Kovacs JM (2012) The application of small unmanned aerial systems for precision agriculture: a review. Precis Agric 13(6):693–712

Zhang YP, Guo JB, Wang S, Wang HG, Ma ZH (2009) Relativity research on near ground and satellite remote sensing reflectance of wheat stripe rust (in Chinese). ActaPhytophylacica Sin 36:119–122

Zhihao Q, Minghua Z, Thomas C, Wenjuan L, Huajun T (2003) Remote sensing analysis of rice disease stresses for farm pest management using wide-band airborne data. International Geosciences and Remote Sensing Symposium, IV: 2215–2217, July 21-25, 2003, Toulouse, France

Biopesticides: Current Status and Future Prospects in India

6

Yashoda Nandan Tripathi, Kumari Divyanshu, Sunil Kumar, Lav Kumar Jaiswal, Atif Khan, Hareram Birla, Ankush Gupta, Surya Pratap Singh, and Ram Sanmukh Upadhyay

Abstract

For over a century, chemical control of pests is a common practice in agriculture. The average reduction in global crop loss due to use of pesticides is around ~39%. The postharvest losses and quality decline caused by storage pests are major problems in a subtropical country like India. So, the farmers have relied heavily on the use of chemical pesticides to improve their crop production, which is now paying a huge toll on the human health and environment. Though the chemical pesticides are very effective, what concerns over their use is their effect on soil and environment and presence of residue in food products. Another major issue is the development of resistance in the pests. Therefore, the use of biopesticides to control pests is now preferred over synthetic pesticides because of their pest control ability and diverse mode of actions which helps in avoiding resistance development in the pests. In a country like India with a huge diversity of plants, there is an urgent need for identifying new biopesticides which can serve the purpose of pest control. India needs to develop its own biocontrol agents (BCA) because it will be cost-effective and also environment-friendly. Major hurdle in the development and use of new biopesticides in India is the commercialization process. The farmers are reluctant to use the new products because of high cost and no practical knowledge.

Y. N. Tripathi (✉) · K. Divyanshu · S. Kumar · R. S. Upadhyay
Department of Botany, Institute of Science, Banaras Hindu University, Varanasi, India
e-mail: yashodan.tripathi3@bhu.ac.in

L. K. Jaiswal · H. Birla · A. Gupta · S. P. Singh
Department of Biochemistry, Institute of Science, Banaras Hindu University, Varanasi, India

A. Khan
Water & Steam Chemistry Division, BARC Facilities, Kalpakkam, Tamil Nadu, India

Homi Bhabha National Institute, Mumbai, India

C. Keswani (ed.), *Bioeconomy for Sustainable Development*,
https://doi.org/10.1007/978-981-13-9431-7_6

Keywords
Biopesticides · Biocontrol agents · Agriculturally important microorganisms · Green revolution · Sustainable agriculture

6.1 Introduction

Indian agriculture sector accounts for 18% of India's gross domestic product (GDP), and about 58% population of India is dependent upon agriculture for its livelihood. India has the tenth largest arable land resources in the world and is among the 15 leading exporters of agricultural products in the world. Agricultural exports from India reached USD 38.21 billion in FY18 and USD 21.61 billion between April and October 2018.The food grain production during 2017–2018 was estimated at record 284.83 million tonnes, and the government is targeting to increase it to 285.2 million tonnes in 2018–2019. The gross value added by agriculture, forestry and fishing is estimated at Rs. 17.67 trillion (USD 274.23 billion) in FY18. The above data clearly depicts the reliance of Indian economy over agriculture. The Indian government is aiming to double the farm income by 2022. Impetus to this daunting task of doubling the farmers' income can be provided only by increasing investments in agricultural infrastructure such as irrigation facilities, warehousing and cold storage and by using genetically modified crops (APEDA, Union Budget 2018–2019). Innovative solutions are required to meet the ever-increasing demands for food and fibre by fast-growing population of India. These demands can only be met by protecting crops from pest losses while conserving limited natural resources and maintaining quality of the environment (Naranjo et al. 2015). Usage of biopesticides has seen rise in recent years, and according to a data presented by the Union Ministry of Agriculture, between 2010–2011 and 2016–2017, the all-India consumption of biopesticides increased from 5151 to 6340 tonnes (by 23%), while that of chemical pesticides grew from 55,540 to 57,000 tonnes (only 2%) for the period under review.

In the early periods of the nineteenth century, Agostine Bassi was the first one to use spores of the white muscardine fungus (*Beauveria bassiana*) to protect silkworms from insect pests and diseases (Xiao et al. 2012). But till date, the biopesticide market has not evolved as per the expectations, and its market is still very restricted when compared to synthetic pesticides. The use of biopesticides to control pests is preferred over synthetic pesticides just because of their mode of action. On one hand, where synthetic insecticides are neurotoxic to pests, biopesticides perform their action by inducing mating disruption, anti-feeding, suffocation and desiccation of the pests. Unlike chemical pesticides, which are made from industrial chemicals, biopesticides are derived from plant extracts, fungi, bacteria, protozoans and minerals. They are used for crop protection and are found to be benign for both humans and the environment (Olson 2015). In the presently followed pest control strategy, the biopesticides have grown importance over the conventional synthetic pesticides. Biopesticides have become popular because of the variety of mode of actions they offer; hence, resistance development in pests is slower (Bisen et al. 2015; Fraceto et al. 2018; Ram et al. 2018; Singh et al. 2017). Therefore, there is substantial scope in identifying and developing biopesticides as alternative pest management resource (Rajapakse 2016). To feed the ever-growing population of India, more crop production is required in the limited amount of land which is available for cultivation. To support the fast-growing population of the nation, along with a high emphasis on achieving food grain self-sufficiency, Indian farmers were compelled to make considerable use of pesticides. Undoubtedly, the use of chemical pesticides has provided a valuable aid to agricultural production, increasing crop protection and yield, but their uncontrolled use has traded heavily on the environment. The overall benefit derived from the pesticides has been overshadowed by the discovery of pesticide residues in various dimensions of the environment (Yadav et al. 2015). Their nontarget effects on beneficial microbes are critical for soil health; their residues are left in feed and fodder and also cause environmental pollution (Singh et al. 2015).

The average production loss due to pests in India has been reported to be as high as USD 42.66 million (Subash et al. 2017). Very few chemical pesticides are available in the Indian pesticide market. This is because some products have been withdrawn from the market for regulatory or commercial reasons or due to the development of resistance by the pests against these products. So, there is an urgent need to look for some better technologies and products based on biological processes to control the pests (Kumar 2012). Biopesticides have been found to be the most suitable candidate for this purpose, but in a developing nation like India, there are several hurdles in the path of commercializing biopesticides. Major constraint being the slow action and low persistence of biopesticides when exposed to solar UV, high cost of production and lack of awareness among the Indian farmers. Apart from this, the poor awareness of decision-makers about opportunities offered by biopesticides, lack of multidisciplinary expertise in the crucial later stages of development, difficulty in conducting toxicological tests and the long testing period of bioactive compounds before registration and commercialization are the other challenges that need basic attention (Keswani 2015a, b; Keswani et al. 2016).

According to a report, the pesticide consumption in India increased by almost 50% from 2009 to 2015. However, per hectare use of pesticide in India was 0.29 kg/ha in 2015 which was far less compared to other countries like China (13.06 kg/ha), Japan (11.85 kg/ha), Brazil (4.57 kg/ha) and other Latin American countries (Subash et al. 2017).

6.2 Classification of Biopesticides

The three different classes of biopesticides that the US Environmental Protection Agency (EPA) has identified are microbial, biochemical and plant-incorporated protectants (PIPs). In microbial pesticides, whole microorganism such as bacteria, fungi, viruses and others are used as pesticides as the active ingredient and have been used efficiently to control insect pests. Despite being specific for its target pest, each active microbial ingredient can check the growth of several different kinds of pests.

In biochemical pesticides, microbial extracts or natural products from other sources like plant extracts or yeast fermentation products are used. These pesticides perform their action by nontoxic mechanisms. The biochemical pesticides often include the following: (a) semiochemicals (hormone mimics), insect sex pheromones that interfere with their mating and population build-up; (b) hormones, moult hormones (ecdysteroids) and juvenile hormones (IGR); (c) natural plant regulators, auxins, gibberellins, cytokinins and inhibitors; and (d) scented extracts (attractants) that are very small molecules and are used as traps and vegetable oil (Rajapakse 2016).

Plant-incorporated protectants (PIPs) are those substances that are produced naturally in genetically modified (GM) crops and are typically macromolecular in nature (Parker and Sander 2017). Such examples of producing transgenic plants may include incorporation of Bt gene, kinase inhibitor gene, lectins, chitinase, etc. into the plant genome. No harmful effect on either humans or animals is observed by the protein products which are produced by these transgenic plants to develop resistance against the pests (Kumar 2012).

All the prototype biopesticides have evolved from the bacteria *Bacillus thuringiensis* (Bt) that produce a toxin (Bt toxin) which binds to insect gut receptor protein and disrupt the gut upon ingestion. All the three varieties of biopesticides (microbial, biochemical and PIP varieties) are derived from this bacteria and its toxin.

6.3 Growth Drivers for India Biopesticide Market

The Indian agriculture has seen a sharp increase in the use of chemical pesticides in the last few years posing serious implications on human health, environment and groundwater. Hence, it has become imperative to find an environment-friendly substitute for the chemical pesticides. Biopesticides have been seen as next generation pesticides and are also believed to have a huge potential to promote sustainable agriculture in this country. On account of presence of higher pesticide residues in food crops and increasing pest resistance, strict regulations have been imposed on use and sale of synthetic pesticides by some developed countries. Around 500 biopesticides, duly registered by the Central Insecticide Board (CIB), are available in the Indian market. The efficacies of these products have been demonstrated in many laboratories, but the major problem is their quality control. The factors driving the growth of the Indian biopesticide market are the environment-friendly nature of biopesticides, encouraging public support policies, increasing public awareness and lesser development of pest resistance.

6.3.1 Government Efforts to Promote Use of Biopesticides

In India, the pesticides are used as the most important tool to protect public health and help in agricultural development. However, the overuse of pesticides is now showing the reverse effect. Pesticides constitute an integral part of the present-day agriculture, but there are some serious problems which are associated with it and these problems must be tackled with strong policies. There are several government agencies which regulate the use of pesticides taking into consideration their residues in food and water (Yadav et al. 2015). According to an annual global report, there are three million cases of acute, severe poisoning which occur due to intoxication of pesticides (Gunnell and Eddleston 2003; WHO 1990). It is a significant health issue in India as well, but the government has till date failed to produce any national data for it.

However, the Government of India (GOI) has shown its concerns about the ill-effects of pesticide use on human health by implementing measures, such as integrated pest management (IPM), prohibition of highly hazardous pesticides, restricting the use of toxic compounds and the development of a National Implementation Plan (NIP). In addition, the use of biopesticides has been strongly recommended by the National Farmer Policy. IPM, which is an eco-friendly approach to manage pests, has been used by GOI to strengthen and modernize pest management approaches. In IPM, various cultural, mechanical and biological methods and need-based use of chemical pesticides, preferably biopesticides and biocontrol agents (BCA), are employed. The 31 central IPM centres operating across the nation perform the tasks of monitoring the pests and diseases, production and release of biocontrol agents and biopesticides and conservation of biocontrol agents apart from providing training to farmers at the basic level (Yadav et al. 2015).

The health risks associated with persistent organic pesticides has greatly perturbed international community. All the nations are now seeking mutual help in order to reduce and minimise the production, use and release of persistent organic pesticides. India has also signed and ratified three international legally binding instruments out of six which have been negotiated and concluded till date. India signed the Stockholm Convention on Persistent Organic Pollutants (SCPOPs), which is one of the most important legally binding international agreements to protect human health and the environment from some of the most dangerous chemical on earth on 14 May 2002, which was later ratified on 13 January 2006.

Being a member nation of SCPOPs, India had to develop a National Implementation Plan (NIP) which recommends certain priorities to the government in order to ensure the implementation of the obligations of the convention. Some of the recommendations of NIP for the GOI include the following: to encourage and develop nonpersistent organic pesticide alternatives to DDT, encouragement for production and demonstration of neem-based biopesticides, designing distinct mechanism to deal with new persistent organic pesticides and strengthening of institutions involved in pesticides research and capacity building.

The manufacturing, sale, transport and distribution, export, import and use of pesticides are regulated by the Ministry of Agriculture through the ‘Insecticides Act 1968’. The central and state governments are assisted on technical matters by Central Insecticide Board and Registration Committee of India (CIB&RC). The data for all kinds of pesticides are collected by CIBRC.

6.3.2 Biopesticides in Integrated Pest Management

One of the biggest growth drivers of biopesticides in the pesticide market is their suitability in integrated pest management (IPM) programs, in combination with other biological, cultural and pesticide approaches. Some advantages of using biopesticide are that they are less toxic to pollinators and are compatible with natural enemies such as hymenopteran parasitoids. Also, in addition, they help in delaying pest resistance when used in rotation with synthetic pesticides (Birch et al. 2011). The biopesticides are now being made more efficient by combining them with 'soft' biological options. For example, the biopesticides which are used for controlling codling moth are based on a granulosis virus and have been available for many years but because of their poor performance were not in use but became highly useful on integration with a codling moth female sex pheromone to disrupt mating and thus helping in control (Krawczyk et al. 2010).

6.3.3 Increased Health Awareness

Exposure of humans to the pesticides present in the environment can be through different routes such as by inhalation, ingestion or dermal contact. The effects of pesticide poisoning are sometimes devastating. Several diseases like cancer, kidney failure, immunosuppression, sterility, etc. are attributed to pesticide poisoning. During the last two decades, public concern over pesticides' adverse effect has been clearly visible. After successful implementation of green revolution, India is now facing another bigger challenge of conserving its ecosystem from the toxic chemical pesticides (Abhilash and Singh 2009). The people have now become more health conscious, and now the demand for organic products are on rise in the Indian market.

6.4 Market of Biopesticides in India

Presently, biopesticides encompasses a minor portion of entire crop fortification market which accounts for 5% in total and is supposed to contribute up to 50% of the overall pesticide market by 2050 (Parker and Sander 2017). Presently, the estimated global market of biopesticides is of ~USD 3–4 billion which corresponds to compound annual growth rate (CAGR) of 14.1%. The Indian biopesticide market is also growing rapidly and is centred across a few companies which dominate the organised sector of the Indian biopesticide market. Pest Control India, International Panaacea Ltd., T Stanes and Biotech International are the major organized players (Table 6.1). Other companies dominating Indian market for their products, services and continuous product developments are Camson Bio Technologies Ltd., Sri Biotech Laboratories India Ltd., Valent Biosciences Corp., Eid Parry, etc.

The major share of the Indian biopesticide market is contributed by these few companies because of their sustained R&D procedures and deep-rooted distribution

Table 6.1 List of some Indian biopesticide companies and their products

S. No.	Company name	Product name	Properties	Reference
1.	Pest Control India	Niprot®	Controls the soil-borne fungal pathogen and protects the crops against root and collar rot. It also protects the plants against damping off and root-knot nematode and uses formulation of antagonistic fungi	https://www.pestcontrolindia.com
		Su-Mona®	Controls and tackles soil-borne pathogens affecting horticulture crops and is also effective against bacterial root knot and lesion nematodes. It increases systemic resistance and also plant growth	
		Bio-Jaal™	It is the first oil-based biopesticide in India, enhances field stability, adheres on insect body in a better way and kills the pathogen at low exposure. It does not affect beneficial insect	
2.	Biotech International Ltd.	Biolep	Contains strain of *Bacillus thuringiensis* var. *kurstaki,* serotype H-3 a, 3b, strain Z-52. Available in wettable powder form. Highly active up to 3rd instar larva of insect. It does not harm human beings, animals, fishes, birds, etc	https://www.biotech-int.com/biosepticides.html
		Bioline/Verti-Star	Contain strain of *Verticillium lecanii.* It is highly active on sucking pests like scale insects, *Coccus viridis*, aphids, thrips, mealybugs, jassids, hoppers, etc. on crops like mustard, sugarcane, mango, sorghum, rice, cotton, tomato, grapes, pomegranates, etc. infects all stages of insects including eggs, larvae, pupae, nymphs and adults and is safe for environment, and human beings	
		Biomet/Ankush	Contains strain of *Metarhizium anisopliae.* Available in wettable powder form and aqueous suspension. Highly active on white grubs, beetle grubs, caterpillars, semiloopers, cutworms and sucking pests. It infects all stages of insects including eggs, larvae, pupae, nymphs and adults. The spores attach to insect cuticle/integuments; they germinate and penetrate insect body	

(continued)

Table 6.1 (continued)

S. No.	Company name	Product name	Properties	Reference
		Biorin/Kargar	Contains strain of *Beauveria bassiana*. Available in wettable powder form as well as aqueous suspension. It is highly active on lepidopteran caterpillars including *Helicoverpa, Spodoptera, Plutella*, borers, hairy caterpillars, pests of vegetables and fruit plants and sucking pests like scale insect, mealy bug, etc.	
		Biovirus-H	Contains strain of nuclear polyhedrosis virus (NPV) of *Helicoverpa armigera*. Available in aqueous suspension. Highly active on *Helicoverpa armigera* pest of cotton, pigeon pea, gram, tomato, okra, brinjal, chilli, cabbage, pea, groundnut, tobacco, millets, oilseed crops, roses etc.	
		Biovirus-S	Contains strain of nuclear polyhedrosis virus (NPV) of *Spodoptera litura*. Available in aqueous suspension. Highly active on *Spodoptera litura* pest of cotton, groundnut, pulses, cabbage, chillies, tobacco, oil seeds crops, roses, etc.	
3.	International Panaacea Ltd.	Milgo™ (fungicide)	Contain *Ampelomyces quisqualis,* it controls the powdery mildew and is also effective against *Botrytis cinerea, Alternaria solani, Colletotrichum, Coccodes, Cladosporium cucumerinum*, etc. at early growth stage of pathogens. The target crops include cucurbits, grapes, apple, peas, beans, tomato, pulses, cumin, chillies, coriander, mango, ber, peas, strawberry, medicinal and aromatic crops and roses that are infected by powdery mildew	https://www.iplbiologicals.com
		Varunastra™, Vertifire-L (insecticide)	It contains *Verticillium lecanii* and infects all the stages of insects. Aphids, thrips, mealybugs, white flies and all types of mites are its target insects. The target crops include banana, grapes, guava, citrus, mango, sapota, apple, coconut, paddy, cotton, tomato, chilly, brinjal, onion, okra, tea, cardamom, coffee, aromatic and medicinal crops	
		Nematofree (Nematocide)	It effectively controls plant parasitic nematodes specifically root-knot nematodes infecting tomato, brinjal, carrot and okra crop. It is effective against nematode (*Meloidogyne* spp.), reniform nematode (*Rotylenchulus* spp.) etc.	
		Suraksha™ (growth promoter)	It contains product of *Rhizobacteria* and enhances the growth of crop plants like grapes, banana, pomegranate, guava, mango, citrus, chilly, tomato, brinjal, okra, tea, coffee, plantation crop, medicinal and aromatic crops, cotton, maize, paddy, wheat, etc.	

4.	T. Stanes	Bio-Nematon	Bio-Nematon contains spores and mycelial fragments of a selective strain of naturally occurring entomopathogenic fungus *Paecilomyces lilacinus*. The spore of this fungus acts by infecting, parasitizing and killing eggs, juveniles and young adults of most phytophagous nematode species. It is compatible with neem and other biopesticide products	http://www.tstanes.com
		Nimbecidine EC	It is a neem oil-based botanical insecticide containing azadirachtin and other limonoids including meliantriol, salannin, nimbin and various other terpenoids. It controls pests such as whitefly, aphids, thrips, mealybugs, caterpillars and leafhoppers in a wide range of crops. It exhibits its mode of action by acting as an antifeedant, repellent, oviposition deterrent, insect growth regulator and sterilant	
		Bio-Power	Bio-Power contains spores and mycelial fragments of a selective strain of naturally occurring entomopathogenic fungus *Beauveria bassiana*. It is a biological insecticide; it effectively controls pests such as borers, cutworms, root grubs, leafhoppers. Whitefly, aphids, thrips and mealybug. It increases the productivity by improving the crop health and is eco-friendly	
5.	Fortune Biotech	Fortune Aza	It is a 3% azadirachtin-formulated insect growth regulator (IGR). It controls more than 300 species of insect and prevents moulting between larval, pupal and nymphal stages	http://www.fortunebiotech.com/products.php
		Fortune Aza Technical	Most potent tetranortriterpenoid isolated till date from neem seeds. It is effective against more than 300 insect species as antifeedant, repellent and as an insect growth regulator. It provides a broad-spectrum pest control for fruits, vegetables and plantation crops, greenhouses, turf, outdoor ornamentals and agricultural crops such as sugarcane, paddy, cotton and tea. It is stable at cool temperature in the absence of water and light and also stable in organic solvents	
		Fortuneem Cake	It controls nematode population and other soil-borne diseases and is rich in NPK and micronutrients. It controls nematode population by increasing the metabolism of microbes, which proves to be toxic for the nematodes	

(continued)

Table 6.1 (continued)

S. No.	Company name	Product name	Properties	Reference
6.	Excel Crop Care	Sudocel	Sudocel contains a prominent soil-inhabiting bacterium *Pseudomonas fluorescence*. The bacterium acts as an excellent bioagent by coiling around and penetrating and killing fungi that are pathogenic to crops. They also secrete enzymes and antibiotics which can digest their cell walls	http://www.excelcropcare.com/product
		Tricho HR	It contains a beneficial fungi *Trichoderma harzianum* 1% and is highly effective in controlling wide range of soil-borne crop diseases. It is biodegradable and target specific and leaves no toxic residues	

channel spread across several regions of India. Due to their sincere efforts, the consumption of biopesticides increased from 219 tonnes in 1996–1997 to 683 tonnes in 2015–2016 (Subash et al. 2017). The pace of development of Indian biopesticide market is not very impressive. Its market can be only expanded with government's aid not only in monetary terms but also by the development of storage facilities at different level of supply chain, which require special training and skills.

6.5 Analysis of Indian Biopesticide Industry

While considering legislative bodies governing biopesticide industries in India, two critical apprehensions should be taken into account. Firstly, guidelines must be coherent to guarantee human and environmental safety and, secondly, to illustrate constant and dependable quality of biopesticide products (Chandler et al. 2011). According to (Chandler et al. 2011), only sanctioned biopesticide products can be legally used for crop protection in most of the developed and developing countries. In India, production of biopesticides is principally structured by legal agendas originally considered for chemical pesticides and insecticides. The Insecticides Act, 1968, and Insecticides Rules, 1971, legalize import, registration process, manufacture, sale, transport, circulation and utilization of insecticides with a view to overcome risk to human beings and animals, as well as all connected matters (Abhilash and Singh 2009) (Fig. 6.1).

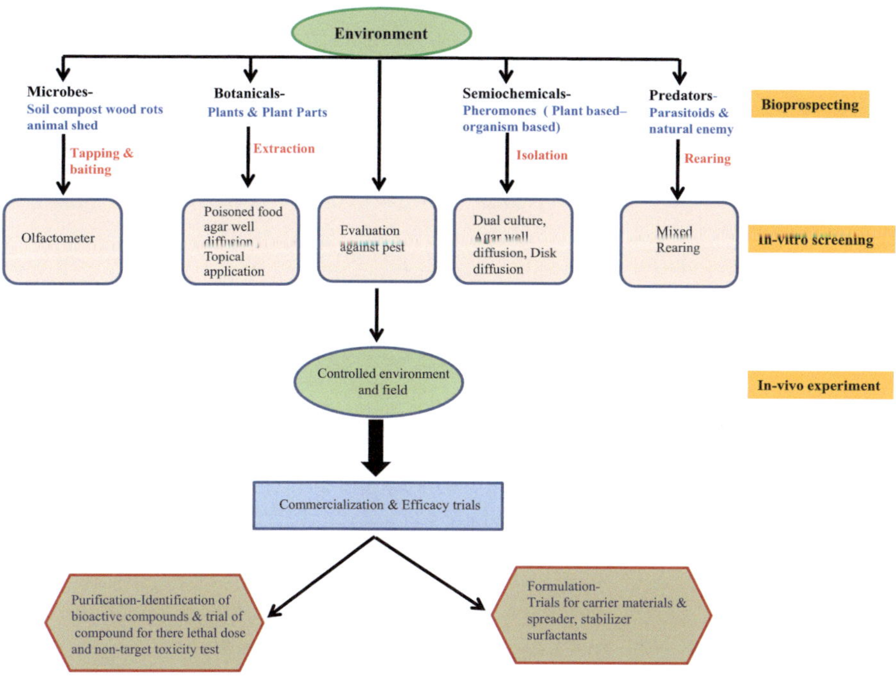

Fig. 6.1 Steps in production and commercialization of different types of biopesticides

Pesticide industry in India has been largely divided into six groups comprising herbicides, insecticides, fungicides, plant growth regulators, biopesticides and rodenticides (Bharatbhai 2017). There is an enormous prospective of development of biopesticide market in India in future years depending on the government support and increasing consciousness on the use of nontoxic and environmental friendly pesticide in country. The Indian biopesticide market has witnessed an unexpected growth in FY 2007–2012 in which western and southern regions of India have been a major sink where biopesticides are chiefly expended. The biopesticide market has observed a growth and revenue contribution of 26.4% compared to 10.2% growth recorded in total pesticide market during the same period. So far, more than 500 types of biopesticides are available in the Indian market, which are suitably registered by the Central Insecticide Board (CIB). Widespread research on these biopesticides in national laboratories has confirmed the effectiveness of these biopesticides (Singh et al. 2016), but maintaining their quality requires considerable R&D approach which is a monotonous assignment and not all the companies are efficient towards this technology (Table 6.2). Furthermore, their high prices have restricted their consumption by poor Indian farmers which demote the superiority of these pesticides in Indian market.

6.6 Comparison of Indian and Global Biopesticide Market

According to an estimate, there will be around nine billion people on the globe by 2050. In order to increase agriculture yield, pesticides are used, and their present estimated market value is of USD 50–60 billion. The global pesticide use has increased by 50-fold in the last 70 years (Vendan 2016). The average reduction in global crop loss due to use of pesticides is around ~39% (Oerke 2006). The current global biopesticide market is approx. 6% of the total pesticide market and is estimated to be of ~USD 3–4 billion. The compound annual growth rate (CAGR) of biopesticides is 14.1% which is relatively higher compared to CAGR of synthetic pesticide (4.8%).

The Indian population is expected to exceed 1.5 billion by 2050. There will be a huge pressure on the agricultural sector to feed such a gigantic population by increasing food production in an environmentally sustainable manner. To achieve this objective, use of pesticides is preferred in Indian agricultural system. India is not only among the biggest consumers of pesticides, but also it ranks 12th in production of it. Vendan (2016) reported that per hectare consumption of pesticides is 280 g/ha in India, but an interesting thing is that per hectare consumption of chemical pesticides is significantly low compared to global data. Of the total pesticides consumed in India, insecticides contribute to more than 50%, while herbicides' and fungicides' share are low. Due to limited number of chemical pesticides available in the Indian market and also because of development of pest resistance and health hazards associated with synthetic pesticides, the demand for biopesticides has been growing in India as well. According to a report of the Department of Agriculture Cooperation and Ministry of Agriculture and Farmers Welfare, the consumption of

Table 6.2 Indigenous biopesticides (list of representative biopesticides registered in India under section 9 (3) of the Insecticides Act, 1968)

Biopesticides	Taxus	Formulations	Targets	Trade name
Bacillus thuringiensis subsp. *israelensis*	Bacterium	5.0% WP, 5.0% AS	Lepidopteran pests	Tacibio
B. thuringiensis subsp. *Kurstaki*		5.0% WP, 7.5% WP	Lepidopteran pests	Bio Dart
Pseudomonas fluorescens		0.5%, 1.0% WP	Soil-borne diseases	Bio Dart
Bacillus subtilis		2.0% AS		
Bacillus thuringiensis subsp. *sphaericus*		1.3% FC	Mosquito larvae	VectoLex
Bacillus thuringiensis subsp. *galleriae*		1.3% FC		
Trichoderma viride	Fungus	1.0% WP	Soil-borne pathogens	Bioderma
Beauveria bassiana		2.15% WP, 10% SC or 1.0%, 1.15%	Coffee berry borer, diamondback moth, grasshoppers, whiteflies, aphids	Myco-Jaal
Ampelomyces quisqualis		2.0% WP	Powdery mildew	Bio-Dewcon
Trichoderma harzianum		0.5%, 1.0%, 2.0% WP	Soil-borne pathogens	Biozim
Metarhizium anisopliae		1.0%, 1.5% WP	Coleoptera, Lepidoptera, termites, mosquitoes, leafhoppers, beetles, grubs	Biomet
Paecilomyces lilacinus		1.0%	Whitefly	Yorker
Verticillium lecanii		1.15%	Whitefly, coffee green bug, homopteran pests	Verisoft
Verticillium Chlamydosporium		1.0% WP	Nematodes	
Nuclear polyhedrosis virus of *Helicoverpa armigera*	Virus	0.43%, 0.5%, 0.64%, 2.0%	*Helicoverpa armigera*	Helicide
Nuclear polyhedrosis virus of *Spodoptera litura*		0.5%, 2.0%	*Spodoptera litura*	Spodocide

Source: Ministry of Agriculture and Farmers Welfare, Government of India
Note: WP wettable powder, *AS* aqueous solution or aqueous suspension, *SC* suspension concentrate, *FC* flowable concentrate

biopesticides has increased from 219 tonnes in 1996–1997 to 683 tonnes in 2000–2001 and further to around 3000 tonnes in 2015–2016. In 2016, the Indian biopesticide market was USD 70.45 million, and it is growing with a CAGR of 17.08% (https://www.mordorintelligence.com/industry-reports/indian-biopesticides-market).

6.7 Biopesticide Commercialization and Regulatory Barrier

Of the overall pesticide market in India, biopesticides constitute around 3% only. Till date, only 14 biopesticides have been registered under the Insecticide Act 1968 in India (Subash 2017). According to a forecast, the 5-year compound annual growth rate of Indian biopesticide industry will be of 17.08% between 2017 and 2022 (https://www.mordorintelligence.com/industry-reports/indian-biopesticides-market). However, the overreliance of Indian farmers on synthetic chemical pesticides can be removed only by substantial increment of biopesticide industry. Many factors that have been hampering the potential growth of biopesticides in the Indian market are their unavailability, lesser reach and also continued disappearance of mixed/multiple cropping system. Development of new biopesticides is difficult because of the several features of the agricultural sectors. The farmers find it hard to use new products because they compare the cost involved and the profit earned in exactly the same manner as the companies which will develop any biopesticide product only if it earns a good profit:

Lack of profit from niche market products – a lot of biopesticides are highly selective like the bioinsecticides based on baculoviruses, such as the CpGV. They are typically selective against a particular insect; thus, they have low profit potential as the size of their market is limited.

Fixed costs – the farmers who use biopesticides face a large fixed cost compared to other farmers who use conventional chemical pesticides because the fixed cost associated with biopesticides is not distributed among a large number of farmers, and hence early users find it disadvantageous.

Farmers risk aversion – A large number of farmers who have been using chemical pesticides for a very long time have gained substantial experience and confidence in their effectiveness, while there are a lot of uncertainties in farmers' mind when it comes to using new products for which they don't have even practical knowledge.

On the basis of several characteristics of the living and nonliving entities that make up the biopesticides, every government tries to regulate their authorization and use. These regulations are meant for human and environmental safety and also to ensure that reliable biopesticide products enter into the market. In EU, only those biopesticides are used for crop protection whose efficacy has been quantified and proved and the guidance of OECD in this respect is even more strict; only those biopesticide that gets authorization pose minimal or zero risk. The major hurdles in the path of commercialization of new biopesticides are the regulatory authorities who sometimes don't have enough expertise in biopesticides, and they tend to delay in the process of authorization. In addition, another obstacle is that the regulatory system for biopesticides is based on chemical pesticide model. Several of the government regulators have been treating the biopesticides in the same manner as chemical pesticides and are thus slow in recognizing the fact that for biopesticides a separate regulatory process is needed (Chandler et al. 2011).

Even if India has a great diversity of botanicals, the commercialization of botanical pesticides is difficult here due to quality control and product standardization issues. Like synthetic pesticides, the improper and excessive use of botanical pesticides leads to development of pest resistance. The phytotoxicity of botanical pesticide is also a matter of concern. For example, neem oil-based biopesticide is often phytotoxic to tomato, brinjal and ornamental plants at high levels (Nawaz 2016).

6.8 Pest Management Strategies

6.8.1 Microbial Biopesticides

Microbial biopesticides are also known as biological control agents (BCAs). In this category of pesticide, the active component is a microorganism that is naturally occurring or genetically modified bacteria, fungi, algae, viruses or protozoans. The advantages of microbial pesticide are their higher selectivity and less or no toxicity in comparison with conventional chemical pesticides present in market (MacGregor 2006). The commonly used microbial biopesticides are living organisms which are pathogenic or toxic for the target pest. They broadly include biofungicides (*Trichoderma, Pseudomonas, Bacillus*), bioherbicides (*Phytophthora*) and bioinsecticides (Bt) (Gupta and Dikshit 2010). Microbial pesticides contain a variety of microorganisms of different genera and species of bacterium, fungus, virus, protozoan or alga, rickettsia, mycoplasma and nematodes. They dismantle, kill or inhibit their target pests either by producing toxic metabolites or by causing diseases that leads to death of the pest. They also prevent their target pest by establishment of other microorganisms through competition or various other modes of action (Clemson 2007). Microbial biopesticides may be delivered to crops in many forms including live organisms, dead organisms or spores (Fig. 6.2).

Out of the total global biopesticide present in market for all crop types, the contribution of bacterial biopesticides is about 74%, fungal biopesticides 10%, viral biopesticides 5%, predator biopesticides 8% and other biopesticides 3% (Mishra 2015). By 2008, there were approximately 73 microbial active ingredients that were registered by the USEPA.

6.8.1.1 Bacterial

Bacteria are prokaryotic, unicellular organism of varying length and shape. The maximum number of the insect pathogenic bacteria belongs to the families of *Bacillaceae*, *Pseudomonadaceae*, *Enterobacteriaceae*, *Streptococcaceae* and

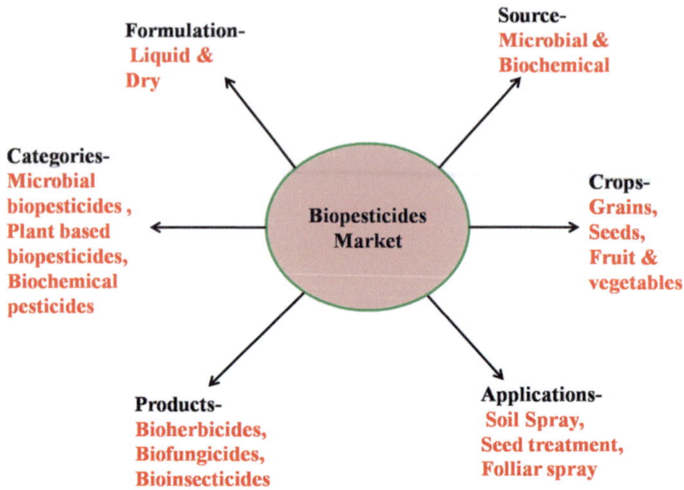

Fig. 6.2 Types of biopesticide market on the basis of their origin and application

Table 6.3 Bacterial species used as biopesticide

Bt variety	Target pest	References
Serratia entomophila	Grass grub	Johnson (2001)
B. thuringiensis subsp. *Israelensis*	Mosquito and blackflies	Kabaluk (2010)
B. thuringiensis subsp. *Kurstaki*	Lepidopteran larvae	
B. thuringiensis subsp. *Galleriae*	Colorado potato beetle	
Lysinibacillus sphaericus	Mosquito larvae	Berry (2012)
Bacillus moritai	*Diptera*	Kunimi (2007)
Burkholderia spp.	Chewing and sucking insects and mites; nematodes	Ruiu (2018)
Saccharopolyspora spinose	Insects	

Micrococcaceae (Kachhawa 2017). Among all types of biopesticides, the most common is bacterial pesticides. The major volume of commercially produced microbial control agent are bacteria, including both gram-positive, spore-forming *B. thuringiensis* and *Lysinibacillus sphaericus*, and gram-negative, non-spore-forming *Serratia entomophila*. According to the most recent data, *Chromobacterium subtsugae and B. thuringiensis* subspecies are the most commonly used microbial pesticides at global level (Glare et al. 2012; Lacey et al. 2015). The major species of *B. thuringiensis* species are used for suppression of different types of lepidopteran pests, forest pests, mosquitoes and black flies (Table 6.3). The insecticidal activity of *B. thuringiensis* and *L. sphaericus* is mainly due to protein toxins (δ endotoxins or Cry proteins) within parasporal inclusions, which must be ingested to become larvicidal. This endotoxin is soluble in the high pH of the stomach and is cleaved to the toxic moieties by gut proteolytic enzymes. These toxins bind to the receptors and create pores on enterocytes followed by osmotic imbalance resulting in cell rupture, compromising integrity of the gut epithelium. Invasion of the main body cavity and subsequent septicemia caused by gut-resident bacteria leads to death of the target pest (Lacey et al. 2015). Due to their high specificity and safety in the environment, *B. thuringiensis* and Cry proteins are efficient, safe and sustainable alternatives to chemical pesticides for the control of insect pests. Bt has been widely used to control insect pests important in agriculture, forestry and medicine.

6.8.1.2 Viral

Like bacteria and fungi, the entomogenous viruses also play a significant role in crop protection. These viruses are host-specific, infecting only one or a few closely related species; thus, the nontargeted insects are not infected by them. A US-based company, Omnilytics, has developed different types and variety of phage products used for the control of *Xanthomonas campestris* pv. *vesicatoria* and *P. syringae* pv. *tomato* for the treatment of a disease mainly caused in tomatoes [bacterial spot and bacterial speck on tomatoes] (Frampton et al. 2012). Among the insect viruses found in nature, those belonging to the baculoviruses family (*Baculoviridae*) were considered for the development of most commercial viral biopesticides (Kachhawa 2017). Baculoviruses are particularly attractive for use as biopesticides due to their high host specificity. They have been shown to have no negative impacts on plants, mammals, birds, fish, or nontarget insects (D'Amico 2007). Baculoviruses are host-specific, rod-shaped enveloped viruses having circular, supercoiled double-stranded DNA (80–180 kbp), which are 230–385 nm in length and 40–60 nm in diameter (Rohrmann 2011). Around 600 baculoviruses have been isolated from *Lepidoptera* (butterflies and moths), *Hymenoptera* (sawflies) and *Diptera* (mosquitoes) (Biagini et al. 2011). These viruses are mainly used as phage pesticides, and they exhibit efficient horizontal transmission. When ingested by the host insect, infectious virus coat gets dissolved, and it is liberated internally and become active. The budded virus initiates infection to other tissues in the haemolymph, i.e. fat bodies, nerve cells, haemocytes, etc. The cell infected in the second round of virus replicate in the insect larva also produce budded virus but in addition occlude virus particles within polyhedral in the nucleus. The accumulation of polyhedral within the insect proceeds until the host consists almost entirely of a bag of virus. In the severe stage of infection, the insect liquefies and thus releases polyhedral, which can infect other insects upon ingestion. The infected larvae show symptoms of negative geotropism before succumbing to the virus infection, thereby facilitating widespread dissemination. Within a few days, the host larvae are unable to digest food and so weaken and die (Thakore 2006). Under favourable conditions, the virus kills the pest within a week (Kachhawa 2017).

6.8.1.3 Nematodes

Entomopathogenic nematodes are non-segmented, soft-bodied roundworms. The maximum members are obligate parasites, and some are facultative parasites of insects. The natural habitat of these nematodes are soil and recognize their specific host by means of carbon dioxide release, motion, vibration and other chemical substances secreted by the host (Kachhawa 2017). Heterorhabditidae and

Steinernematidae are the two families that have been effectively used as biopesticide (Jess et al. 2005). Entomopathogenic nematodes are considered as nontoxic to humans and as effective biopesticide because of host-specific characteristic. The adult nematodes are not free-living; only juvenile stage is free-living and infective to the host. Both family of nematodes *Heterorhabditis* and *Steinernema* are present in symbiotic association with bacterial species *Photorhabdus* and *Xenorhabdus*, respectively (Ferreira and Malan 2014). The immature juvenile stage release cells of their symbiotic bacteria from their intestines into the haemocoel. The bacteria multiply in the insect haemolymph, and the infected host usually dies within 24–48 h of infection. After the death of the host, nematodes continuously feed on the host body tissue and get mature. The progeny nematodes develop through four juvenile stages to the adult. Depending on the available resources, one or more generations may occur within the host dead body, and a large number of infective juveniles are often released into environment to infect other hosts and continue their life cycle (Kachhawa 2017). Some important nematodes used as biopesticides are shown in Table 6.4.

6.8.1.4 Protozoan

Entomopathogenic protozoans are extremely diverse group of organisms comprising almost 1000 protozoan species mainly attacking numerous insect species like grasshoppers and heliothine moths (Senthil-Nathan 2015) and are commonly

Table 6.4 Nematode strains used as biopesticide

Name of nematodes	Host	References
S. glaseri	White grubs (scarabs, especially Japanese beetle, *Popillia* sp.), banana root borer	Poinar and Grewal (2012)
S. kraussei	Black vine weevil, *Otiorhynchus sulcatus*	Ansari (2010)
S. carpocapsae	Turf grass pests – billbugs, cutworms, armyworms, sod webworms, chinch bugs, crane flies. Orchard pests, ornamental and vegetable pests – codling moth, banana moth, cranberry girdler, dogwood borer and other clearwing borer species, black vine weevil, peach tree borer, shore flies (*Scatella* spp.)	Cruz-Martínez (2017)
S. feltiae	Fungus gnats (*Bradysia* spp.), shore flies, western flower thrips	
S. riobrave	Citrus root weevils (*Diaprepes* spp.) mole crickets	Bender (2014)
H. bacteriophora	White grubs (scarabs), cutworms, black vine weevil, flea beetles, corn rootworm, citrus root weevils (*Diaprepes* spp.)	Goudarzi (2015)
H. indica	Fungus gnats, root mealybug, grubs	van Niekerk and Malan (2012)
H. marelatus	White grubs (scarabs), cutworms, black vine weevil	Miles (2012)
Phasmarhabditis hermaphrodita	Molluscs	Ruiu (2018)
Heterorhabditis downesi	Black vine weevil *Otiorhynchus sulcatus*	Williams (2013)

referred as microsporidians (Kachhawa 2017). Microsporidia such as *Nosema* spp. and *Vairimorpha necatrix* are generally host-specific and slow-acting, producing chronic infections with general weakening of the host. *Nosema pyrausta* generally infects European corn borer (Kachhawa 2017) and *Nosema locustae* infects the grasshoppers (Senthil-Nathan 2015). The biological activities of these protozoan species are complex. The spore formed by these protozoans is the infectious stage and has to be ingested by the insect host for pathogenicity. The spore germinates in the midgut and sporoplasm is released invading the target cells causing infection of the host. The infection results in reduced feeding, vigour, fecundity and longevity of the insect host as inundated applied microbial control agents. There are many benefits like persistence and recycling in host populations and their debilitating effect on reproduction and effect overall fitness of target insects. Naturally, parasitoids and insect predators commonly play role as vectors distributing the disease.

6.8.2 Botanical Biopesticides

Since ancient time, India is the richest source of different varieties and diversity of plant species. India possesses the largest diversity of plant species having 47,000 plant species and total 7–8% of the world (Ghosh 2017). Botanical pesticides are potential alternative sources over chemical pesticide and are not harmful to the environment and other nontargeted pests. It is also known as 'phytochemical insecticides' and 'green chemical insecticides'. New bioactive chemicals are being isolated and characterized every day from different varieties of plant species; more than 6000 species of plants have been screened against various types of pests. There are approximately 1005 species of plants having insecticidal properties, 384 with antifeedant properties, 297 with repellent properties and 31 with growth inhibiting properties (Vendan 2016).

Botanical pesticides are made by using some parts, plant extract or whole plant. They have the ability of insect killing, sterilization, weed control and plant growth-regulating activities. The application of botanical pesticides for the crop and stored products' protection from insect pests has been a part of traditional agriculture form many years. In insect pest management, a number of plant products derived from neem, custard apple, tobacco, pyrethrum, etc. have been used as effective and safer insecticides (Nawaz et al. 2016). Azadirachtin compounds derived from the neem tree is sold under various trade names and used for several food crops and ornamental plants for controlling whitefly, thrips, scale and other pests. It affects the reproductive and digestive process of target pests (Dutta 2015) and does not affect other biocontrol agents. Neem products are effective against more than 350 species of arthropods, 12 species of nematodes, 15 species of fungi, 3 viruses, 2 species of snails and 1 crustacean species (Kandpal 2014). Several components of its leaves and seeds show marked insect control potential, and due to their relative selectivity, neem products can be recommended for many programmes on crop pest management. The major advantages of neem products over conventional chemical pesticides are its biodegradable and nontoxic nature to nontarget organisms

Table 6.5 Some plant products used as biopesticides

Plant products use as biopesticide	Target pests	References
Linalool	Peach potato aphid (*Myzus persicae*)	Gabryś (2005)
Neem	A variety of sucking and chewing insect	Cloyd (2009)
Pyrethrins	Flowerbugs and lacewings	Pezzini and Koch (2015)
Rotenone	Leaf-feeding insects, such as aphids, certain beetles (asparagus beetle, bean leaf beetle, Colorado potato beetle, cucumber beetle, flea beetle, strawberry leaf beetle, and others) and caterpillars, as well as fleas and lice on animals	Nawaz et al. (2016)

Table 6.6 Potential biopesticides (from plant extract) used in India

Plant extract	Effective against (target pests)	References
Adathodakashayam and Pudhina kashayam	Leaf folder, bacterial leaf blight, *Helminthosporium* leaf spot	Chauhan (2018)
Thriphala kashayam	Bacterial leaf blight and *Helminthosporium* leaf spot,	
Cow's urine arkam and sweet flag arkam	Bacterial leaf blight, *Helminthosporium* leaf spot, vein clearing disease, fusarium wilt	Kandpal (2014)
Garlic arkam	Leaf folder, bacterial leaf blight, *Helminthosporium* leaf spot	
Neem seed extract (for all crops)	Leaf folder, aphids, jassids, fruit borer and stem borer	
Andrographis kashayam and Sida kashayam	Aphids and borers in brinjal, ladies finger	Balasubramaniam (2008)
Extract of the species *Clitoria ternatea* (butterfly pea)	*Helicoverpa* spp.	Damalas and Koutroubas (2018)
Stilbenes isolated from grapevine extracts	*S. littoralis*	Akacha (2017)

(Senthil-Nathan 2015). Some important botanical biopesticides are shown in Table 6.5 and Table 6.6.

6.8.3 Biofungicides

Every year, Indian agriculture faces a huge loss due to deteriorating pathogens and detrimental pests (Savary et al. 2012). Some of these pathogens are gradually acting, while others act recklessly on eating the crops whose symptoms are sometime not easily noticeable to unaided eye (Yang et al. 2017). The pathogen which illustrates graphical symptoms after their full puffed infection comprises fungus. Fungi causes blight disease, damping of roots and seedling and endorses tanning of soft tissues (Broders et al. 2007). Till now, humans have established diverse chemical compounds as fungicides like strobilurin, alkyl phosphonate and etridiazole to defend the plant from these damage and upsurge the agricultural efficiency, but massive use of these compounds ensued in its build-up in different trophic levels and divisions of ecosystem and lastly getting and depositing inside human body causing injurious and sometime incorrigible diseases (Borges et al. 2015). To avoid this deposition, a substitute method of exclusion of these fungi was revealed which can be used in gigantic amount without any damage to any organism of ecosystem comprising humans (Cuthbertson and Murchie 2010). These replacement technique elaborated use of biofungicides which were isolated using biological source and can be used on different isolates of pathogenic fungi to exterminate the infecting fungus at a bulky scale (Heydari and Pessarakli 2010). These biological preparations comprise biofungicides which are isolated from fungi like *Trichoderma harzianum, Bacillus subtilis* and *Gliocladium virens* (Abbey 2018). Irrespective of quality of biocontrol agent, the course involved in its development, formulation and its delivery to the target system also regulates its effectiveness in the challenging area.

A thought-provoking part of a biofungicide against *Botrytis cinerea* on the microorganisms involved in alcoholic fermentation and on the grape biofilm. The usefulness was confirmed in the grape vineyard with some yeast like *Candida sake* or any bacteria like *Bacillus subtilis* to diminish the grapevine disease. Some of the biofungicides used was found to be of great advantage equated to its chemical equivalents. Based on the above submission and its exploitation in numerous arenas studied by Kumar and Singh (2014), the mounting attention in biofungicides accompanying with the product includes the following: (a) they are fundamentally less toxic than conventionally used chemical pesticides; (b) biofungicides distress only the target organism in broad-spectrum chemical fungicides which disturb unrelated organisms like birds, other insects and mammals; (c) by using as a constituent of integrated pest management, these biofungicides significantly limit the use of conventional pesticides; and (d) low cost of biofungicides paralleled to their synthetic counterparts.

Trichoderma spp. is an additional biopesticide technology established in recent years and found to be an operative fungicide against soil-borne diseases (Sharma et al. 2014). *Trichoderma* safeguards plant from pathogens by augmenting plant growth and protection under certain circumstances and by parasitizing detrimental fungi around the rhizosphere (Mukherjee et al. 2012). They are competent to parasitize plant pathogenic fungi in soil and yield antibiotics and fungal cell wall mortifying enzymes. They also strive with soil-borne pathogens for carbon and nitrogen and able to encourage plant growth by improving production of auxin like compounds (Vinale et al. 2008). The product of *Trichoderma* spp. is applied on seed coatings before seeding to soil. This mode improves the nutrient absorption by seeds and also substitutes fertilizer constraint of plant by up to 50%. One-time application of this product is sufficient to curtail inhibiting off disease caused by *Pythium* spp., *Sclerotium rolfsii* and *Rhizocotina solani*. Presence of these fungi is also interrelated with amplified sprouting and endurance of seeds (Keswani et al. 2013, 2014).

It is noted that around 250,000 higher plant species existing on this earth that are reflected as a reservoir of bioactive compounds with uncountable uses, comprising their use as pharmaceuticals and fungicides (Rungsung et al. 2015). Out of various fungicides isolated from several sources and verified for its usefulness, thymol and carvacrol are established as an operative ingredient against harmful fungal species whose projected mechanism comprises disbanding of fungal cell wall and cell membranes (Nazzaro et al. 2017). These products are extremely dynamic against *Botryodiplodia theobromae* and *Colletotrichum acutatum* which are the contributing agents of stem-end rot and anthracnose. They are the main components of essential oils of *Laminaceae* members like oregano, thyme and savoury. These mediators cause modifications in accumulation and morphology of hyphae which settles in lessening of hyphal diameter and lysis of their walls (Soylu et al. 2007). They are reasonably vigorous than other biofungicides due to their hydrophobicity which consents them to pass through fungal cell membrane which further distresses pH homeostasis and equilibration of inorganic ions resulting in disorganized cell structure and lysis (Cristani et al. 2007).

Isothiocyanates resulting from glucosinolates in the plant cells of family Crucifereae are found to retain compelling antifungal activity (Dufour et al. 2015). These isothiocyantaes comprises allyl, ethyl, methyl and benzyl. They impede fungal cells which encompass enzymes countering with the disulphide bonds or thiocyanate anion ensuing in inactivation of sulfhydryl enzymes. Furthermore, they are also involved in disengagement of oxidative phosphorylation reactions (Calmes et al. 2015). Purified metabolic extract of carrots are also found to contain antifungal properties due to occurrence of dodecanoic acid and pentadonoic acids which is active against yeast *Candida lambica* ensuing in inhibition of sporulation (Martínez 2012).

6.8.4 Bioinsecticides

Bioinsecticides are naturally formed insecticides which are produced or secreted as a by-product of organisms such as nematodes, bacteria and plants (Gašić and Tanović 2013). Microorganisms and plants are supposed to be a chief basis of biopesticides due to existence of higher constituents of antimicrobial agents and bioactive compounds (Nefzi et al. 2016). Natural enemies that comprise predator, pathogen and some birds are also sometimes used as a form of bioinsecticide in controlling insect pests (Knutson and Ruberson 2005). Bioinsecticides are also accessible in the form of plant extracts or crucial oils. They are acquired from leaves, flower, roots, and bark which are fresh or dried in nature (Chougule and Andoji 2016). Bioinsecticides are also mined from microorganisms including bacteria, virus, fungi, nematode and protozoans. Around 100 bacteria have been recognized as insect pathogens among which *Bacillus thuringiensis* is found to be most powerful microbial control agent. About 1000 virus which are specific to insects like *Nuclear polyhedrosis virus* (NPV) has been isolated which are competent to infest around 525 types of insects worldwide (Koul 2011). There are certain constructive features of utilising microbial biopesticides as described by Jindal et al. (2013): (a) These are bioactive compounds which are normally nontoxic to nontarget organisms including humans. (b) They are able to show synergistic effect with synthetic chemical pesticides. (c) The deposits left after their use do not show any antagonistic effect on humans or any other animals. (d) They improve the count of advantageous soil microflora which in turn progresses plant growth.

Most of the fungi existing in nature are damaging to plant health by triggering serious disease, but certain groups of fungi like entomopathogenic fungi which are natural managers of insect population are active against diverse agricultural pests. They act by penetrating the insect cuticle and producing toxins inside haemolymph which circumvents immune response of insects (Hajek and Leger 1994). The use of a fungus *Metarhizium anisopliae* which are pathogenic towards *Aedes aegyptii* mosquitoes has also been testified and the consequences designated a deterioration in this mosquito species due to their augmented vulnerability towards infection from this entomopathogen (Scholte et al. 2007).

The first achievement of viral bioinsecticide as baculovirus into the environment also resulted in improved destruction of spruce sawfly *Diprion hercyniae* during World War II. NPVs are largest group of virus which are used as bioinsecticide which blights many species of insects belonging to *Lepidoptera*, *Hymenoptera*, *Diptera* and *Trichoptera*. Their genomes are composed of double-stranded circular DNA and are able to reproduce inside the nucleus of host cells (Rohrmann 2008). Bacterial bioinsecticides has been reflected as the inexpensive and extensively used approaches of insect management (Chattopadhyay et al. 2017). In this scheme, convinced species of bacteria are used for poisoning the insects, but genus *Bacillus* are most broadly used pesticides (Ruiu 2015). One of the *Bacillus* species, *Bacillus thuringiensis*, has established molecular mechanisms to harvest endotoxins which produce transmembrane pores in the walls of insect gut that results in cell lysis due to osmotic discrepancy (Roh et al. 2007). Certain reports recommend that *B. thuringiensis* necessitates participation of commensal gut bacteria of insects to be wholly pathogenic (Broderick et al. 2009). Utilization of these toxins encompasses integration of toxin producing genes into the plant genome which is expressed as inactivated cry protein in designated part of plant. These proteins are cleaved and activated inside the insect gut once it is ingested causing pore formation and cell lysis of insects (Palma et al. 2014). Some commonly used bioinsecticides are given in the Table 6.7.

Table 6.7 Commonly used bioinsecticides and their target insects

S. No.	Bioinsecticide	Target organism	References
1.	Azadirachtin	*Amrasca devastans, Myzus persicae, Sitobion avenae, Lipaphis erysimi*	Akbar (2010)
2.	Thymol	*Megalurothrips sjostedti, Eloidogyne incognita, Helicotylenchus dihystera, Pratylenchus brachyurus*	Pumnuan and Insung (2016)
3.	*Phytoseiulus* spp. (predator organism)	*Tetranychus urticae, Tetranychus evansi*	Wekesa (2007)
4.	*Neoseiulus* spp. (predator organism)	*Tetranychus urticae, Oligonychus perseae*	McMurtry (2015)
5.	Baculovirus	Insects of genera *Lepidoptera, Hymenoptera, Coleoptera, Diptera*	Herniou (2004)

Pheromones are chemical signals emanated by living organisms to interconnect organisms of opposite sex (Yew and Chung 2015). Most of them are discriminating and produce no toxic residues, and production costs are significantly lower than that of synthetic chemical pesticides (Hajek and Eilenberg 2018). Pheromones are also of major importance in pest management when used in combination with traps as 'pheromone traps' using 'attract and kill' technique. This technique is called as mating disruption and is found effective in regulating a number of pests (Campos and Phillips 2014). This method is found operative in interruption of grape moth, grapevine moth and codling moth.

Certain bioinsecticides are active at every stage, while some are effective at definite stages of pest life cycle to reduce their population below a threshold level (Singh et al. 2010). Some biopesticides have a superiority in application due to their change in modes of action which permits them to overcome resistance from conventional pesticides and their influence on nontarget organisms (Spence and Lewis 2010). Synthetic female pheromones are used as attractants for males into traps which is used for mating disruption (Samietz et al. 2012). It is of utmost importance in development of biopesticide to overcome the problem of low shelf life, improper formulation as well as economic feasibility (Chandler et al. 2011).

6.9 Prospects of Biopesticides and Plant Diseases in India

Management of pest in an eco-friendly is no more a dream. Biopesticides can now be produced using the tools of molecular biology, biotechnology and nanotechnology in crop plants itself in a sustainable manner. They have been viewed as a safer alternative to chemical pesticides for a very long time. Their branding or commercialization can be done only by cooperation of the public and private sectors. Discovery of new substances and extensive research on formulation and delivery can help facilitate the development, commercialization and consumption of biopesticides.

India is an agriculture-based economy as around 58% of its population depends upon agriculture for its livelihood. Thus, it is very right to say that still agriculture serves as the backbone of Indian economy. With such a huge population getting employment due to this sector and also 18% GDP being contributed by it, agriculture sector needs proper attention of the government. The pest management practice in India is very complicated as a large number of farmers, crops and pests are involved. To control the pests, insecticides, fungicides and herbicides are commonly used, and they all are synthetic in nature. However, the share of insecticides is the highest in total pesticide use in India. According to a report (2016–2017) of the Ministry of Chemicals and Fertilizers and the Government of India, pesticides contributed to 3% in cotton, 1.9% in paddy, further lower in wheat (0.7%) and 0.3% in sugarcane of the total cost of cultivation. Insecticides contribute maximum of all the pesticides produced in India, its share being 39% in 2016–2017. Mancozeb, 2-4-D, acephate and profenofos are the major pesticides produced in India. Mancozeb and acephate are also among the top 5 pesticides exported from India, while glyphosate and atrazine are the major pesticides imported from China.

The Indian pesticide industry is facing tough challenges, the major being the stringent global environmental laws, little attention on R&D by domestic manufacturers due to high costs, need for innovation and product diversification, lack of awareness about safe use of pesticides among farmers, long gestation period for new products and product quality assurance. Thus, these negative environmental externalities can only be ruled out by using biopesticides, which can play a pivotal role in moving the focus from chemical pesticides to reliable, sustainable and environment-friendly alternative. The speed of development of biopesticide industry in India is not very impressive; it contributes only 3% of the pesticide market in India.

Commercialization and availability of biopesticides in the Indian market can be facilitated not only by maintaining low cost to farmers but also by the regulations that help in speedy registration of low-risk compounds with provision of incentives under the insecticides act. More field research of the presently available biopesticides needs to be done along with that recombinant DNA technology can also be employed to enhance efficacy of biopesticides in different cropping systems, which in turn may lower the continuous search for new substances. Some nanotechnology like microencapsulation of the available biopesticides can also be done to improve their residual action and hence their field use (Damalas and Koutroubas 2018). Information about the genes from the microbes and crop plants can be exploited to isolate genes beneficial against a particular pest, and these genes can be utilised to control insect pests and diseases (Kumar 2015).

Not only from pests and weeds but plant diseases caused by various groups of plant pathogens are often a big challenge in agriculture in India. Currently, chemical pesticides are being used in India to control plant diseases caused by several pathogens. Since, the use of these synthetic pesticides can severely damage the environment, so the reduction or elimination of the pesticide is inevitable.

The goal of having a clean and safe environment can be achieved by using new tools that are based on biological control agents (BCAs) and natural antimicrobial chemicals for disease control. A lot of research laboratories in India are now focussing on identification of the microorganisms and antimicrobial botanicals which can either inhibit the growth or kill the plant pathogens. All the countries need to develop their own BCAs because of the imposition of several quarantine procedures by the other countries. The mode of action of each BCA is different in different condition, so the Indian scientist must focus on developing indigenous BCAs as foreign BCAs may not work in Indian climate efficiently.

References

Abbey JA, Percival D, Abbey L, Asiedu SK, Prithiviraj B, Schilder A (2018) Biofungicides as alternative to synthetic fungicide control of grey mould (*Botrytis cinerea*) – prospects and challenges. Biocontrol Sci Technol:1–22

Abhilash P, Singh N (2009) Pesticide use and application: an Indian scenario. J Hazard Mater 165(1–3):1–12

Akacha M, Chaieb I, Laarif A, Haouala R, Boughanmi N (2017) Effects of Melia azedarach leaf extracts on nutritional behavior and growth of Spodoptera littoralis. Tunis J Plant Prot 12(Special Issue):61–70

Akbar MF, Haq MA, Parveen F, Yasmin N, Khan MFU (2010) Comparative management of cabbage aphid (Myzus persicae (Sulzer) (Aphididae: Hemiptera) through bio-and synthetic-insecticides. Pak Entomol 32(1):12–17

Ansari MA, Shah FA, Butt TM (2010) The entomopathogenic nematode Steinernema kraussei and Metarhizium anisopliae work synergistically in controlling overwintering larvae of the black vine weevil, Otiorhynchus sulcatus, in strawberry growbags. Biocontrol Sci Technol 20(1):99–105

Balasubramanian AV, Arumugasamy S, Vijayalakshmi K, Subhashini S (2008) Plant products as biopesticides: building on traditional knowledge of Vrikshayurveda: traditional Indian plant science

Bender GS, Bates LM, Bethke JA, Lewis E, Tanizaki G, Morse JG, Godfrey KE (2014) Evaluation of insecticides, entomopathogenic nematodes, and physical soil barriers for control of Diaprepes abbreviatus (Coleoptera: Curculionidae) in citrus. J Econ Entomol 107(6):2137–2146

Berry C (2012) The bacterium, Lysinibacillus sphaericus, as an insect pathogen. J Invertebr Pathol 109(1):1–10

Bharatbhai PP (2017) Market potential and awareness of different fungicides for control of diseases in tomato in Anand district (Doctoral dissertation, AAU, Anand)

Biagini P, Bendinelli M, Hino S, Kakkola L, Mankertz A, Niel C, Okamoto H, Raidal S, Teo CG, Todd D (2011) Family Circoviridae. In: AMQ K, Adams MJ, Carstens EB, Lefkowitz EJ (eds) Virus taxonomy: ninth report of the International Committee on Taxonomy of Viruses. Elsevler Academic, San Diego, pp 343–349

Birch E, Nicholas A, Begg GS, Squire GR (2011) How agro-ecological research helps to address food security issues under new IPM and pesticide reduction policies for global crop production systems. J Exp Bot 62(10):3251–3261

Bisen K, Keswani C, Mishra S, Saxena A, Rakshit A, Singh HB (2015) Unrealized potential of seed biopriming for versatile agriculture. In: Rakshit A, Singh HB, Sen A (eds) Nutrient use efficiency: from basics to advances. Springer, New Delhi, pp 193–206

Borges, L. D., Garcia, L. A., Fabri, C. E., Lima, A. M., de Godoy, R. D. F., Werlang, R. C. (2015). U.S. Patent application no. 13/876,613

Broderick NA, Robinson CJ, McMahon MD, Holt J, Handelsman J, Raffa KF (2009) Contributions of gut bacteria to Bacillus thuringiensis-induced mortality vary across a range of Lepidoptera. BMC Biol 7(1):11

Broders KD, Lipps PE, Paul PA, Dorrance AE (2007) Characterization of Pythium spp. associated with corn and soybean seed and seedling disease in Ohio. Plant Dis 91(6):727–735

Calmes B, N'Guyen G, Dumur J, Brisach CA, Campion C, Iacomi B, Pigné S, Dias E, Macherel D, Guillemette T, Simoneau P (2015) Glucosinolate-derived isothiocyanates impact mitochondrial function in fungal cells and elicit an oxidative stress response necessary for growth recovery. Front Plant Sci 6:414

Campos M, Phillips TW (2014) Attract-and-kill and other pheromone-based methods to suppress populations of the Indian meal moth (Lepidoptera: Pyralidae). J Econ Entomol 107(1):473–480

Chandler D, Bailey AS, Tatchell GM, Davidson G, Greaves J, Grant WP (2011) The development, regulation and use of biopesticides for integrated pest management. Philos Trans R Soc Lond Ser B Biol Sci 366(1573):1987–1998

Chattopadhyay P, Banerjee G, Mukherjee S (2017) Recent trends of modern bacterial insecticides for pest control practice in integrated crop management system. 3 Biotech 7(1):60–60
Chauhan A, Ranjan A, Jindal T (2018) Biological control agents for sustainable agriculture, safe water and soil health. In: Paradigms in pollution prevention. Springer, Cham, pp 71–83
Chougule PM, Andoji YS (2016) Antifungal activity of some common medicinal plant extracts against soil borne phytopathogenic fungi Fusarium oxysporum causing wilt of tomato. Int J Dev Res 6(3):7030–7033
Clemson HGIC (2007) Organic pesticides and biopesticides, Clemson extension, home and garden information center. Clemson University, Clemson
Cloyd RA, Galle CL, Keith SR, Kalscheur NA, Kemp KE (2009) Effect of commercially available plant-derived essential oil products on arthropod pests. J Econ Entomol 102(4):1567–1579
Cristani M, D'Arrigo M, Mandalari G, Castelli F, Sarpietro MG, Micieli D, Venuti V, Bisignano G, Saija A, Trombetta D (2007) Interaction of four monoterpenes contained in essential oils with model membranes: implications for their antibacterial activity. J Agric Food Chem 55(15):6300–6308
Cruz-Martínez H, Ruiz-Vega J, Matadamas-Ortíz PT, Cortés-Martínez CI, Rosas-Diaz J (2017) Formulation of entomopathogenic nematodes for crop pest control-a review. Plant Protect Sci 53(1):15–24
Cuthbertson AGS, Murchie AK (2010) Ecological benefits of Anystis baccarum in an orchard ecosystem and the need for its conservation. J Environ Sci Technol 7(4):807–813
D'Amico V (2007) Baculoviruses in biological control: a guide to natural enemies in North America. Cornell University, Ithaca. http://www.nysaes.cornell.edu/ent/biocontrol/pathogen/baculoviruses
Damalas CA, Koutroubas SD (2018) Current status and recent developments in biopesticide use. Agriculture 8:1–6
Dufour V, Stahl M, Baysse C (2015) The antibacterial properties of isothiocyanates. J Microbiol 161(2):229–243
Dutta S (2015) Biopesticides: an eco-friendly approach for pest control. World J Pharm Pharm Sci 4(6):250–265
Ferreira T, Malan A (2014) Xenorhabdus and Photorhabdus, bacterial symbionts of the entomopathogenic nematodes Steinernema and Heterorhabditis and their in vitro liquid mass culture: a review. Afr Entomol 22(1):1–14
Fraceto LF, Maruyama CR, Guilger M, Mishra S, Keswani C, Singh HB, deLima R (2018) *Trichoderma harzianum* based novel formulations: potential applications for management of Next-Gen agricultural challenges. J Chem Technol Biotechnol 93:2056–2063. https://doi.org/10.1002/jctb.5613
Frampton RA, Pitman AR, Fineran PC (2012) Advances in bacteriophage-mediated control of plant pathogens. Int J Microbiol 2012:326452
Gabryś B, Dancewicz K, Halarewicz-Pacan A, Janusz E (2005) Effect of natural monoterpenes on the behaviour of the peach potato aphid Myzus persicae (Sulz.). IOBC WPRS Bulletin 28(10):29–34
Gašić S, Tanović B (2013) Biopesticide formulations, possibility of application and future trends. Pestic Fitomed 28(2):97–102
Ghosh AK (2017) Shivendu K. Srivastava: commercial use of biodiversity: resolving the access and benefit sharing issues. Proc Zool Soc, Springer India 70(2):206–206
Glare T, Caradus J, Gelernter W, Jackson T, Keyhani N, Köhl J, Marrone P, Morin L, Stewart A (2012) Have biopesticides come of age? Trends Biotechnol 30(5):250–258
Goudarzi M, Moosavi MR, Asadi R (2015) Effects of entomopathogenic nematodes, Heterorhabditis bacteriophora (Poinar) and Steinernema carpocapsae (Weiser), in biological control of Agrotis segetum (Denis & Schiffermuller) (Lepidoptera: Noctuidae). Turk Entomol Derg 39(3):239–250
Gunnell D, Eddleston M (2003) Suicide by intentional ingestion of pesticides: a continuing tragedy in developing countries. Oxford University Press, Oxford

Gupta S, Dikshit A (2010) Biopesticides: an eco-friendly approach for pest control. J Biopest 3(Special Issue):186
Hajek AE, Eilenberg J (2018) Natural enemies: an introduction to biological control. Cambridge University Press, Cambridge
Hajek AE, St. Leger RJ (1994) Interactions between fungal pathogens and insect hosts. Annu Rev Entomol 39(1):293–322
Herniou EA, Olszewski JA, O'reilly DR, Cory JS (2004) Ancient coevolution of baculoviruses and their insect hosts. J Virol 78(7):3244–3251
Heydari A, Pessarakli M (2010) A review on biological control of fungal plant pathogens using microbial antagonists. J Biol Sci 10(4):273–290
https://www.mordorintelligence.com/industry-reports/indian-biopesticides-market
Jess S, Schweizer H, Kilpatrick M, Grewal P, Ehlers R, Shapiro-Ilan D (2005) Mushroom applications. In: Nematodes as biocontrol agents. CABI Publishing, New York, pp 191–213
Jindal V, Dhaliwal GS, Koul O (2013) Pest management in 21st century: roadmap for future. Biopest Int 9(1):1–22
Johnson VW, Pearson JF, Jackson TA (2001) Formulation of Serratia entomophila for biological control of grass grub. In Proceedings of the New Zealand plant protection conference. N Z Plant Protect 54:125–127
Kabaluk JT, Svircev AM, Goettel MS, Woo SG (Eds.) (2010) The use and regulation of microbial pesticides in representative jurisdictions worldwide. International Organization for Biological Control of Noxious Animals and Plants (IOBC), p 99
Kachhawa D (2017) Microorganisms as a biopesticides. J Entomol Zool Stud 5(3):468–473
Kandpal V (2014) Biopesticides. Int J Environ Res Dev 4(2):191–196
Keswani C (2015a) Eco-friendly management of plant diseases by biosynthesised secondary metabolites of *Trichoderma* spp. J Brief Ideas. https://doi.org/10.5281/zenodo.15571
Keswani C (2015b) Proteomic studies of thermotolerant strains of *Trichoderma* spp. Ph.D. thesis. Banaras Hindu University, Varanasi, India, p 126
Keswani C, Singh SP, Singh HB (2013) A superstar in biocontrol enterprise: *Trichoderma* spp. Biotechnol Today 3(2):27–30
Keswani C, Mishra S, Sarma BK, Singh SP, Singh HB (2014) Unravelling the efficient applications of secondary metabolites of various *Trichoderma* spp. Appl Microbiol Biotechnol 98:533–544
Keswani C, Sarma BK, Singh HB (2016) Synthesis of policy support, quality control, and regulatory management of biopesticides in sustainable agriculture. In: Singh HB, Sarma BK, Keswani C (eds) Agriculturally important microorganisms: commercial and regulatory requirement in Asia. Springer, Singapore, pp 167–181
Knutson AE, Ruberson J (2005) Field guide to predators, parasites and pathogens attacking insect and mite pests of cotton: recognizing the good bugs in cotton. Texas FARMER Collection
Koul O (2011) Microbial biopesticides: opportunities and challenges. CAB Rev 6:1–26
Krawczyk G, Hull L, Bohnenblust E (2010) Utilization of mating disruption and codling moth granulosis virus (CMGV) in conventional commercial apple orchards in Pennsylvania, USA. IOBC WPRS Bull 54:71–74
Kumar S (2012) Biopesticides: a need for food and environmental safety. J Biofertil Biopestic 3(4):1–3
Kumar S (2015) Biopesticide: an environment friendly pest management strategy. J Biofertil Biopestic 6:1
Kumar S, Singh A (2014) Biopesticides for integrated crop management: environmental and regulatory aspects. J Biofertil Biopestic 5:e121
Kunimi Y (2007) Current status and prospects on microbial control in Japan. J Invertebr Pathol 95(3):181–186
Lacey L, Grzywacz D, Shapiro-Ilan D, Frutos R, Brownbridge M, Goettel M (2015) Insect pathogens as biological control agents: back to the future. J Invertebr Pathol 132:1–41
MacGregor JT (2006) Genetic toxicity assessment of microbial pesticides: needs and recommended approaches. Int Assoc Environ Mutagen Soc 1–17

Martínez JA (2012) Natural fungicides obtained from plants fungicides for plant and animal diseases: InTech 3–28

McMurtry JA, Sourassou NF, Demite PR (2015) The Phytoseiidae (Acari: Mesostigmata) as biological control agents. In: Prospects for biological control of plant feeding mites and other harmful organisms. Springer, Cham, pp 133–149

Miles C, Blethen C, Gaugler R, Shapiro-Ilan D, Murray T (2012) Using entomopathogenic nematodes for crop insect pest control. Pacif NW ext publs 1–9

Ministry of Agriculture & Farmers Welfare Department of Agriculture, Cooperation & Farmers Welfare Directorate of Plant Protection, Quarantine &Storage, Government of India. http://ppqs.gov.in/

Mishra J, Tewari S, Singh S, Arora NK (2015) Biopesticides: where we stand? In: Plant microbes symbiosis: applied facets. Springer, New Delhi, pp 37–75

Mukherjee M, Mukherjee PK, Horwitz BA, Zachow C, Berg G, Zeilinger S (2012) Trichoderma–plant–pathogen interactions: advances in genetics of biological control. Indian J Microbiol 52(4):522–529

Naranjo SE, Ellsworth PC, Frisvold GB (2015) Economic value of biological control in integrated pest management of managed plant systems. Annu Rev Entomol 60:621–645

Nawaz M, Mabubu JI, Hua H (2016) Current status and advancement of biopesticides: microbial and botanical pesticides. J Entomol Zool Stud 4(2):241–246

Nazzaro F, Fratianni F, Coppola R, Feo VD (2017) Essential oils and antifungal activity. Pharmaceuticals (Basel, Switzerland) 10(4):86

Nefzi A, Abdallah BAR, Jabnoun-Khiareddine H, Saidiana-Medimagh S, Haouala R, Danmi-Remadi M (2016) Antifungal activity of aqueous and organic extracts from Withania somnifera L. against Fusarium oxysporum f. sp. radicis-lycopersici. J Microb Biochem Technol 8:144–150

Oerke E-C (2006) Crop losses to pests. J Agric Sci 144(1):31–43

Olson S (2015) An analysis of the biopesticide market now and where it is going. Outlooks Pest Manag 26(5):203–206

Palma L, Muñoz D, Berry C, Murillo J, Caballero PJT (2014) Bacillus thuringiensis toxins: an overview of their biocidal activity. Toxins (Basel) 6(12):3296–3325

Parker KM, Sander M (2017, April) Environmental fate of double-stranded RNA (dsRNA) biopesticides from RNA interference (RNAi)-based crop protection. In: ACS national meeting 2017

Pezzini DT, Koch RL (2015) Compatibility of flonicamid and a formulated mixture of pyrethrins and azadirachtin with predators for soybean aphid (Hemiptera: Aphididae) management. Biocontrol SciTechnol 25(9):1024–1035

Poinar GO Jr, Grewal PS (2012) History of entomopathogenic nematology. J Nematol 44(2):153

Pumnuan J, Insung A (2016) Fumigant toxicity of plant essential oils in controlling thrips, Frankliniella schultzei (Thysanoptera: Thripidae) and mealybug, Pseudococcus jackbeardsleyi (Hemiptera: Pseudococcidae). J Entomol Res 40:1–10

Rajapakse RHS, Ratnasekera D, Abeysinghe S (2016) Biopesticides research: current status and future trends in Sri Lanka. In: Agriculturally important microorganisms. Springer, Singapore, pp 219–234

Ram RM, Keswani C, Bisen K, Tripathi R, Singh SP, Singh HB (2018) Biocontrol technology: eco-friendly approaches for sustainable agriculture. In: Brah D, Azevedo V (eds) Omics technologies and bio-engineering: towards improving quality of life volume II microbial, plant, environmental and industrial technologies. Academic, London, pp 177–190

Roh JY, Choi JY, Li MS, Jin BR, Je YH (2007) Bacillus thuringiensis as a specific, safe, and effective tool for insect pest control. J MicrobiolTechnol 17(4):547

Rohrmann GF (2008) Baculovirus molecular biology. National Library of Medicine (US), National Center for Biotechnology Information, Bethesda, 154 pp

Rohrmann GF (2011) Baculovirus molecular biology, 2nd edn. http://www.ncbi.nlm.nih.gov/books/NBK49500/. NCBI, Bethesda

Ruiu L (2015) Insect pathogenic bacteria in integrated pest management. Insects 6(2):352–367

Ruiu L (2018) Microbial biopesticides in agroecosystems. Agronomy 8(11):235

Rungsung W, Ratha KK, Dutta S, Dixit AK, Hazra J (2015) Secondary metabolites of plants in drugs discovery. World J Pharm Res 4(7):604–613

Samietz J, Baur R, Hillbur Y (2012) Potential of synthetic sex pheromone blend for mating disruption of the swede midge, Contarinia nasturtii. J Chem Ecol 38(9):1171–1177

Savary S, Ficke A, Aubertot JN, Hollier C (2012) Crop losses due to diseases and their implications for global food production losses and food security. Food Secur Springer 4:519–537

Scholte EJ, Takken W, Knols BG (2007) Infection of adult Aedes aegypti and Ae. albopictus mosquitoes with the entomopathogenic fungus Metarhizium anisopliae. Acta Trop 102(3):151–158

Senthil-Nathan S (2015) A review of biopesticides and their mode of action against insect pests. In: Environmental sustainability. Springer, New Delhi, pp 49–63

Sharma P, Sharma M, Raja M, Shanmugam V (2014) Status of Trichoderma research in India: a review. Indian Phytopathol 67(1):1–19

Singh RK, Sanyal PK, Patel NK, Sarkar AK, Santra AK, Pal S, Mandal SC (2010) Fungus–benzimidazole interactions: a prerequisite to deploying egg-parasitic fungi Paecilomyces lilacinus and Verticillium chlamydosporium as biocontrol agents against fascioliasis and amphistomiasis in ruminant livestock. J Helminthol 84(2):123–131

Singh S, Gupta R, Sharma S (2015) Effects of chemical and biological pesticides on plant growth parameters and rhizospheric bacterial community structure in Vigna radiata. J Hazard Mater 291:102–110

Singh HB, Keswani C, Bisen K, Sarma BK, Chakrabarty PK (2016) Development and application of agriculturally important microorganisms in India. In: Agriculturally important microorganisms. Springer, Singapore, pp 167–181

Singh HB, Sarma BK, Keswani C (eds) (2017) Advances in PGPR research. CABI, Wallingford, 408 pages, ISBN-9781786390325

Soylu S, Yigitbas H, Soylu EM, Kurt Ş (2007) Antifungal effects of essential oils from oregano and fennel on Sclerotinia sclerotiorum. J Appl Microbiol 103(4):1021–1030

Spence KO, Lewis EE (2010) Biopesticides with complex modes of action: direct and indirect effects of DiTera® on Meloidogyne incognita. Nematology 12(6):835–846

Subash SP, Chand P, Pavithra S, Balaji SJ, Pal S (2017) Pesticide use in Indian agriculture: trends, market structure and policy issues. In: Policy brief. ICAR-National Centre for Agricultural Economics and Policy Research, New Delhi, India, p 43. Available on: http://www.ncap.res.in/upload_files/policy_brief/pb43.pdf.

Thakore Y (2006) The biopesticide market for global agricultural use. Ind Biotechnol 2(3): 194–208

van Niekerk S, Malan AP (2012) Potential of South African entomopathogenic nematodes (Heterorhabditidae and Steinernematidae) for control of the citrus mealybug, Planococcus citri (Pseudococcidae). J Invertebr Pathol 111(2):166–174

Vendan SE (2016) Current scenario of biopesticides and eco-friendly insect pest management in India. South Indian J Biol Sci 2(2):268–271

Vinale F, Sivasithamparam K, Ghisalberti EL, Marra R, Woo SL, Lorito M (2008) Trichoderma–plant–pathogen interactions. Soil Biol Biochem 40(1):1–10

Wekesa VW, Moraes GD, Knapp M, Delalibera I Jr (2007) Interactions of two natural enemies of Tetranychus evansi, the fungal pathogen Neozygites floridana (Zygomycetes: Entomophthorales) and the predatory mite, Phytoseiulus longipes (Acari: Phytoseiidae). Biol Control 41(3):408–414

Williams CD, Dillon AB, Harvey CD, Hennessy R, Mc Namara L, Griffin CT (2013) Control of a major pest of forestry, Hylobius abietis, with entomopathogenic nematodes and fungi using eradicant and prophylactic strategies. For Ecol Manag 305:212–222

World Health Organization, & United Nations Environment Programme (1990) Public health impact of pesticides used in agriculture. World Health Organization, Geneva

Xiao G, Ying SH, Zheng P, Wang ZL, Zhang S, Xie XQ, Shang Y, Leger RJ, Zhao GP, Wang C, Feng MG (2012) Genomic perspectives on the evolution of fungal entomopathogenicity in Beauveria bassiana. Sci Rep 2:483

Yadav IC, Devi NL, Syed JH, Cheng Z, Li J, Zhang G, Jones KC (2015) Current status of persistent organic pesticides residues in air, water, and soil, and their possible effect on neighboring countries: a comprehensive review of India. Sci Total Environ 511:123–137

Yang J, Hsiang T, Bhadauria V, Chen X-L, Li G (2017) Plant fungal pathogenesis. Biomed Res In 2017:9724283

Yew JY, Chung H (2015) Insect pheromones: an overview of function, form, and discovery. Prog Lipid Res 59:88–105

From Genetic Engineering to Gene Editing: Harnessing Advances in Biology for National Economic Development

7

Chandra Sekhara Rao Nuthalapati

Abstract

This chapter has examined the nature and adoption of biotechnologies, socio-economic impacts, regulatory frameworks and concerns for rising farm incomes in a cross-country perspective. The product development in biotech has been moving from just insect/herbicide resistance to breaking yield barriers, drought tolerance and quality enhancing traits, just from 3 to 31 crops, a large share of acreage in developing countries and increasing penetration of public sector. The frontiers have been moving forward with the fundamental breakthrough in the form of CRISPR-Cas 9 technique with wide-ranging applications. A rigorous study of peer-reviewed literature shows that GE crop cultivation has increased yields and net income, reduced pesticide usage and helped conserve tillage. Biosafety laws have been stifling product development, and therefore harnessing biotechnologies necessitate enabling policies like a legal framework for bio-safety, labelling and trans-boundary movement. Developing countries need to put in place regulations for the new plant breeding techniques on par with the conventional plant breeding techniques. The policy implications have been then drawn for utilization of opportunities in advancement of biotechnology for developing country agriculture.

Keywords

Yield effect · Selection bias · Halo effect · Employment · Regulatory framework · IPRs · Drought tolerance · Labelling · Consolidation

JEL Classification

J4 · K19 · L1 · O3 · Q10 · Q16

C. S. R. Nuthalapati (✉)
Institute of Economic Growth, University of Delhi Enclave, New Delhi, India

C. Keswani (ed.), *Bioeconomy for Sustainable Development*,
https://doi.org/10.1007/978-981-13-9431-7_7

7.1 Introduction

Theory of growth attributes unequal adoption of technologies due to lack of institutions as one of the major reasons for differential growth of countries in the world (Acemoglu and Robinson 2012). Technological change, according to Schultz, brings in new factors of production and act as 'low-priced sources of permanent income streams' for economic growth. Slow growth in traditional agriculture is explained by the dependency upon a particular set of factors of production, the profitability of which has been exhausted. The way forward then lies in the farmers acquiring, adopting and learning how to use effectively a profitable new set of factors (Schultz 1964). Biotechnology is both a general-purpose enabling technology and a source of radical innovation. It presents an opportunity to introduce a variety of genetic traits into farming systems around the world that replace, compete with or otherwise affect the value of existing production techniques and products. The recent advances in biology increase our understanding of life so much that experts say these discoveries are likely to define changes in the way we live in the twenty-first century. In fact, the twenty-first century is predicted to be the century of biology. The biotech sector got a big boost from deciphering of human genome in the new century, which is being widely regarded as the 'biological equivalent of landing on Moon' (Rao and Dev 2010).

Food crisis of 2007 brought back the 'classical development paradigm' which views agriculture as an engine of economic growth, industrialization and structural transformation and stresses on uni-modal strategy of modernizing the entire agricultural sector, including the smallholder sector rather than just the high-value segment (Durr 2016). Many developing countries are passing through the 'Schultz' stage, where rising agricultural incomes fall behind the rapidly growing non-farm incomes, exacerbating rural-urban disparities (Barrett et al. 2010: 451). While China and India are the striking examples of this phenomenon, countries in Africa continue to be food insecure, and East Asian countries suffer from large food imports (Otuska 2013: 7–8). Lack of modernization of agriculture is one of the reasons for 'middle income trap' that has been haunting countries like Brazil, Mexico, Malaysia, Argentina, South Africa, China and India (Eichengreen et al. 2013; Armstrong and Westland 2016).

The decline in agricultural productivity due to climate change is estimated to be to the tune of 10–38% in individual crops by 2050, and spatial spread is likely to be adverse to developing countries and regions (Muller and Robertson 2014; Rao 2015). Further, the scope of agriculture is expanding in the world to cater to the rising demands in non-food applications like fuels, fine chemicals and other products (Zilberman et al. 2013). Concerted efforts are needed to counter the reversal of secular decline in food prices after the 1990s in most countries of the world including India (Dev and Rao 2010; Rao et al. 2015). Apart from the level of prices, excessive volatility and spikes are one of the most critical economic and food security challenges (Swinnen and Riera 2013). To sum up, there is a pressing need to modernize small farm agriculture and raise agricultural productivity in view of the need to put back agriculture as engine of growth in line with 'classical development paradigm' as well as issues arising out of climate change, expanding role of agriculture to non-food requirements, raising food prices and price volatility. Then, the issue to be addressed is whether and how the rapidly diffusing biotechnologies can serve this purpose. We present a framework here to follow in this chapter.

7.2 Conceptual Framework

Theoretically, there can be both positive and negative impacts of any technology, including agricultural technologies, which can both be direct and indirect. It is noteworthy that technology has impacts on adopters, on non-adopters as well as on populations unrelated directly to the production process of the sector. However, the actual extent of these impacts is moderated by the available infrastructure, political, socio-economic contexts of regions as well as the characteristics of the adopters along with asset distribution patterns (Adato et al. 2007).

There is a consensus in the extant literature on poverty-reducing effects of agricultural growth (Ahluwalia 1978; Mellor 2006). The experience of poverty reduction in poor agrarian societies reveals that raising the productivity of small-scale farming is the key requirement to overcome poverty, because the poor are concentrated in the rural areas and their livelihoods are based on agriculture (Lewis 1954; Rao and Dev 2010). Beyond the obvious effects, technologies can increase growth and employment opportunities in rural non-farm sector and thereby contribute to poverty reduction (Mellor 2006). This in turn will have an upward pressure on wages. However, the poverty-reducing effects of technology depend on the nature of technology, nature of poverty and type of institutions in the adopting region (de Janvry and Sadoulet 2002). The predominantly rural nature of poverty and hunger in the contemporary world makes productivity-enhancing technologies exceedingly important (WB 2007). Technologies can also play big role in enhancing the nutrients required to alleviate hidden hunger and achieve sustainable development goals (Bouis et al. 2019).

Several studies have shown that seed-fertilizer technologies of the 1960s made a positive impact on agricultural growth, helped in diversifying to high-value crops and made a dent on poverty in Asia and Latin America, while the African continent could not derive significant gains for lack of necessary policy support and unavailability of improvements in crops of local interest (Hazell 2009; Pingali 2012). It is clear from Green Revolution experience that new agricultural technologies cannot be harnessed without enabling policy framework.

This chapter looks, in a cross-country perspective, at nature of biotechnologies and their diffusion patterns, provides a critical evaluation of the impact of the genetically engineered (GE) crops on farm incomes, analyses evolving regulatory frameworks and examines consolidation in seed and agricultural biotechnology and emerging countervailing forces for smallholder agriculture. This chapter does not go into the biosafety issues and remains confined to agronomic and socio-economic impacts and policy-related issues.

7.3 Changing Landscape of Biotechnologies

Standard narrative in development literature posits that predominantly multinational-developed biotechnologies will be tailor-made to the cultivation requirements of industrial agriculture of developed countries and the crops and traits of importance to resource-poor farmers in developing countries will be bypassed (Rao 2004; Rao and Dev 2009). However, the recent shifts in both technology development and adoption across developing countries allay these fears to some extent, though issues arising out of concentration in the seed industry continue to be of concern. These issues are dealt with later in this chapter.

The foremost among the recent shifts is moving of technology frontiers from genetic engineering to gene editing (Hefferon and Herring 2017), leaving out many of the unintended consequences of introducing a foreign gene through the development and adoption of SU (sulphonylurea)-tolerant canola in the USA. It uses a new gene editing method called CRISPR (clustered regularly interspaced short palindromic repeat). The past few years have witnessed a higher share of developing countries in the total area covered under the GE crops, viz. 53% of the 189.8 million hectares (Table 7.1), and this contrasts with the early years of commercialization. Brazil (26%), Argentina (12%) and India (6%) occupied nearly 85% of the area under these crops in developing countries in 2017.

Commercialized crops moving beyond four crops (soybean, maize, cotton and canola) and public sector, despite diminished funding and regulatory and IPR hurdles, moving ahead and bringing out GE products in several crops are further indicators of moving frontiers. The portfolio of technologies encompassed 31 crops in 2018, and all of them were being commercialized in different countries (Table 7.2). Most prominent among them are drought-tolerant (DT) soybean in Argentina; DT sugarcane in Indonesia; Bt brinjal in Bangladesh; Bt cotton in China, Pakistan and India; virus-resistant (VT) bean in Brazil; VT potato and VT papaya in Argentina; and VT papaya, petunia, sweet pepper and poplar in China. There are approved events now that break yield barriers, afford protection against abiotic stresses like droughts and enhance quality of product (Fig. 7.1). The DT maize was commercialized and is grown in 12 lakh hectares in 2016, and the Water Efficient Maize for Africa (WEMA), donated by the private sector, is likely to be commercialized soon in Kenya. There has been a move towards GE food crops with white maize in S. Africa; non-browning apples, late-blight-resistant potato, sweet corn, sugar beet and papaya in the USA; and Bt brinjal in Bangladesh. However, the private sector still has bulk of these new biotech crop products, despite some progress by the public sector.

7.3.1 Gene-Edited Crops

Notwithstanding the advantages of genetic modification via transgenesis for the introduction of a wider range of traits not available through conventional or mutational breeding, its scope is limited for traits that depend on a small number of

Table 7.1 Country-wise area under genetically engineered crops and approved events in developing countries

Country	Total area under GE crops in Mha	Soybean		Maize		Cotton		Other crops with commercialized biotechnology
		Area in Mha	GM events commercialized (numbers)	Area in Mha	GM events commercialized (numbers)	Area in Mha	GM events commercialized (numbers)	
Brazil	50.2	33.7 (97%)	11	15.6 (90%)	106	0.94 (84%)	12	Mosaic virus-resistant bean; fast-growing eucalyptus; borer-resistant sugarcane
Argentina	23.6	18.1 (78%)	13	5.2 (97%)	77	0.25 (98%)	3	Drought-tolerant soybean; virus-resistant potato
India	11.4	–	–	–	11	11.4 (93%)	6	DMH-11 mustard hybrid cleared by GEAC is waiting for government clearance
China	2.9	–	–	–	73	2.9 (95%)	8	Virus-resistant papaya, petunia, sweet pepper, poplar
Paraguay	2.96	2.68 (96%)	1	0.27 (42%)	22	0.01 (100%)	3	–
Pakistan	3.0	–	–	–	6	3.0 (96%)	2	–
S. Africa	2.73	0.74 (75%)	1	1.96 (85%)	72	0.004 (100%)	6	–
Uruguay	1.1	1.23 (98%)	5	0.06 (86%)	17	–	–	–
Bolivia	1.3	1.283 (100%)		–	1	–		–
Philippines	0.6	–		0.812 (65%)	105	–		–
Total	99.7	57.73	31	23.90	490	18.50	47	–

Source: ISAAA (2017) and Brookes and Barfoot (2018a) and other reports
Note: Figures within the parentheses represent adoption rates of GE crops

Table 7.2 Approved GE technologies (events) in public and private domains

Sl. No.	Crop	Public	Private	Total
1	Maize	0	231	231
2	Cotton	4	59	63
3	Potato	23	46	69
4	Argentine canola	0	41	41
5	Soybean	0	41	41
6	Carnation	0	19	19
7	Tomato	3	8	11
8	Rice	5	3	8
9	Alfalfa	0	5	5
10	Sugarcane	2	3	5
11	Papaya	4	0	4
12	Polish canola	3	1	4
13	Apple	0	3	3
14	Chicory	0	3	3
15	Sugar beet	0	3	3
16	Melon	0	2	2
17	Poplar	2	0	2
18	Rose	0	2	2
19	Safflower	0	2	2
20	Squash	0	2	2
21	Tobacco	1	1	2
22	Bean	1	0	1
23	Creeping bentgrass	0	1	1
24	Eggplant	0	1	1
25	Eucalyptus	0	1	1
26	Flax	1	0	1
27	Petunia	1	0	1
28	Plum	1	0	1
29	Sweet pepper	1	0	1
30	Wheat	0	1	1
31	Cowpea	1	1	2
	Total	53	480	533

Source: isaaa.org

known genes. In reality, many desirable traits are the result of complex interactions of several gene products (Jander et al. 2003; Till et al. 2007). Further, the integration of transgenes into the host genome is non-specific, sometimes unstable and is a matter of public concern when it comes to edible crop species (Stephens and Barakate 2017; Jaganathan et al. 2018). Moreover, the insertion of foreign gene into the crop cultivar is the central point for the unending controversies on their use. Even after occupying nearly 12% of the world cultivated area, more than 90% of the GM crop acreage is under insect resistance and herbicide tolerance, without any solution for breaking yield barriers or addressing abiotic stress. The gene editing methods have

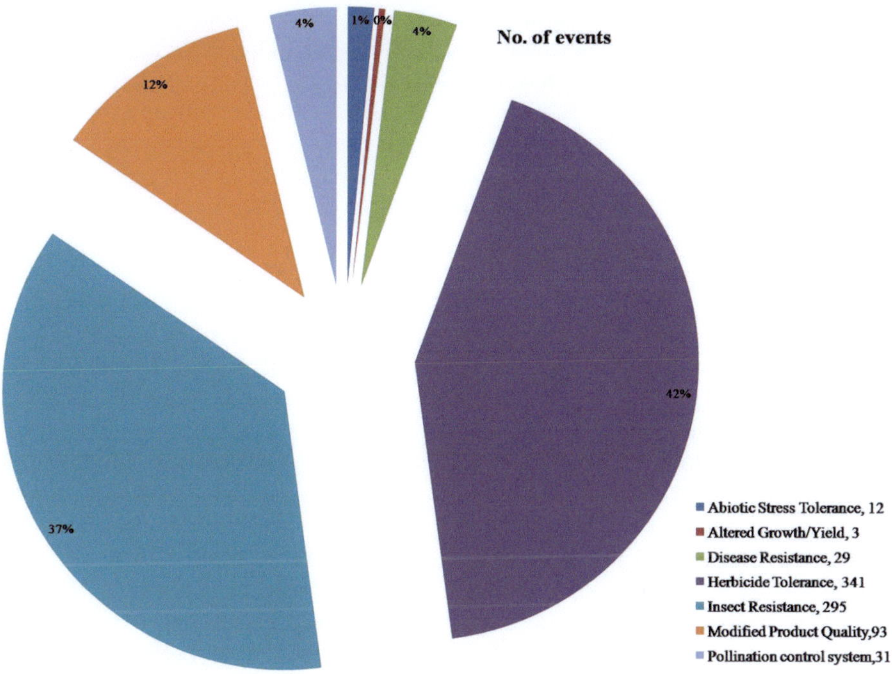

Fig. 7.1 Commercialized genetically engineered (GE) traits

arrived in this background that simplified the whole process of varietal development with desired traits.

A diverse set of gene editing tools called new breeding technologies (NBTs) are revolution arising basic molecular biology research and taking it to an entirely new level (Hefferon and Herring 2017; Adenle et al. 2018). The new breeding techniques (NBTs) make it possible to bring about genetic changes more precisely by targeting specific sites in the genome and allows clear-cut and reliable mutations, setting them apart from genetically engineered crops (Zaidi et al. 2019). Though the first-generation gene editing techniques like zinc-finger nucleases (ZFNs) and transcription activator-like effector nucleases (TALENs) have this potential, they are time-consuming and expensive. The development of the fundamental breakthrough gene editing technique called CRISPR-Cas9 altered the landscape so much and involves simple designing and cloning methods for precise changes in the genome in crop plants (Jaganathan et al. 2018). Other possible modalities include precise DNA sequence editing, gene replacement and simultaneous enhancement of multiple traits (stacking), as well as promoter and regulatory element engineering for altered gene expression patterns (Zaidi et al. 2019). Gene editing technologies or NBTs are rapid, precise and efficient compared to other means of developing desired characteristics in plants, viz. transgenesis, chemical- or radiation-induced mutagenesis and conventional breeding.

NBTs do not involve introduction of new gene sequences and may direct only one or a few nucleotide changes within a plant genome. The gene editing methods enable the plant breeders to know exactly where a change has been made in genome, leave no trace of that process and allow simultaneous editing of all copies of a gene. Consumer benefits such as enhanced nutrient content, prolonged shelf life or improved colour, odour, flavour and texture of the plant can be incorporated simultaneously with producer benefits, such as improved pest resistance or yield (Bortesi and Fischer 2015).

NBTs such as genome editing will be able to contribute substantially to global food and nutritional security with judicious applications and scientifically informed regulation (Zaidi et al. 2019). It should be kept in mind that breeding techniques are generally complementary and not mutually exclusive, that all are essential tools in addressing the challenges of agriculture (Schaart and Visser 2009).

Two applications of gene editing that are commercially approved for production are bruising- and browning-resistant potato using RNA interference technique (by reducing polyphenol oxidase levels) and SU-tolerant canola developed using Rapid Trait Development System, a more precise form of mutagenesis (Seyran and Craig 2018). The USA has so far declared five applications of CRISPR-Cas 9 to not to regulate under the purview of GMOs. These are non-browning button mushroom, waxy corn with enriched amylopectin, green bristlegrass with delayed flowering time, camelina for increased oil content and drought-tolerant soybean.

7.4 Empirical Evidence on Impacts

Several rigorous research studies have focused on the agronomic, environmental and socio-economic impacts of GE crops, just as the debates and controversies have been leading to intense scrutiny of biosafety-related issues of agricultural biotechnologies (Qaim 2009, 2016). Both meta-analyses of studies on impacts (Table 7.3) and crop-wise individual studies (Table 7.4) show higher yields, lower pesticide use and better net returns.

Meta-analysis by Klumper and Qaim (2014) has found a 22% yield increase associated with 68% profit gain and 38% reduction in pesticide expenditure. The longitudinal studies over the past 19 years show that GE crop cultivation created additional gains of USD 186 billion, conserved biodiversity by saving cultivation of 152 Mha of land (Brookes and Barfoot 2015, 2018a). Evidence from Tables 7.3 and 7.4 point to higher yield gains in developing countries as pest attacks are not effectively controlled in the absence of these technologies. The causative mechanism can be expressed in a damage control framework, following Litchenberg and Zilberman (1986) as Eq. 7.1:

$$Y = F(x)\left[1 - D(z,;\mathrm{Bt},;N)\right] \tag{7.1}$$

where Y is the effective crop yield; $F(.)$ is the potential yield without insect/weed damage, which depends on variable inputs, x; $D(.)$ is the damage function

Table 7.3 Results of meta-analyses on performance of genetically engineered crops

Study	Number of studies covered	Region or crop	Yield gain over conventional	Profit over conventional	Costs over conventional	Pesticide cost over conventional
Klumper and Qaim (2014)	147	All crops	21.57%	68.21%	NS	−39.15%
		IR crops	24.85%	68.78%	5.24%	−43.43%
		HT crops	9.29%	NS	NS	−25.29%
Areal et al. (2013)	133	Developed countries	NS	16 euros/ha	11 euros/ha	
		Developing countries	0.35 tonne/ha	188 euros/ha	−25	
		All countries	0.28 tonne/ha	166 euros/ha		
		Bt corn	0.55 tonne/ha	523 euros/ha		
		Bt cotton	0.30 tonne/ha	84 euros/ha		
		HT soybean	0.03 tonne/ha	16 euros/ha		
Hall et al. (2013)		All countries		66%	23%	
Finger et al. (2011)	177 –maize	Maize – all countries	3.9%		−66.6%	
	454 –cotton	Cotton – all countries	46.3%	86.3%		−48.2%
		India	50.8%	32.5%		−30.0%
		China		−120%		−71.7%
		S. Africa	NS	114%		−51.7%
		Australia	NS			−22.0%
		USA	NS			NS
Gruere and Sengupta (2011)		Cotton	36.2%	58.1%	16%	42%

(continued)

Table 7.3 (continued)

Study	Number of studies covered	Region or crop	Yield gain over conventional	Profit over conventional	Costs over conventional	Pesticide cost over conventional
Carpenter (2010)	168	Developed countries				
		All crops	6%			
		Corn	7%			
		Cotton	7%			
		Soybean	7%			
		Developing countries				
		All crops	29%			
		Maize	85%			
		Cotton	30%			
		Soybean	21%			

Note: NS indicates not significant

Table 7.4 Impacts of genetically engineered crops in different countries

Country	Studies	Percent change in			
		Crop/trait	Physical yield	Pesticide/ herbicide cost	Gross margin in monetary value
India	Kathage and Qaim (2012);	Bt cotton	24		50
	Rao and Dev (2009)		32–47	−13 to −56	70–251
China	Pray et al. (2002)	Bt cotton	19	−67	340
	Qiao (2015)	Bt cotton	34	−50	NA
South Africa	Thirtle et al. (2003)	Bt cotton	22	−36	28
Mexico	Traxler et al. (2003)	Bt cotton	11	−77	12
Argentina	Qaim and de Janvry (2003)	Bt cotton	33	−47	42
USA	Falck-Zepeda et al. (2000);	Bt cotton	10	36	NA
	Carpenter et al. (2002)				
Australia	Fitt (2003)	Bt cotton	0	−48	NA
Argentina	Qaim and Traxler (2005)	HT soybean	0	−42	9
USA	Marra et al. (2004)	HT soybean	0	−33	–
Canada	Brewin and Malla (2012)	Ht canola	10	−54	4
South Africa	Ghouse et al. (2009)	Ht maize	85	79	440
		Bt maize	6	−41	124
Philippines	Yorobe and Smale (2012)	Bt maize	34	−52	54
USA	Fernandez-Cornejo et al. (2005)	Bt maize	9	NA	NA
USA	Kniss (2008)	Bt sugar beet	–	−68	–

Note: NA not available in the study

determining the fraction of potential output being lost to insect pests (it can take values in the 0–1 interval); and N is the exogenous pest pressure and can be reduced by either pesticide applications (z) or Bt technology adoption. Bt technology will reduce insecticide use if farmers use lots of insecticides in conventional crop. On the other hand, this technology can help in reaching potential yield $F(.)$ by reducing $D(.)$, if they were not using chemical insecticides for effective control of pests in the conventional crop. Similar finding of higher yield gains in the developing countries was observed in the case of weed control through use of GE crops by Brookes (2005) in Romania on herbicide-tolerant (HT) soybean (29–33% increase); Smale et al. (2012) in Bolivia on HT soybean (30% increase); and Kalaitzandonakes et al. (2015) on HT maize in Kenya.

The positive yield effects have been noticed in all the Bt cotton growing countries, except in Australia, where the reduction in pesticide expenditure led to benefits by increasing gross margin to the tune of 79 Australian dollars per hectare (Table 7.4). Apart from that, cultivating herbicide-tolerant soybean and sugar beet enabled the cultivators to raise another crop in the same field and additionally led to conservation of tillage (Marra et al. 2004; Kniss 2008), apart from enabling the farmers to spend time on non-farm activities through reduced time in weed management in soybean (Fernandez-Cornejo et al. 2005; Qaim and Traxler 2005).

Huge welfare gains from adoption of GE crops are shown to people of developing countries in the economy-wide models, despite trade barriers in the EU countries (Anderson et al. 2008; Anderson 2010). On the other hand, there are studies that show positive indirect effects of adopting GE crops. Bt cotton adoption in India increased household employment and income (Subramanian and Qaim 2009), especially for the hired female workers (Subramanian and Qaim 2010; Rao and Dev 2009), as well as calorie intake (Kouser and Qaim 2013). Adopting women farmers valued labour-saving benefit more in Bt corn grown in S. Africa, while men preferred yield-enhancing benefit, signalling gender perspectives in looking at new technologies (Ghouse et al. 2016). Globally, the reduction in pesticide sprays is estimated to have saved 671 million kilograms (8.2% reduction) of active ingredient, and the environmental impact associated with herbicide and insecticide use on these crops fell by 18.5% (Brookes and Barfoot 2018b).

On the downside, Bollgard II cotton in India developed resistance to pink bollworm in Western India (Fabrick et al. 2014), while Bt cotton worked without resistance in China (Qiao 2015) and USA (Carriere et al. 2003). However, resistance to American bollworm continues in India, and resistance to pink bollworm can be delayed by mixing refugia with biotech seeds (Kranthi 2015). Weeds developed resistance of Bt cotton in places where HT crops are grown and higher quantities of glyphosate are applied. NASEM (2016) has concluded that this is not because of GE crops per se and variations in applied herbicides can prolong resistance. Several technologies like Bt brinjal, HT cotton, HT maize and virus-resistant cassava are in the pipeline with documented evidence of huge benefits to farmers and to the national exchequer (Rao et al. 2018; Ashok et al. 2017). The recent application of biotech potato with bruise and late blight resistance and cold storage is found to reduce grower costs by 28%, apart from environmental benefits including a reduction of 2.5 million acre applications of pesticides, 740 million fewer pounds of carbon dioxide emitted and 84 billion gallons less of water used (Guenthner 2017).

7.4.1 Isolating Technology Effect

Higher yields and net returns in new technologies could be either due to technology effect or because better and motivated farmers self-select themselves for adoption. Therefore, isolating the technology effect by separating the confounding factors is critical in evaluating technologies, as otherwise ‘farmer effect’ can be wrongly attributed to technologies (Rao 2013). Several studies have applied econometric tools to separate technology effect and found higher yields in insect-resistant cotton (Kathage and Qaim 2012; Rao and Dev 2009; Stone 2011; Gruere and Sun 2012; Morse et al. 2012), herbicide-tolerant soybean (Smale et al. 2012; Fernandez-Cornejo and McBride 2002; Fernandez-Cornejo et al. 2005) and insect-resistant maize (Yorobe and Smale 2012). Meta-analysis by Witjaksono et al. (2014) found positive yield and revenue impacts after controlling for selection bias, estimation and measurement bias. Further, using panel data models, Kouser and Qaim (2011) and Krishna and Qaim (2012) have shown that there were significant pesticide reductions in Bt cotton cultivation and that these are sustainable for adopters apart from helping the non-adopters with declined pest population, resulting in a *halo effect*. This corroborates research findings, on the benefits to non-adopters, in the realm of biological science reported in *Science* by Wu et al. (2008) in China, Hutchison et al. (2010) and Carriere et al. (2003) in the USA.

7.5 Policy Framework for Harnessing Biotechnology

The current logjam in commercializing biotech crops stems for excessive and faulty regulation giving credence to uninformed and speculative fears (Adenle et al. 2018; Zaidi et al. 2019). Harnessing potential of biotechnologies is conditioned on putting in place an elaborate institutional mechanism to scrutinize technologies for biosafety, labelling, trans-boundary movement of GE foods in concurrence with Cartagena Protocol on Biosafety (CPB) and strengthening property rights through patent laws, apart from developing institutional framework for risk management and risk communication (Craig et al. 2017). This is a tall order for developing country governments, and most of them, especially those in the African continent and Caribbean region countries, are not equipped to do this, stoking fears of recurrence of the Green Revolution experience (of bypassing poor countries), though there has been some improvement in recent years (Morris 2017; Rosado and Craig 2017). The countries with relatively stronger agricultural research capabilities are moving ahead in this trajectory, and their regulatory frameworks have been analysed in this section (Table 7.5). Countries having commercial and postcolonial ties to EU in Africa, the Middle East and South and Southeast Asia adopted more precautionary approach, while those having closer ties to the USA, including most in the Western Hemisphere plus the Philippines, have generally adopted a less precautionary approach (Herring and Paarlberg 2016).

Overarching legal framework with exclusive personnel for regulation was put in place in very few countries like Brazil, S. Africa, Mexico and very recently in the Philippines, while the same has been in the process in Argentina and India, as well as in Bangladesh and Pakistan. As could be seen from Table 7.5, either Ministry of Agriculture (Argentina, China, S. Africa) or Ministry of Science and Technology (Brazil, Mexico and the Philippines) handle this except in India where Ministry of Environment, Forest and Climate Change takes a final decision. Reforms to the framework in Brazil in 2008 making National Technical Commission on Biosafety (CTNBio) as the single agency for taking decisions on approvals quickened the

Table 7.5 Policy framework in selected developing countries

		Arrangement in countries				
Regulatory issue	Brazil	Argentina	India	China	S. Africa	Mexico
Overarching law	Available	Not in place	Not in place	Not in place	Available	Available
Controlling ministry	Office of the President (CNBS) and Ministry of Science and Technology (CNBio)	Central Food Ministry	Ministry of Environment, Forest and Climate Change	Ministry of Agriculture	Ministry of Agriculture	Executive secretary nominated from Ministry of Science and Technology and approved by the president
Regulatory approach	Precautionary	Precautionary	Precautionary	Precautionary	Precautionary	Precautionary
Current stage of commercialization	Outlook very positive with quick approvals after 2008	On the whole, quite positive	Moratorium from Feb 2010 on approvals. Field trials, put on hold since early 2012, are revoked in 2013	Came to a standstill amid resistance	Process of approvals moving quickly	11-year moratorium ended in 2009
		Approvals faster after 2012 change of policy				Now, approval process moving quickly
Purpose of harnessing agri-biotechnology	National development and exports	National development and exports	Domestic food security and exports	Domestic food security	Domestic food security and exports	Domestic food security
Separate permissions	Not needed	Not needed	Need 'No Objection certificate'	Needed	Not needed	Not needed
Approval of staked events	Treated as new events	Allows applications for transgenic combining two approved events without full analysis	Treated as new event, though the consisting events are approved	No clear policy	Treated as new events	Evaluates them as different than the parental one
Type of labelling law	Mandatory (process based)	Voluntary	Mandatory (process based)	Mandatory (process based)	Voluntary so far (product based)	Voluntary

Labelling law	Mandatory labelling for >1%	No labelling regime. Does not differentiate GM and non-GM	Ministry of Consumer Affairs, Food and Public Distribution from 2013 for packaged form	Compulsory for soybean, maize, cotton, canola and tomato	By Ministry of Health. Only when allergens or human/animal proteins are present	No mandatory labelling. But, labelling of GM content of seeds
Enforcement of labelling	Not enforced	Not applicable	Not done presently	Enforced	Does not arise	Does not arise
UPOV Treaty, 1978	Joined in 1999	Joined in 1994	Not a member. But, enacted PPV&FR Act in consonance with UPOV '78	Joined in 1999	Joined in 1977	Joined in 1997
Patent laws	Not strong patents Plant and animals not patentable Microorganisms patentable with conditions	Strong protection to animals, plants, plant varieties, microorganisms, biological processes and genes	Microorganisms patentable	Patent law, 2008 Regulation on protection of new plant varieties, 1997	NA	Strong protection to plants, plant varieties and microorganisms Biological processes not patentable

Source: Compiled by the author from various sources

process of harnessing technology and made it the leading GE crop cultivator, overtaking Argentina. To retain the edge, Argentina centralized all biotech-related decision-making by forming an exclusive Biotechnology Directorate in the Ministry of Agriculture, Fisheries and Livestock since 2009 and further reformed in 2012 to take decisions within 24 months by reducing from 42 months (USDA 2015).

Most countries take decisions on commercialization at the federal level, except in India and China. The Punjab Seed Council gave approvals to Bt cotton varieties in Pakistan until 2014, and there is uncertainty on the competent authority at the moment (Spielaman et al. 2015). In India, permissions from respective state governments are required to undertake trials since 2010 (Gupta 2011), and only eight of them have allowed trials since then. The moratorium imposed in 2010 continues in India and probably subject to the verdict of a case in the Supreme Court. Mexico came out of an 11-year moratorium in 2009 and accelerated approvals since then. Though approvals have stopped after Bt cotton in China, there was a shift in policy in 2016, which was witnessed in Chinese government acquiring biotech major company, Syngenta that aimed at allaying fears of foreign domination in technology and pushing forward transgenic crops to overcome imports and legitimize widespread illegally grown GE maize and rice crops (Economist 2016).

Development of biotechnological products in private domain is also conditioned on IPR protection either through UPOV (the Union for the Protection of New Varieties of Pulses) route or sui generis system. While Brazil, Argentina, China, S. Africa and Mexico joined UPOV 1978, India followed sui generis system and formulated Protection of Plant Varieties and Farmers Rights' Act, 2011, to enable protection to traditional knowledge by farmers. In the past few years, several African countries have joined UPOV, and several others are in consultation to do so (Jefferson and Padmanabhan 2016).

Labelling GE products has become a major issue of contention in recent times with demands for consumers' choice. While Brazil, China and India follow the mandatory process-based labelling methods, Argentina, South Africa and Mexico follow the voluntary product-based method. Though there are divergent views on which of these two methods of labelling helps consumers make an informed decision, published academic research concludes that voluntary labelling serves the purpose better than mandatory labelling (Bansal and Gruere 2012). However, a study from the USA concluded that mandatory labels led to a 19% reduction in opposition to GE food (Kolodinsky and Lusk 2018).

Most of the developing countries have signed and ratified Cartagena Protocol on Biosafety (CPB) except Argentina, though Brazil continues to have reservations about strict liability regime. The compliance to the CPB requires traceability arrangements on the source of GE product and also specific guidelines on coexistence of conventional, organic and GE crops (Bailey 2002; Wilson et al. 2008). None of the developing countries has the system for traceability and coexistence, except Brazil which has put in place rules for coexistence (Table 7.5). Stricter policies might end up with increased segregation costs for value chain actors, reduced international competitiveness (Boccaletti et al. 2017). Mexico is unique in that the argument that the places of primary source of origin of crops should be left GE free has forced them to keep GE free zones in some states. The issue of liability and compensation is another contentious matter which the developing countries will have to address in the years to come (Vigani and Olper 2012; Punt et al. 2017). Free flow of crop products from developing countries would require harmonization of GMO standards (de Faria and Wieck 2015) (Table 7.6).

7.5.1 Regulation of Gene-Edited Crops

Gene-edited crops, prima facie, seem less susceptible to stigmatization on issues of biosafety, as the technology leaves no sign of transgenics, and the ensuing products are like those from conventional plant breeding (Hefferon and Herring 2017; Adenle et al. 2018; Zaidi et al. 2019). These technologies are new, and their products might carry unintended effects like conventional plant breeding and transgenic technologies (EFSA 2012), depending on the kind of NBT used. Extensive development of NBTs over the last decade creates a situation where every country has to decide on the appropriate way to classify them for handling applications of potential trials and commercialization. The issue that arises in this regard concerns their similarity or difference to the GMOs, as they follow the latter in biological research pipeline. The first mover is Argentina by legislating to regulate NBTs on a product to product basis, using the concept of novel combination of genetic material (Srinivas 2018). The unfolding regulatory trajectories across countries show that those with product-based approach to GMOs treat NBTs like any other products of conventional plant breeding, while countries using process-based approach consider NBTs as GMOs (Seyran and Craig 2018). The former countries regulate them on a case-by-case assessment and include the USA, Canada, Argentina, Brazil, Chile, Columbia, China, Sweden and Australia (Lassoued et al. 2018). Japan also framed rules in early 2019 allowing gene-edited crops like those from conventional plant breeding. On the other hand, European Union Court of Justice decided that any organism obtained through an NBT that applies mutagenesis will be classified from now on with a retrospective effect from the day that the directive went into force as a GMO within the scope of the directive (Purnhagen et al. 2018).

NBTs are in limbo in several countries in view of the politically sensitive nature of biotechnology, difficulty of detecting NBTs from products of conventional plant breeding and wide spectrum of technologies in gene editing. These technologies range from simple refinement of conventional plant breeding without altering the genome in RNA-dependent DNA methylation (RdDM), transgene-free products with site-specific genome changes in clustered regularly interspersed short palindromic repeats (CRISPR), transcription activator-like effector nuclease (TALEN) and zinc-finger nucleases (ZFN) to gene insertions in SDN3. Therefore, the problem arises in classifying products of different NBTs either as GMOs or non-GMOs. This is a unique challenge that emanates with the advent of NBTs only. Excessive regulation as in case of GMOs can stifle investments, product development and also adversely effects consumer confidence.

Table 7.6 Policy framework in selected developing countries – *Contd*

Regulatory issue	Arrangement in countries					
	Brazil	Argentina	India	China	S. Africa	Mexico
Cartagena protocol	Ratified. But opposed to strict liability. Incorporated precautionary principle	Signed, but not ratified	Ratified and a biosafety clearing house is set up in the ministry of envt and forests	Signed and ratified	Signed and ratified. Department of Agriculture, Forestry and Fisheries (DAFF) is looking instead of DEF	Signed and ratified
GE imports	Only GE events approved for commercial production can be imported	Does not differentiate GE and non-GE	Soybean oil from Brazil, Argentina and the USA is allowed	LLP with 0% tolerance for unapproved events	1% tolerance. But, processed product allowed	Allowed under NAFTA. However, the imports should not be used for production but only for consumption. 2% level tolerance by the Secretariat of Agriculture, Livestock, Rural Development, Fisheries and Food (SAGARPA)
	0% tolerance for unapproved events		Zero-tolerance policy for unapproved events	Permitted 10 events of soybean from the USA, Brazil, Argentina and 17 events of GM corn	Only approved events allowed. 54 events in five crops – soybean, maize, cotton, canola and rice	
Coexistence	Rules exist	No policy	No specific regulation	No rules	No rules	Biosafety law provision 90 established GEO free zones by SAGARPA
Traceability	No system	No official system	No system	No system	No system	No system

Animals	GE dairy cattle produced. Recombinant proteins in pipeline. Cloning done. No GE animals so far. Same regulatory framework	Transgenics in pipeline for growth hormones. Cloning allowed. Regulation same as agriculture	In its infancy. Two buffaloes are cloned successfully. No regulations on production or marketing of cloned animals	Transgenic animals being developed, but none are approved so far	No animals so far. Regulation is same	No GE animals or products. Covered under same regulatory framework as plants

Source: Compiled by the author

7.6 Rising Investment Requirements with New Technologies and Consolidation in Seed Industry

The recently completed mergers and acquisitions in the seed industry indicate continuation of the ongoing trend in the seed industry that entails huge investments with the advent of new technologies and their convergence with information and communication technologies. However, they have raised concerns on improving small farm agriculture, especially in developing countries like India through rising seed prices (Bryant et al. 2016). To mention the top three, these are 130 billion USD merger of Dow and Dupont, 66 billion USD takeover of Monsanto by Bayer and 43 billion USD acquisition of Syngenta by China Chemical Corporation, all in the past 2 years. Even before these big ticket consolidations, 'the big six' corporations collectively controlled more than 75% of global agrochemical market, 63% of the commercial seed market and almost three-fourths of R&D expenses in the seeds and pesticides sector, and the sector has been witnessing transformation to oligopoly (Lianos et al. 2016). These recent consolidations are continuation of long-term trend in the industry of agrochemical companies taking over seed companies and likely to persist for some time to come (Rao 2004; Lianos et al. 2016; Howard 2015), if the countervailing forces discussed below do not grow strong enough to counter the trend. Besides the secular trend of agrichemical companies taking over seed companies, these corporations have been acquiring software start-ups for precision farming (Gullickson 2018). The erstwhile Monsanto established Climate Corporation as a subsidiary, and AGCO bought Precision Planting from Climate Corporation. DowDupont acquired granular to connect farmers with big data analytics. John Deere invested 305 million USD to acquire Blue River Technology that designed and integrated computer vision and machine learning technology in lettuce fields to reduce herbicide use and potential applications in other crops.

The public sector breeders quite often get stonewalled with patent hurdles in their effort to develop varieties even in orphan crops. The patent thickets create a situation referred to as 'tragedy of the anti-commons' by Heller and Eisenberg (1998), in which no one will be able to assemble a product overcoming the maze of patents and results in underuse of (or non-use) of resources. Golden rice is a classic example of this phenomenon, as its development was stalled for a long time to overcome the 40 odd patents from different owners (Jefferson and Padmanabhan 2016). The challenge of ever-rising share of the private sector in global food and agriculture R&D, which stood at 44% in 2009, is another big concern, as private sector cannot compensate for the decline of public research in view of its focus on technology development, while public universities and institutes continue to be the source of upstream research (Pardey et al. 2015). The private sector research however can have high social benefits (to farmers) relative to private benefits (to companies) (Fig. 7.2) and can be utilized for the societal gains, with a clear understanding that public research can only create agricultural public goods (Dalrymple 2008).

These disturbing developments mask another set of developments rising as countervailing forces to protect small farm agriculture. The locus of R&D expenditures in the world is now slowly shifting towards developing countries. In 2009, about

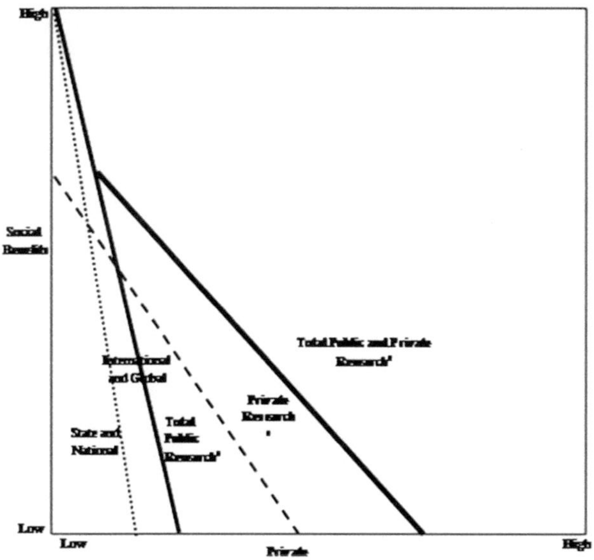

Fig. 7.2 Hypothetical relationship between the social and private benefits from public and private research. (*Source*: Dalrymple 2008)

42% of global R&D investment was done in the middle-income countries including China, Brazil and India, though low-income countries continue to have a miniscule of this total (Pardey et al. 2015). Analyses of the changing landscape of biotechnology across developing countries show that they have realized the need to be proactive to save resource-poor farmers through energizing public sector research. China, for example, acquired Switzerland-based Syngenta at a price of 43 billion USD through its National Chemical Corporation.

Several developing countries like Brazil, Argentina, China, the Philippines, Bangladesh and Pakistan have been racketing up public sector research in biotechnology and have brought out crop products in recent years. Not surprisingly, these crop products possess traits of importance to resource-poor farmers like drought tolerance, examples of which are given earlier. Another significant positive development is in the realm of legal framework of property rights, whereby gene patents have either recently been invalidated (in the USA and Australia) or likely to be done in the near future in many other countries. Also, subsequent to the Nagoya Protocol on access and benefit sharing, negotiations are underway in the Intergovernmental Committee to negotiate an international legal instrument to protect traditional knowledge and access and benefit sharing (Jefferson and Padmanabhan 2016). In India, new 2013 patent guidelines, if enforced, might change the scenario away from strong patents (Ravi 2013).

Beyond IPR protection, stringent regulations through biosafety laws dampen research and product development by the public sector as well as small investors as happened in India and Argentina, and this is referred to as 'IP-regulatory' complex

(Graff and Zilberman 2016). Despite not having patent protection, Bollgard I event of Monsanto enjoyed monopoly rights in India because of the arduous process of getting event approval leaving other companies dependent on Monsanto for seed development (Graff et al. 2015). Gene-edited crops might be less controversial on issues of bio-property as the relatively less cost, time and infrastructural requirements do not enable first-mover advantages to the developers, apart from lower chances for geographical concentration of the industry (Hefferon and Herring 2017; Lassoued et al. 2018).

7.7 Conclusions and Policy Implications

Any technology, including agricultural technologies, can in principle have both positive and negative impacts in varying degrees and only one of them in a certain magnitude. Empirical evidence in the specific socio-economic, cultural and institutional milieu is necessary to evaluate technologies. This chapter has examined the diffusion of genetically engineered crops in developing countries in a cross-country comparative perspective in regard to their nature and adoption, impacts, necessary supplementary policies and challenges associated with rising investment needs in the seed sector.

The pace of discovery in biological sciences has been rapid, and biotechnologies are moving beyond genetic engineering, and the first non-GE biotechnological crop using gene editing was released in 2016 in the form of sulphonylurea (SU)-tolerant canola (mustard) in the USA followed by bruising- and browning-resistant potato. Within the GE crops, technology has deepened to commercialize several (31) GE crops in place of just 3 crops earlier, viz. soybean, maize and cotton, and from single gene expressions like insect/herbicide resistance to second-generation products like drought tolerance and improved quality attributes. The developing countries accounted for 54% of the 189.8 million hectares of area under these crops in 2017 negating fears that resource-poor farmers in these countries would not be benefited from these technologies.

Economic and agronomic impacts of GE crops have been rigorously studied, as the controversies on their utility continue. The peer-reviewed research findings suggest higher yields, higher net income and lower chemical use with conservation tillage. The most recent meta-analysis estimated 22% yield gain associated with 39% reduction in plant protection expenditure and 68% higher net income. The longitudinal studies have shown that cultivation of these crops over the past 19 years has resulted in gains of 150 billion USD to world agriculture.

Technologies need supplementary policies to optimize social welfare. Excessive and uninformed regulation has been the bane of biotech crop adoption in the contemporary world. Very few of the developing countries could put a legal framework for biosafety, and our study reveals that this is still a work in progress with inadequate efforts to create a professional body that allays the fears of consumers and arrives at decisions based on scientific data. The countries like Brazil and Argentina have been moving fast in diffusion of these technologies, as they could put this mechanism in place. India and China are yet to navigate this process, not to speak of the many low-income developing countries, especially from the African continent. For instance, the Biotechnology Regulatory Authority of India (BRAI) bill has been in the making since 2008 in India without any end in sight. Uncertainty in regulatory approvals can have the adverse impacts of abandonment of investment and thereby innovation. A 2-year delay reduces the net present value of normal returns for a private investment into a new GM crop variety by about one-third that renders the investment loss-making (Smyth et al. 2016). Above all, several developing countries including India are yet to devise regulatory norms for regulating gene-edited crops developed using new breeding technologies (NBTs) unmindful of the ideological opposition by civil society groups as in Shiva (2018). These countries need to classify gene-edited crops as non-GMOs and regulate as those from conventional plant breeding by following the path taken by developing countries like Argentina, Brazil, China, Chile and Columbia, besides those among developed countries, viz. the USA, Canada, Australia, Sweden and Japan.

There has been a dramatic shift in the seed industry with rising investment demands with the availabilities of new technologies as well as convergence of technologies. The consolidation with the mega mergers and acquistions in recent times have to be viewed in this background. This trend is likely to continue for some time to come in view of the rising investment needs with convergence of technologies and falling margins of chemical companies. Foremost among the countervailing forces to this consolidation is the racketing up of public sector research by the national agricultural research system (NARS) in developing countries. Second is the recent trend of reversal of patent protection for DNA sequences that started with the Supreme Court verdict in *Association for Plant Pathology* vs. *Myriad Genetics* in the USA. However, it is premature to foresee the final outcome of this trend. The developing countries will gain by internalizing these technologies into their national agricultural research systems and invest more in both upstream and downstream research, besides proactively participating in the ongoing review process of negotiating international legal instruments for traditional knowledge. Enabling regulatory policies will go a long way in unleashing the huge potential created in the private sector.

The third countervailing force to overcome the asymmetric power of corporation in biotechnological research is to forge public-private partnerships like in case of drought-tolerant (DT) soybean in Argentina, DT maize in Africa and several others. Innovative platforms like Public Intellectual Property Resource for Agriculture (PIPRA) established in the University of California, Davis, and African Technology Foundation in Kenya show the way forward for the governments of developing countries to act in the interest of resource-poor farmers. It should not be forgotten that much of patented technology by companies has their origins from the upstream research done in public universities. Viewed from that angle, it becomes clear that science behind technologies and private sector domination need to be separated in taking decisions about their utilization. Excessive regulation or lack of regulatory mechanism has been stifling technologies developed by the public institutions and small companies that are of interest to developing country agriculture. The developing countries are likely to benefit more from engaging with development discourse on how to harness new opportunities arising out of rapid discoveries in biological sciences for raising incomes and welfare of rural populations with predominantly agriculture-based livelihood and thereby achieve higher national economic development.

References

Acemoglu D, Robinson JA (2012) Why nations fail: origins of power, prosperity and poverty. Profile Books Ltd., London

Adato M, Meinzen-Dick R, Hazell P, Lawrence H (2007) Integrating social and economic analyses to study impacts on livelihoods and poverty: conceptual frameworks and research methods. In: Adato M, Meinzen-Dick R (eds) Agricultural research, livelihoods, and poverty: studies of economic and social impacts in six countries. Johns Hopkins University Press, Baltimore, pp 20–55

Adenle AA, Morris EJ, Murphy DJ, Phillips PWB, Trigo E, Learns P, Quemada Y, Li H, Falck-Zepeda J, Komen J (2018) Rationalizing governance of genetically modified products in developing countries. Nat Biotechnol 36(2):137–139

Ahluwalia MS (1978) Rural poverty and agricultural performance in India. J Dev Stud 14:298–323

Anderson K (2010) Economic impacts of policies affecting crop biotechnology and trade. New Biotechnol 27(5):558–564

Anderson K, Valenzuela E, Jackson LA (2008) Recent and prospective adoption of genetically modified cotton: a global CGE analysis of economic impacts. Econ Dev Cult Chang 56:265–296

Areal FJ, Riesgo L, Rodriguez-Cerezo E (2013) Economic and agronomic impact of commercialized GM crops: a meta-analysis. J Agric Sci 151:7–33

Armstrong S, Westland T (2016) Escaping the middle income trap. East Asia Forum Blog. Retrieved from http://www.eastasiaforum.org/2016/03/28/escaping-the-middle-income-trap

Ashok KR, Giuliani AM, Thilagavathi SV, Raj R, Ramamoorthy MD, Sanjeevi Kumar A (2017) Trait valuation in genetically modified crops: an ex-ante analysis of GM cassava against cassava mosaic disease. Agric Econ Res Rev 30(2):223–234

Bailey R (2002) The looming trade war over plant biotechnology. Cato Trade Policy Analysis, 18. Center for Trade Policy Studies, Cato Institute, Washington, DC

Bansal S, Gruere GP (2012) Implications of mandatory labelling of GM food in India: evidence from the supply side. Food Policy 37(4):467–472

Barrett C, Carter MR, Timmer CP (2010) A century-long perspective on agricultural development. Am J Agric Econ 92(32):447–468

Boccaletti S, Farncesca P, Soregaroli C (2017) Segregation between GM and non-GM inputs and EU feed and food supply chains, future scenarios. AgBioforum 20(1):1–13

Bortesi L, Fischer R (2015) The CRISPR/Cas9 system for plant genome editing and beyond. Biotechnol Adv 33(1):41–52

Bouis HE, Saltzman A, Birol E (2019) Improving nutrition through Biofortification. In: Fan S, Yosef S, Pandya-Lorch R (eds) Agriculture for improved nutrition: seizing the momentum. CAB International

Brewin DG, Malla S (2012) The consequences of biotechnology: a broad view of the changes in the Canadian canola sector, 1969–2012. AgBioforum 15(3):257–275

Brookes G (2005) The farm level impact of using round up ready soybeans in Romania. AgBioforum 8(4):235–241

Brookes G, Barfoot P (2015) GM crops: global socio-economic and environmental impacts 1996–2013. PG Economics Ltd., Dorchester

Brookes G, Barfoot P (2018a) Farm income and production impacts of using GM crop technology 1996– 2016. GM Crops Food 9(1):59–89

Brookes G, Barfoot P (2018b) Environmental impacts of genetically modified (GM) crop use 1996–2016: impacts on pesticide use and carbon emissions. GMCrops Food 9(3):109–139

Bryant H, Maisashvili A, Outlaw J, Richardson J (2016) Effects of proposed mergers and acquisitions among biotechnology firms on seed prices. Working Paper16-2, Agricultural and Food Policy Centre, Department of Agricultural Economics, Texas A&M Agri Life Extension Services, Texas A&M University. Retrieved from http://www.afpc/tamu.edu/

Carpenter JE (2010) Peer-reviewed surveys indicate positive impact of commercialized crops. Nat Biotechnol 28:319–321

Carpenter J, Felsot A, Goode T, Hamming M (2002) Comparative environmental impacts of biotechnology-derived and traditional soybean, corn, and cotton crops. Council for Agricultural Science and Technology, Ames

Carriere Y, Ellers-Kirk C, Sisterson M, Antilla L, Whitlow M, Timothy JD, Tabashinik BE (2003) Long-term regional suppression of pink bollworm by *Bacillus thuringiensis* cotton. Proc Natl Acad Sci 100(4):1519–1523

Craig W, Obonyo DN, Tepfer M (2017) A strategy for integrating science into regulatory decision-making for GMOs: in genetically modified organisms in developing countries. In: Adenle AA, Morris EJ, Murphy D (eds) Risk analysis and governance. Cambridge University Press, Cambridge, pp 26–38

Dalrymple DG (2008) International agricultural research as a global public good: concepts, the CGIAR experience, and policy issues. J Int Dev 20:347–379

De Faria RN, Wieck C (2015) Empirical evidence on the trade impact of asynchronous regulatory approval of new GMO events. Food Policy 53:22–32

de Janvry A, Sadoulet E (2002) World poverty and the role of agricultural technology: direct and indirect effects. J Dev Stud 38(4):1–26

Dev SM, Rao CN (2010) Agricultural price policy, farm profitability and food security. Econ Political Wkly 45(26–27):174–182

Durr J (2016) The political economy of agriculture for development today: the "small versus large" scale debate revisited. Agric Econ 47:1–11

EFSA (2012) Scientific opinion addressing the safety assessment of plants developed using zinc finger nuclease 3 and other site-directed nucleases with similar function. EFSA J 10(10):2943

Eichengreen B, Park D, Shin K (2013) Growth slowdown redux: new evidence on the middle-income trap. Working Paper No. 18673. National Bureau of Economic Research, Cambridge, USA

Fabrick JA et al (2014) Alternative splicing and highly variable cadherin transcripts associated with field-evolved resistance of pink bollworm to *Bt* cotton in India. PLoS One 9(5):1–13

Falck-Zepeda JB, Traxler G, Nelson RG (2000) Surplus distribution from the introduction of a biotechnology innovation. Am J Agric Econ 82(2):360–369

Fernandez-Cornejo J, McBride WD (2002) Adoption of bioengineered crops. Agricultural Economic Report NO.810. US Department of Agriculture

Fernandez-Cornejo, Hendricks C, Mishra A (2005) Technology adoption and Off-Farm household income: the case of herbicide-tolerant soybeans. J Agric Appl Econ 37(2):549–563

Finger R et al (2011) A meta-analysis on farm-level costs and benefits of GM crops. Sustainability 3:743–762

Fitt GP (2003) Implementation and impact of transgenic *Bt* cottons in Australia. ICAC Recorder 21(4):14–19

Ghouse M, Piesse J, Thirtle C, Poulton C (2009) Assessing the performance of GM maize amongst stall holders in KwaZulu-Natal, South Africa. AgBioforum 12(1):78–89

Ghouse M, Sengupta D, Zambrano P, Falck-Zepeda J (2016) Genetically modified: less drudgery for her, more maize for him? Evidence from smallholder maize farmers in South Africa. World Dev 83:27–38

Graff G, Zilberman D (2016) How the 'IP-Regulatory' complex affects incentives to develop socially desirable products from agricultural genomics. In: Marden E, Godfrey RN, Manion R (eds) The intellectual property- regulatory complex: overcoming barriers to innovation in agricultural genomics. University of British Columbia Press, Vancouver

Graff G, Hochman G, Suntharlingam C, Zilberman D (2015) The competing policy paradigms of agricultural biotechnology: implications and opportunities for emerging and developing economies. AgBioforum 18(2):168–181

Gruere G, Sengupta D (2011) *Bt* cotton and farmer suicides in India: an evidence-based assessment. J Dev Stud 47(2):316–337

Gruere G, Sun Y (2012) Measuring the contribution of Bt cotton adoption to India's cotton yields leap. IFPRI Discussion Paper 01170, April. International Food Policy Research Institute, Washington, DC

Guenthner JF (2017) Economic and environmental benefits of biotech potatoes with traits for bruise resistance, late blight resistance, and cold storage. AgBioforum 20(1):37–45

Gullickson G (2018) 10 Ag mergers and acquisitions from 2017. Available at: http://news.agropages.com/News/NewsDetail%2D%2D-24882-e.htm. Accessed on 27 May 2019

Gupta A (2011) An evolving science-society contract in India: the search for legitimacy in anticipatory risk governance. Food Policy 36:736–741

Hall C, Knight B, Ringrose S, Knox O (2013) What have been the farm-level economic impacts of the global cultivation of GM crops? Syst Rev. CEE review11-002, Collaboration for Environmental Evidence Library. Retrieved from www.environmentalevidence.org/SR110

Hazell PBR (2009) The Asian green revolution. IFPRI Discussion Paper No. 00911. International Food Policy Research Institute, Washington, DC

Hefferon KL, Herring RJ (2017) The end of the GMO? Genome editing, gene drives and new frontiers of plant technology. Rev Agrar Stud 7(1):1–32

Heller MA, Eisenberg RS (1998) Can patents deter innovation? The anti commons in biomedical research. Science 280:698–701

Herring RJ, Paarlberg R (2016) The political economy of biotechnology. Ann Rev Resour Econ 8(8):1–8. Retrieved from http://www.annualreviews.org/journal/resource

Howard PH (2015) Intellectual property and consolidation in the seed industry. Crop Sci 55(2015):1–7

Hutchison WD et al (2010) Area wide suppression of European corn borer with *Bt* maize reaps savings to non-*Bt* maize growers. Science 330:222–225

ISAAA (2017) Global status of commercialised Biotech/GM crops in 2017: biotech crop adoption surges as economic benefits accumulate in 22 years. ISAAA Brief 53. International Service for the Acquisition of Agri-Biotech Applications

Jaganathan D, Ramasamy K, Sellamuthu G, Jayabalan S, Venkataraman G (2018) CRISPR for crop improvement, an update review. Front Plant Sci 9:985

Jander G, Baerson SR, Hudak JA, Gonzalez KA, Gruys K, Gomez E et al (2003) Ethylmethanesulfonate saturation mutagenesis in Arabidopsis to determine frequency of herbicide resistance. Plant Physiol 131(1):139–146

Jefferson DJ, Padmanabhan MS (2016) Recent evolutions in intellectual property frameworks for agricultural biotechnology: a worldwide survey. Asian Biotechnol Dev Rev 18(1):47–67

Kalaitzandonakes N, Kruse J, Gouse M (2015) The potential economic impacts of herbicide tolerant maize in developing countries: review and evidence from Kenya. AgBioforum 18(2):221–238

Kathage J, Qaim M (2012) Economic impacts and impact dynamics of *Bt* cotton in India. Proc Natl Acad Sci 109(29):1652–11656

Klumper W, Qaim M (2014) A meta-analysis of the impacts of genetically modified crops. PlosOne 9(11–29):1–7. Retrieved from http://journals.plos.org/plosone/article?id=10.1371/journal.pone.0111629

Kniss A (2008) Farm-scale analysis of glyphosate-resistant sugar beet the year of commercial introduction in Wyoming. International Weed Science Society Annual Meeting

Kolodinsky J, Lusk JL (2018) Mandatory labels can improve toward genetically engineered food. Sci Adv 4(6):1413

Kouser S, Qaim M (2011) Impact of *Bt* cotton on pesticide poisoning in smallholder agriculture: a panel data analysis. Ecol Econ 70(11):2015–2013

Kouser S, Qaim M (2013) Genetically modified crops and food security. PLoS One. open access journal, June

Kranthi KR (2015) Pink bollworm strikes cotton again. Cotton Statistics and News No.35. Retrieved from http://www.cicr.org.in/pdf/Kranthi_art/Pinkbollworm.pdf

Krishna V, Qaim M (2012) *Bt* cotton and sustainability of pesticide reductions in India. Agric Syst 107:47–55

Lassoued R, Stuart JS, Peter WBP, Hayley H (2018) Regulatory uncertainty around new breeding techniques. Front Plant Sci 9:1291

Lewis A (1954) Economic development with unlimited supplies of labor. In: Gersovitz M (ed) Selected economic writings of W. Arthur Lewis. New York University Press, 1983

Lianos I, Katalevsky D, Ivanov A (2016) The global seed market, competition law and intellectual property rights: untying the Gordian knot. Research Paper Series: 2/2016. Centre for Law, Economics and Society, University of London. Retrieved from www.ucl.ac.uk/cles/research-paper-series

Lichtenberg E, Zilberman D (1986) The econometrics of damage control: why specification matters. Am J Agric Econ 68:261–273

Marra MC, Piggott NE, Carlson GA (2004) The net benefits, including convenience, of roundup ready soybean: results from a national survey. Technical Bulletin 2004-3. NSF Centre for IPM, Raleigh

Mellor JW (2006) Pro-poor growth- the relation between agricultural growth and poverty reduction. In: Radhakrishna R, Rao SK, Dev SM, Subbarao K (eds) India in a globalising world: some aspects of macro economy, agriculture and poverty. Academic Foundation, New Delhi

Morris EJ (2017) Biosafety regulatory systems in Africa. In: David PK, Makinde D, Weebadde CK, Maredia K (eds) Biosafety in Africa, experiences and best practices

Morse S, Mannion AM, Evans C (2012) Location, location, location: presenting evidence for genetically modified crops. Appl Geogr 34:274–280

Muller C, Robertson RD (2014) Projecting future crop productivity for global economic modelling. Agric Econ 45:37–50

NASEM (2016) Genetically engineered crops: experiences and prospects. The National Academies Press, Washington, DC. Retrieved fromwww.nap.edu

Otuska K (2013) Food insecurity, income inequality, and the changing comparative advantage in world agriculture. Agric Econ 44:7–18

Pardey PG, Chan-Kang C, Beddow JM, Dehmer SP (2015) Long-run and Global R&D Funding Trajectories: the U.S. farm bill in a changing context. Am J Agric Econ 97(5):1312–1323

Pingali P (2012) Green revolution: impacts, limits and the path ahead. Proc Natl Acad Sci 109(31):12302–12308

Pray CE, Huang JHR, Rozelle S (2002) Five years of *Bt* cotton in China- the benefits continue. Plant J 31:423–430

Punt MJ, Venus TJ, Justus W (2017) The costs of coexistence on farms in Germany. AgBioforum 20(1):24–36

Purnhagen KP, Esther K, Kleter G, Schebesta H, Visser RGG, Justus W (2018) EU court casts new plant breeding techniques into regulatory limbo. Nat Biotechnol 36(9):799–800

Qaim M (2009) The economics of genetically modified crops. Ann Rev Resour Econ 1:665–693

Qaim M (2016) Genetically modified crops and agricultural development. Palgrave Macmillan, New York

Qaim M, de Janvry A (2003) Genetically modified crops, corporate pricing strategies, and farmers' adoption: the case of *Bt* cotton in Argentina. Am J Agric Econ 85(4):814–828

Qaim M, Traxler G (2005) Roundup ready soybeans in Argentina: farm level and aggregate welfare effects. Agric Econ 32:73–86

Qiao F (2015) Fifteen years of *Bt* cotton in China: the economic impact and its dynamics. World Dev 70:177–185

Rao NC (2004) Plant biotechnology and the emerging scenario. Rev Dev Chang 9(1):69–92

Rao NC (2013) *Bt* cotton yields and performance: data and methodological issues. Econ Political Wkly 48(33):66–69

Rao NC (2015) Disadvantaged regions and social groups: is there a way out? Indian J Agric Econ 70(3):438–449

Rao NC, Dev SM (2009) Biotechnology and pro-poor agricultural development. Econ Polit Wkly 44(52):56–56
Rao NC, Dev SM (2010) Biotechnology in Indian agriculture: potential, performance and concerns. Academic Foundation, New Delhi
Rao NC, Pray CE, Herring RJ (2015) Biotechnology for a second green revolution in India: socioeconomic, political and public policy issues. AgBioforum 18(2):126–141. Retrieved from www.agbioforum.org
Rao NC, Carl EP, Ronald JH (2018) Biotechnology for second green revolution in India, overview of issues. In: Rao NCS, Pray CE, Herring RJ (eds) Biotechnology for a second green revolution in India, socioeconomic, political and public policy issues. Academic Foundation, New Delhi
Ravi B (2013) Gene patents in India: gauging policy by an analysis of the Grants made by the Indian patent office. J Intellect Prop Rights 18(4):323–329
Rosado A, Craig W (2017) Biosafety regulatory systems overseeing the use of genetically modified organisms in the Latin America and Caribbean region. AgBioforum 20(2):120–132
Schaart JG, Visser RGF (2009) Novel plant breeding techniques-consequences of new genetic modification-based plant breeding techniques in comparison to conventional plant breeding (Report 2009-02). The Netherlands Commission on Genetic Modification (COGEM), Bilthoven, pp 2785–2794
Schultz TW (1964) Transforming traditional agriculture. Yale University Press, New Haven
Seyran E, Craig W (2018) New breeding techniques and their possible regulation. AgBioforum 21(1):1–12
Shiva V (2018) Biodiversity, GMOs, gene drives and militarised mind. Retrieved from http://www.ipsnews.net/2016/07/biodiversity-gmos-gene-drives-and-the-militarised-mind/
Smale M, Zambrano P, Paz-Ybarnegaray R, Fernandez-Montano W (2012) A case of resistance: herbicide-tolerant soybeans in Bolivia. AgBioforum 15(2):191–205
Smyth SJ, Jose FZ, Karinne L (2016) The costs of regulatory delays for genetically modified crops. Estey J Int Law Trade Policy 17(2):173–195
Spielaman D, Nazli H, Ma A, Zambrano P, Zaidi F (2015) Technological opportunity, regulatory uncertainty, and *Bt* cotton in Pakistan. AgBioforum 18(1):98–112
Srinivas KR (2018) Regulating genome edited crops and European court of justice ruling. Asian Biotechnol Dev Rev 20(1–2):89–97
Stephens, J., Barakate, A. (2017). Gene editing technologies – ZFNs, TALENs, and CRISPR/Cas9, Encyclopedia of applied plant sciences, 2nd B. Thomas, B. G. Murray, D. J. Murphy (Cambridge, MA: Academic), 157–161. https://doi.org/10.1016/B978-0-12-394807-6.00242-2
Stone G (2011) Field *versus* farm in Warangal: *Bt* cotton, higher yields and larger questions. World Dev 39(3):387–398
Subramanian A, Qaim M (2009) Village-wide effects of agricultural biotechnology: the case of *Bt* cotton in India. World Dev 37(1):256–267
Subramanian A, Qaim M (2010) The impact of *Bt* cotton on poor households in rural India. J Dev Stud 46(2):295–311
Swinnen J, Riera O (2013) The global bio-economy. Agric Econ 44:1–5
The Economist (2016) Gene policy transfer. The Economist 23. Retrieved http://www.economist.com/news/china/21697272-china-may-relax-its-almost-total-ban-growing-gm-food-gene-policy-transfer
Thirtle C, Beyers L, Ismael Y, Piesse J (2003) Can GM-technologies help the poor? The impact of *Bt* cotton in Makhathini Flats, Kwa Zulu-Natal. World Dev 31(4):717–732
Till BJ, Cooper J, Tai TH, Colowit P, Greene EA, Henikoff S, Comai L (2007) Discovery of chemically induced mutations in rice by TILLING. BMC Plant Biol 7:19
Traxler G, Godoy-Avila S, Falck-Zepeda J, Espinoza-Arellano J (2003) Transgenic cotton in Mexico: a case study of the Comarca Lagunera. In: Kalaitzandonakes N (ed) The economic and environmental impacts of Agbiotech. Kluwer, New York
USDA (2015) Argentina annual biotechnology report, GAIN report. United States Department of Agriculture Foreign Agricultural Service

Vigani M, Olper A (2012) International trade and endogenous standards: the case of GMO regulations. World Trade Rev 11(3):415–437
WB (2007) World Development Report (2008). Agriculture for development. The World Bank, Washington, DC
Wilson WW, De Vuyst EA, Taylor RD, Koo WW, Dahl BL (2008) Implications of biotech traits with segregation costs and market segments: the case of roundup ready wheat. Eur Rev Agric Econ 35(1):51–73
Witsjaksono J, Wei X, Mao S, Gong W, Li Y, Yuan Y (2014) Yield and economic performance of the use of GM cotton worldwide over time: a review and meta-analysis. China Agric Econ Rev 6(4):616–643
Wu KM, Lu YH, Feng HQ, Jiang YY, Zhao JH (2008) Suppression of cotton bollworm in multiple crops in China in areas with *Bt* toxin-containing cotton. Science 321:1676–1678
Yorobe JM, Smale M (2012) Impacts of Bt maize on smallholder income in the Philippines. AgBioforum 15(2):152–162
Zaidi SSA, Vanderschuren H, Qai M, Mahfouz MM, Kohli A, Mansoor S, Tester M (2019) New plant breeding technologies for food security. Science 363(6434):1390–1391
Zilberman D, Kim E, Kirschner S, Kaplan S, Reeves J (2013) Technology and the future bio economy. Agric Econ 44:95–102

Part II

Industrial Biotechnology

Biotechnology Directive: A Major Step in Biotechnology Patent Law in Europe

8

Li Jiang

Abstract

Biotechnology has become a new crucial technology of increasing economic growth. Nowadays, biotechnology has been widely applied in the fields of agriculture, pharmaceutical industry, medicine, energy, and environment protection. With the development of new processes in biotechnology, new adjustments are needed from established patent rules. Thus the Directive was drafted by the Commission to meet the demands of biotechnology industry. The Biotechnology Directive had successively treated the patentability of gene-related inventions, the exceptions to patent and moral issues. In addition, the Directive generally achieved the goal of harmonization of patent laws among the member states. To some extent, the Directive simplified the uncertainty of the patent law which is benefit to increase the research investment and development funds, but the remaining issue limiting its wide acceptance have been discussed in this chapter.

Keywords

Biotechnology directive · Bioeconomy · Patent laws · Stem cells · Plant varieties · Animal varieties · European Union

8.1 Introduction

Biotechnology has become a new crucial technology of increasing economic growth. Nowadays, biotechnology has been widely applied in the fields of agriculture, pharmaceutical industry, medicine, energy, and environment protection. In 2003, US biotech companies employed approximately 200,000 people and

L. Jiang (✉)
Shanghai International College of Intellectual Property, Tongji University, Shanghai, People's Republic of China

C. Keswani (ed.), *Bioeconomy for Sustainable Development*,
https://doi.org/10.1007/978-981-13-9431-7_8

generated USD 39.2 billion in revenues.[1] The rapid development of biotechnology is also reflected in the increasing of applications of biotechnology patent. According to the European Commission report, in the past 10 years, biotechnology patent at the United States Patent and Trademark Office and the European Patent Office have increased by 13–15% per year on average, compared with the 5% annual growth rates for all patents.[2]

There are three main sectors of biotechnology market: the United States, Europe, and Japan. The European Union, which identifies the importance of biotechnology to its future economic development, is committed to becoming a more competitive participant in this bright prospect market.[3] Inevitably, the patent regulations have received unprecedented challenges and struggled to adjust the system to this new technology.

One of the earliest important patent documents in Europe is the 1973 version of the European Patent Convention (EPC). It is a multilateral treaty which provides a complete system of patent protection for contracted nations. This system guarantees a European patent has an equal influence to a national patent.[4] Then in 1994, the WTO Agreement on Trade-Related Aspects of Intellectual Property Rights (TRIPS) entered into effect. This agreement is consist of a detailed series of minimum international legislative and regulatory standards.[5] As Daniel Gervais evaluated in his book, TRIPS was one of "most significant milestones in the development of intellectual property in the twentieth century."[6]

With the development of new technology especially biotechnology, new adjustments are needed from established patent rules. Thus the Directive was drafted by the Commission to meet the demands of biotechnology industry. There are two basic rationales behind the Biotechnology Directive.

One rationale was a lag of Europe compared with other economic areas.[7] Among three main areas of the United States, Europe, and Japan, they all believe that patent system can increase investment activity and enable a patented technology to be

[1] Hilderth M, Resilience: Americas Biotechnology Report 2003, Emst & Young, July 2003

[2] Commission of the European Communities, Report from the Commission to the European Parliament and Council, "An Assessment of the Implications for Basic Genetic Engineering Research of Failure to Publish, or Late Publication of, Papers on Subjects which could be Patentable as Required under Art.16(b) of Directive 98/44/EC on the legal Protection of Biotechnological Inventions" 7(Brussels 2002)

[3] Communication on Promoting the competitive environment for the industrial activities based on biotechnology within the Community SEC(91) 629 final

[4] Braendli P, 'The future of the European patent system' (1995) International Review of Intellectual Property and Competition Law

[5] Sommer T, 'Patenting the animal kingdom? From cross-breeding to genetic make-up and biomedical research' (2008) International Review of Intellectual Property and Competition Law

[6] Gervals D, The TRIPS Agreement: *Drafting History and Analysis* (2nd edn London Sweet& Maxwell, London2003

[7] MacQueen H, Waelde C and Laurie G, Contemporary Intellectual Property (Oxford 2008, New York)

protected worldwide. Thus they all retain propatenting attitudes and improve their patent system to promote development of economy. The United States Patent Office preferred the equal treatment of biotechnology and other technologies.[8] This led to a liberal attitude of the United States toward patentability of biotechnological inventions. Consequently, if the US patent principles confer an advantage than European patent principles do, it puts pressures to Europe to reform the patent system.[9]Under this structure, the Directive adapted in 1998 aimed to clearly express the patentability of biotechnological inventions.

The other was the nonuniformity of member states. Despite the harmonization of substantive patent law in the contracting states, the interpretation and application of these laws are disagreeable.[10]Such as case Genentech v Wellcome Foundation, the EPO and UK courts have different standards of obvious.[11] In order to reinforce the research capability and promote economy in the whole Europe, the uncertainty of patent law in member states should be avoided. The Directive held the promise of harmonizing the rule of biotechnological patent throughout the EU. All member states have the obligation to transpose or implement in their national law. However, before July 2000, only six member states amend their national law to in line with the Directive. According to the second report of the Commission (2005), 21 member states had apprised the Commission of their instruments implementing the Directive.[12]

Originally, the first draft of Directive 1988 was rejected by the European Parliament. The main reason was that the Commission lacked considerations from ethical dimension. Some animal welfare groups and religious groups strongly protested against the drafted Directive and suggested that ethical considerations should be added.[13]After 10 years, the Directive 98/44 was finally adapted. However, a problem of potential conflict between legal systems arose. Because the EPC which is the rules of an intergovernmental treaty belongs to a non-EC instrument, the European Union (EU) has no jurisdiction over the EPC. For the purpose of releasing this discrepancy, in 1999 the Administrative Council of the European Patent Office decided to make some changes to the rules in the implementing regulations for adjusting the EPC to the Directive. Moreover, Rule 23b(1) EPC provides that: "Directive 98/44/EC of 6 July 1998 on the legal protection of biotechnological

[8] Morneault M, 'Stem Cell Research and Human Cloning: Where Do We Draw The Line?' (2005) New Eng.L.Rev.523

[9] Drahos P, 'Biotechnology patents, markets and morality' (1999) European Intellectual Property Review

[10] Braendli P, 'The future of the European patent system' (1995) International Review of Intellectual Property and Competition Law

[11] *Genentech Inc.* v *Wellcome Foundation Ltd* (1989) 8 RPC 147; (1988) 15 IPR 423

[12] Report from the Commission to the Council and the European Parliament, development and implications of patent law in the field of biotechnology and genetic engineering, at 2

[13] Sommer T, 'Patenting the animal kingdom? From cross-breeding to genetic make-up and biomedical research' (2008) International Review of Intellectual Property and Competition Law

inventions shall be used as a supplementary means of interpretation."[14] These countermeasures basically bridged the gap between the EU Biotechnology Directive and the EPC.

8.2 Discovery or Invention

In tradition patent system, the distinction between discovery and invention is explicit and unambiguous. As Kolle defined, "discovery is the unearthing of causes, properties or phenomena already existing in nature; invention is the application of such knowledge to the satisfaction of social needs,"[15] However, with the development of gene technology, the difference became vague and problematic.

In case Genentech v Wellcome Foundation, the House of Lords held that disclosing the structure of DNA belong to discovery which is excluded by the subject matter of patent.[16]According to Article 52(2)(a) EPC, discoveries which are not considered as inventions seem not to be patentable. As a result of lacking of patent protection, investments of human and nonhuman resource were reduced to some extent. Many people indicated that many efforts need to be thrown in order to obtain the DNA. As Crespi pointed out, "the inventor has not simply discovered or confirmed the existence of a gene but has been the first to characterize it, to define it chemically, and to make it available in a way that serves some useful purpose."[17] In case Relaxin, the Opposition Division held that the claims related to DNA sequence of a natural substance were classified as invention instead of discovery.[18]

The Directive settled the debate and affirmed the patentability of gene-related applications. Article 3(2) stipulated that "Biological material which is isolated from its natural environment or produced by means of a technical process may be the subject of an invention even if it previously occurred in nature." Furthermore, the Article added that "An element isolated from the human body or otherwise produced by means of a technical process, including the sequence or partial sequence of a gene, may constitute a patentable invention, even if the structure of that element is identical to that of a natural element."

Conclusively, due to the adaption of Directive, the patenting of a gene-related biological product was no longer the obstacle to gene technology research and development but the driving force to this area.[19]

[14]Aerts R J, 'Biotechnological patents in Europe-functions of recombination DNA and expressed protein and satisfaction of the industrial applicability requirement' (2008) International Review of Intellectual Property and Competition Law

[15]Kolle, 'For the evolution of this in the EPC' (1974) 5 I.I.C. 140 at 147–148; *IBM/Document retrieval* [1990]

[16]*Genentech Inc.* v *Wellcome Foundation Ltd* (1989) 8 RPC 147; (1988) 15 IPR 423

[17]Crespi, 'Patents on Genes: Can the Issues be Clarified?' (2000) 5(3) Bio-Science Law Review 199–204, at 199/200

[18]Icos Corporation/Seven transmembrane receptor [2002] OJEPO 293.307

[19]Sena G, 'Directive on Biotechnological Inventions: patentability of discoveries' (1999) International Review of Intellectual Property and Competition Law

8.3 Protection of Gene-Related Inventions

The dispute about the validity of gene-related patent mainly focuses on three general criteria – novelty, inventive step, and industrial application. The significant influence of Biotechnology Directive was embodied in the industrial application step.

8.3.1 Novelty

A dilemma exists in patenting genes. From one hand, it should be categorized as discovery because of its natural property. On the other hand, the novelty exam is designed according to the availability of genes.[20] The adaption of the Biotechnology Directive relieved this tension and confirmed the patentability of gene-related inventions under certain requirements.

Furthermore, according to Recital 22 of the Directive, "the granting of a patent for inventions which concern such sequences or partial sequences should be subject to the same criteria of patentability as in all other areas of technology: novelty, inventive step, and industrial application." The exam for novelty mainly rests on the prior art documents and availability to the public. Because the examination involves the individual case, it is difficult to draw a unified standard of novelty. From the previous cases, it is basically definite that isolated DNA natural counterpart cannot influence its novelty in Europe.[21] In the circumstance of the structural identity between the application and the known DNA sequence, the applicant can seek to use a patent as long as it provides a new function.[22]

Another notable issue is that novelty and inventive step are different from each other. In investigating novelty, items of prior art should be left separate. Besides, examination of inventive step only happen after novelty is met.[23]

8.3.2 Inventive Step

The request that the invention must not be "obvious to the person skilled in the art" should be satisfied in tests of inventive step. Three main factors can be distinguished during examination. First, the degree of proximity to the prior art and "near

[20] Zekos G I, 'Nanotechnology and biotechnology patents' (2006) Journal of Law & Information Technology

[21] Amanda Warren-Jones, 'Patenting DNA: a lot of controversy over a little intangibility' (2004) Medical Law Review.

[22] Oser A, 'Patenting (partial) gene sequence taking particular account of the EST issue' (1999) International Review of Intellectual Property and Competition Law

[23] Howlett M J and Christie A F, 'An analysis of the approach of the European, Japanese and United States Patent Offices to patenting partial DNA sequences (ESTs)' (2003) Review of Intellectual Property and Competition Law

anticipations" should be considered.[24] As Lindley stated, the inventive step is not "a mere analogy, or on the mere application of a principle"; instead it calls for "some ingenuity to overcome a practical difficulty in the adaptation or application."[25] Second, many situations of appeal addressed to the argument that it is "obvious to try."[26]But before finding a new technology from known techniques, almost all researches begin with simple trial-and-error methods.[27]So the requirement of inventive step needs something to mark out the line between the claimed research and the known research.[28]Also if the notional research group is in a strong expectation that there is no commercial reason to do it, taking the step could be inventive. Third, if the invention is the reason of commercial success, this achievement should be considered in the test of inventive.[29]

During the course of the Human Genome Project, the problem that "once sequencing methodology became routine no patent protection should attach" had arisen.[30] EPO held that the homologous DNA sequence was no inventiveness.[31]

8.3.3 Industrial Application

The industrial applicability requirement is the most controversial in patentability of gene-related invention. Article 52(1) EPC said that the patentable invention is "susceptible of industrial application." The Article further specified that "an invention shall be considered as susceptible of industrial application if it can be made or used in any kind of industry, including agriculture." Rule 27(1)(f) EPC explained that description shall "indicate explicitly, when it is not obvious from the description or nature of the invention, the way in which the invention is capable of exploitation in industry." A borderline seemed to be made to distinct between qualified and unqualified.

Nevertheless, many problems were still unsettled. For an instance pointed out by Amanda Jones, "the problem with gaining protection to probes has been to ensure that the claim is drafted to an actual use, rather than attaching a convenient use to a desirable invention of no more real value than simple knowledge."[32]

The effectiveness of Biotechnology Directive changed it. Article 5(3) of the Directive enlarges the standard of industrial applicability by providing that "the

[24] *Seller's Application* [1980] R.P.C. 103

[25] Lindley L J, Mutoh's application [1984] R.P.C. 85

[26] Paterson, *The European Patent System* (2nd ed, 2001) para 12-42a

[27] Brugger v Medicaid [1996] R.P.C. 635 at 661

[28] *Pharmacia v Merck* [2002] R.P.C. 775, paras 123–124

[29] *Martin v Millwood* [1956] R.P.C. 125 at 139

[30] The Nuffield Council's objection to protecting probes, op.cit.n. 8 at para.3.22

[31] Jones A W, 'Patenting DNA: a lot of controversy over a little intangibility' (2004) Medical Law Review

[32] Jones A W, 'Patenting DNA: a lot of controversy over a little intangibility' (2004) Medical Law Review

industrial application of a sequence or a partial sequence of a gene must be disclosed in the patent application." Recitals 23 and 24 of the Directive further state "(23) Whereas a mere DNA sequence without indication of a function does not contain any technical information and is therefore not a patentable invention; (24) Whereas, in order to comply with the industrial application criterion, it is necessary in cases where a sequence or partial sequence of a gene is used to produce a protein or part of a protein, to specify which protein or part of a protein is produced or what function it performs."

The Directive makes clear that the teaching of a mere reproduction of genetic information is excluded from patentability. However, the teaching of a function of a gene which is to some extent industrially applicable is a patentable invention.[33] Article 5(3) of the Directive has specific requirement on disclosure. The problem is what knowledge is actually needed and the degree of satisfaction of industrial application requirement.

Basically, under the new rule, the standard of industrial application has a more general purport but only a repeat of general criteria which mainly contain the description of the function, use, or purpose of a claimed product.[34] In a recent case Eli Lilly, the appellant applied to invalidate a patent of gene sequence for the reason of lacking of industrial application. This is the first time for the high court to make decision on the patentability of human genes. The Chancery Division revoked the patent and held that "whatever the merit of the discovery of Neutrokine-a, the specification contains no more than speculation about how it might be useful. It does not teach the person skilled in the art how to solve any technical problem."[35] This decision means that a large quantity of similar achievements of Human Genome Science is precluded by patent. As Irvine commented, partner of M&C, "You can't just leave it to clinical research carried out at a later date to get the technical contribution to the art necessary to have a valid patent."[36] Edward Nodder, another partner of Bristow evaluated that "this decision is very important for the UK and European pharmaceutical industry. Biotech products are an increasing source of revenue and patent protection is vital to safeguard the huge investments made by the industry in this area. The judgment provides strong guidance on what the threshold should be for a valid patent in this field of technology which had previously received relatively little judicial consideration."[37]

---

[33] Aerts R J, 'Biotechnological patents in Europe-functions of recombinant DNA and expressed protein and satisfaction of the industrial applicability requirement' (2008) International Review of Intellectual Property and Competition Law

[34] Aerts R J, 'Biotechnological patents in Europe-functions of recombinant DNA and expressed protein and satisfaction of the industrial applicability requirement' (2008) International Review of Intellectual Property and Competition Law

[35] Eli Lilly and Co v Human Genome Sciences Inc., [2008] EWHC 2511(Pat)

[36] First High Court Ruling on Gene Patens, LexisNexis UK legal New Analysis

[37] Press releases, Bristows, 31 July 2008 http://www.brietows.com/?pid=48&level=2&nid=97

8.4 Exception of Patentability

Before the Directive, the general prohibition of patentability is Article 53(b) of EPC. In order to incorporate the EPC into the national law, the European Parliament and the Council accepted the Directive on July 6, 1998. As a result of the Biotechnology Directive, many provisions were amended, and the national patent law was basically uniformed with the EU.[38]

8.4.1 Plant Varieties

The International Convention for the Protection of New Varieties of Plants (UPOV) established in 1961 grants property right in new plant varieties. According to a ban on dual protection given by a sui generis plant breeder's right and patent, plant varieties should be excluded from patentability. For this purpose, the Article 53(b) of EPC precluded plant varieties from patentability.[39]

The following question is how to define plant variety in Article 53(b). In case Plant Genetic Systems,[40] the Board of Appeal specifically stated that "plant variety is characterized by at least one single transmissible characteristic distinguishing it from other plant grouping and which is sufficiently homogeneous and stable in its relevant characteristics." The Board of Appeal broadly explained the exception and held that claim rights over the plant varieties were not patentable.

This decision was overturned in Article 4(2) of the Biotechnology Directive. Article 4(2) further states that "Inventions which concern plants or animals shall be patentable if the technical feasibility is not confined to a particular plant or animal variety." This means if a claim covers two or more varieties, it should be patentable. In order to clarify, Recital 31 of the Directive provides that "Whereas a plant grouping which is characterized by a particular gene (and not its whole genome) is not covered by the protection of new varieties and is therefore not excluded from patentability even if it comprises new varieties of plants."

The principle was reconfirmed in case Novartis.[41] The Enlarged Board of Appeal mentioned that "a claim wherein specific plant varieties are not individually claimed is not excluded from patentability under Article 53(b), even though it may embrace plant varieties."[42] Following to this decision, the Administrative Council of the EPO issued the Implementing Regulations on June 16, 1999. In this document, the rule 23c (b) further states that an invention related to plant varieties should be patentable "if the technical feasibility is not confined to a particular plant or animal variety." In

[38] Schertenleib D, 'The patentability and protection of living organisms in European Union' (2004) European Intellectual Property Review

[39] Bently L & Sherman B, *Intellectual Property Law* (2[nd]edn Oxford, Oxford 2004) 426

[40] *Plant Genetic System/Glutamine synthetase inhibitors*, T356/93 (1993) 24 IIC 618; [1995] EPOR 357; [1995] OJEPO 545 (TBA)

[41] *Novartis/Transgenic plant* G1/98 [1999] EPOR 123, 137 (TBA)

[42] *Novartis/Transgenic plant* G1/98 [2000] EPOR 303, 319 (EBA)

general, this new regulations end to the debate of patentability of genetically modified plant. However, the continuing problems highlighted by the challenge to the Directive still remain unsolved.[43]

8.4.2 Animal Varieties

Unlike plant varieties which are specifically protected by a sui generis system, there is no equivalent regime to protect animal varieties. Originally, Article 53(b) of EPC explicitly precluded animal varieties from patentability.[44] However, this became controversial with the development of biotechnology.

In OncoMouse case,[45] the Board of Appeal stated that Article 53(b) should be narrowly explained and not contained animals in general. The TBA therefore used "species" as a borderline of animal varieties in Article 53(b). Because OncoMouse was not a new species, the exception did not apply. However, the Examining Division said that "animal variety either meant a species or a subunit of a species." As a result, the subject matter of this patent was irrelevant with an animal variety and the exclusion of Article 53(b). However, after the adaption of the Directive, the Opposition Division redefined the meaning of animal variety.

A substantive advance was brought by the Directive. Article 4(2) explains that "inventions which concern animals shall be patentable if the technical feasibility of the invention is not confined to a particular plant or animal variety." That is to say, as long as invention does not limit to a specific animal variety, it can be patentable.

The subsequent cases move forward this principle. In amended case OncoMouse, the Opposition Division noted that "living matter and in particular plants and animals could be patentable."[46] Meanwhile, the Division emphasized that the exclusion confined to the varieties which should not contain the animals in general.

8.4.3 Essential Biological Process

Historically, biological processes of breeding plant and animal are unpatentable, because they belong to natural phenomena. Under Article 53(b) of EPC, European patents shall not be granted in respect of "essentially biological processes for the production of plants or animals."

The subsequent question was the degree of technical intervention to satisfy non-essential biological process. The Technical Board of Appeal didn't give clear answer to this question. The board noted that there were three possible approaches: (a) In the first approach, if any part of process invention is biological, it is excluded from

[43] Llewelyn M, 'The patentability of biological material: continuing contradiction and confusion' (2000) European Intellectual Property Review.

[44] Bently L & Sherman B, *Intellectual Property Law* (2ndedn Oxford, Oxford 2004) 424

[45] *Harvard/Onco-Mouse* [1990] EPOR 501

[46] *Harvard/Onco-mouse* [2003] OJEPO 473, 499

patentability[47]; (b)The second way requires that the decision considers the overall degree of human intervention in the process, which was given in case Lubrizol.[48] TBA held that quality rather than quantity of human intervention was significant. The criterion was "on the basis of the essence of the invention taking into account the totality of human intervention and its impact on the result achieved"[49]; (c) Under the last option, if any technical element exist, the invention is patentable. This method is reflected in the case Novartis[50] and reconfirmed in the Biotechnology Directive. Article 2(2) states that "a process for the production of plants or animals is essentially biological if it consists entirely of natural phenomena such as crossing or selection."

However, The Technical Board of Appeal takes the view that Article 2(2) is somewhat self-contradictory, because crossing and selection which are classified as entirely natural phenomena would not happen without human intervention.[51] In the recent case Broccoli, two questions have been referred to the EBA: (1) "whether a nonmicrobiological process for the production of plants which contained the steps of crossing and selecting plants escaped the exclusion of Article 53(b) merely because it contained, as a further step or as part of any of the steps of crossing and selection, an additional feature of a technical nature"; (2) "if not, what were the relevant criteria for distinguishing nonmicrobiological plant production processes excluded from patent protection under Article 53(b) from nonexcluded ones?"[52] These questions refer to the explanation of the scope of exception from patentability. The case is pending, and the correct approach for interpretation is still to be determined.

According to Article 53(b), which stipulates that the exception of this provision does not apply to "microbiological processes or the products thereof," microbiological process and microorganism could be patentable. In case Plant Genetic Systems,[53] the Board of Appeal defined "microbiological process" as a process "in which microorganisms or their parts are used to make or to modify products or where new microorganisms are developed for specific uses." Simultaneously, the Board clarified the meaning of "microorganism" which should be "generally unicellular organisms with dimensions beneath the limits of vision which can be propagated and manipulated in a laboratory." However, the problem is whether the products which are genetically modified varieties from microbiological process could be patentable. This issue was discussed in case Plant Genetic Systems.[54] The Board of Appeal rejected the patent of genetically modified plant. The reason is that

---

[47] *Hospital/Contraceptive methods*, T820/94 [1995] EPOR 446

[48] *Lubrizol/Hybrid plant*, T320/I87 [1990] OJEPO 71.

[49] *Plant Genetic Systems/Glutamine synthetase inhibitors*, T356/93 [1995] EPOR 357.

[50] *Novartis/Transgenic plant*, T1054/96 [1999] EPOR 123,135(TBA).

[51] Sommer T, 'Patenting the animal kingdom? From cross-breeding to genetic make-up and biomedical research' (2008) International Review of Intellectual Property and Competition Law

[52] *Plant Bioscience/Broccoli*, T83/05 [2008] EPOR 14

[53] *Plant Genetics Systems/Glutamine synthetase inhibitors* [1995] OJEPO 545 (Board of Appeal)

[54] *Plant Genetics Systems/Glutamine synthetase inhibitors* [1995] OJEPO 545 (Board of Appeal)

"technical processes including a microbiological step may not simply be equated with microbiological processes." Because the process as a whole could not be viewed as microbiological, the resulting product would belong to the scope of exception. The problem was not distinctly and clearly answered in this case.

This was changed due to the implementation of the Biotechnology Directive. First, Article 2 of the Directive provides the definition of microbiological process as "any process involving or performed upon or resulting in microbiological material." Literally, in comparison with the concept given by the Board of Appeal, it was extended. Secondly, Article 4(3) of the Directive added the qualification "technical process" to Article 53(b) of EPC. Obviously, the scope of patentable subject matter expands into genetically manipulated process or the product of such a process.[55]Consequentially, the impact of the exclusion will be minimized. Thirdly, Article 4 (3) of the Directive expressly noted that the exception paragraph "shall be without prejudice to the patentability of inventions which concern a microbiological or other technical process or a product obtained by means of such a process." Therefore, this provision solves the problem as to the patentability of plant varieties of microbiological or technical process. The Directive clearly said that plant or animal varieties shall not be patentable only because they are the result of microbiological or technical process.

8.5 Morality Issues

It seems to be easy to conclude that patent has a closer connection with economy than morality. Historically, the morality provision existed for a long time but was rarely used. Until recently, the development of Biotechnology raised many ethical issues. The morality objection became one of the strongest forces by the Greens, animal rights campaigners, and others to reject the Biotechnological patent.

In Article 53(a) of the European Patent Convention 1973, it states that European patents shall not be granted to "inventions the commercial exploitation of which would be contrary to 'ordre public' or morality; such exploitation shall not be deemed to be so contrary merely because it is prohibited by law or regulation in some or all of the Contracting States". However, neither the definition of "ordre public" nor the benchmark of morality was provided by this provision.

As Gitter notes, "morality is an exceedingly complex standard to implement as a criterion of patentability."[56] In dealing with the increasing applications of biotechnology patent, the European Patent Office had established two standards: one is "abhorrence" standard, and the other is "unacceptability" standard. Respectively, there are two methodologies relevant to the moral standard: the "balancing exercise"

[55] Bostyn S J R, 'The patentability of genetic information carrier' (1999) Intellectual Property Quarterly

[56] Donna M Gitter, 'Led Astray by the Moral Compass: Incorporating Morality into European Union Biotechnology Patent Law' (2001) 19 *Berkeley Journal of International Law* 1, 21

approach followed "unacceptability" standard; the "rebuttable presumption" approach followed "abhorrence" standard. According to Amanda Warren Jones, this distinction is significant because "under the 'balancing exercise' all of the issues considered form part of the reason why the invention is patentable or not: whereas the 'rebuttable presumption' approach identifies a single issue upon which the decision rests."[57]

The first instance of adapting "abhorrence" standard is Lubrizol[58] case. The Opposition Division held that "an invention will be excluded from patent protection only where the public in general would regard the invention as so abhorrent that the grant of a patent would be inconceivable." In reference to the benefit, the invention of hybrid transgenic plant might bring to the solution of food crisis, it should be excluded from Article 53(a) of EPC.

The second case Relaxin[59] further confirmed the "abhorrence" standard. In this case, the European Patent Office stated that DNA was not life, and the use of pregnancy had not offended human dignity. "An overwhelming consensus" which the invention was abhorrent would be required to fall within the scope of immorality under Article 53(a). Thus, the Division rejected the oppositions of the Green Party. This decision accorded with the general principle that the exceptions to patentability should be constructed narrowly.[60]

Compared with the "abhorrence" standard, the "unacceptability" standard is stricter and higher. The approach which adheres to this standard is called the utilitarian approach. As Shawn H.E. Harmon said, "this approach weighs risks/harms against benefits such as individual financial reward, economic development, and scientific advancement which may promote better healthcare and greater health."[61]

This approach was first used in OncoMouse case.[62] The Technical Appeal Board mainly put the emphasis "on a careful weighing up of the suffering of animals and possible risks to the environment on the one hand and the invention's usefulness to mankind on the other." After balancing animal suffering, environment risks, and usefulness to mankind, the decision was made.[63] The benefit to cancer research outweighed other aspects, the patent would be maintained.

The formalistic treatment of the morality criterion in OncoMouse case was referred to but not fully spread in case Plant Genetic Systems. In this case, the Technical Appeal Board applied the utilitarian analysis through weighing benefit

---

[57] Jones A W, 'Finding a common morality codex for biotech- a question of substance' (2008) International Review of Intellectual Property and Competition Law

[58] *Lubrizol/Hybrid plants*, T320/87 [1990] EPOR 173

[59] *Howard Florey/Relaxin*, T74/91 [1995] EPOR 541

[60] Crespi R S, 'The human embryo and patent law-a major challenge ahead' (2006) European Intellectual Property Review

[61] Harmaon S H E, 'From engagement to reengagement: the expression of moral values in European patent proceedings, past and future' (2006) European Law Review

[62] *HARVARD/Onco-mouse,*T19/90, 1990, O.J. EPO 12/1990, 476, and 1992 O.J. EPO 110/1992, 588

[63] *Harvard/Transgenic animal* (T-315/03) [2005] E.P.O.R. 31 at 10.5–10.8 [161]--[164]

against disadvantage. However, the Opposition Division in the same case stated that "its function has to be seen as a measure to ensure that patents would not be granted for inventions which would universally be regarded as outrageous."[64] Furthermore, the Opposition Division indicated that the possibility of genetically modified plants disturbing the ecological balance "has no bearing on whether a patent is granted or not."[65] This represented the uncertainty in relation to the morality assessment.

An important change was made by the implementation of the Biotechnological Directive. Although the first proposal of the Directive was rejected by the European Parliament because it lacked ethical considerations, the 10-year revision added many public concerns about the morality of biotechnology inventions. Article 6(1) of the Directive said "Inventions shall be considered unpatentable where exploitation or publication would be contrary to public policy or morality; however, exploitation shall not be deemed to be so contrary merely because it is prohibited by law or regulation." This reflected Article 53(a) of EPC. Significantly, Article 6(2) gives some direct examples of immoral inventions: "a. Processes for cloning human beings; b. Processes for modifying the germ line genetic identity of human beings; c. Uses of human embryos for industrial or commercial purposes; d. Processes for modifying the genetic identity of animals which are likely to cause them suffering without any substantial medical benefit to man or animal, and also animals resulting from such processes."

In the Directive, a series of definitions and interpretation was laid down to serve the aim "to provide an equal level of patent protection for biotechnological inventions in all the Member States."[66]Anomalously, the morality provision seemed not to be stricter. According to Amanda Warren-Jones, "there being a discernable intention in the Biotech Directive to leave the morality provision outside of this overarching aim at clarity."[67] He indicated that the list of unpatentable subject matter was nothing but a "general guideline." The individual nations can choose what moral standard being applied.[68]

One of exclusions in Article 6(2) is the use of human embryos for industrial or commercial purpose. However, the provisions do not provide a distinct explanation of "uses of human embryos." In case University of Edinburgh/Stem Cell Isolation, two interruptions of Rule 23d(c) EPC which was equal to Article 6(2) of the Directive were discussed: one is narrowly explained as "uses of human embryos as such"; the other is broadly explained as "uses of human embryos together with the

[64] Plant Genetic System/Glutamine synthetase inhibitors, T356/93 (1993) 24 IIC 618; [1995] EPOR 357; [1995] OJEPO 545 (TBA)

[65] Plant Genetic System/Glutamine synthetase inhibitors, T356/93 (1993) 24 IIC 618; [1995] EPOR 357; [1995] OJEPO 545 (TBA)

[66] The "Council's Reasons" for amending the Commission's 1992 version of the proposed Directive (1992 version, OJ EC C44, at 36 (16 February 1993); OJ EC C101, at 71 (9 April 1994)

[67] Jones A W, 'Finding a common morality codex for biotech- a question of substance' (2008) International Review of Intellectual Property and Competition Law

[68] Jones A W, 'Finding a common morality codex for biotech- a question of substance' (2008) International Review of Intellectual Property and Competition Law

cells being retrieved there from by destruction of the embryos, namely, human ES cells."[69] The Opposition Division stated that broadly interruption should be applied. The Division further explained that "if the patenting of a product is ethically unacceptable, it is hardly conceivable that the patenting of "uses" of this product can be judged differently."[70] But Claudio Germinario, a former EPO Appeal Board member, has an opposite view. He indicated that this decision was contrary to the previous findings of the Appeal Boards that the interruption of exclusions from patentability should be narrow.[71]It should be noted that the UK patent Office agrees the Germinario's opinion.[72]

The following case WARF seamlessly addressed to the problem of the correct approach to Rule 23(d)(c). The filed application did not direct to the method which involved the destruction of the human embryo. Instead, they claimed the product which was derived from those methods. The Enlarged Board of Appeal finally rejected the appeal on November 25, 2008. The decision clearly said that "Rule 28(c) EPC forbids the patenting of claims directed to products which at the filing date could be prepared exclusively by a method which necessarily involved the destruction of the human embryos from which the products are derived, even if the said method is not part of the claim."[73] This means the Europe Patent Office refused to permit any patent which involving the destruction of human embryos or their use for industrial or commercial purposes.

However, the controversy of patentability of human stem cell will be prolonged. As Dr. Paul Chapman pointed out, the decision leaves a backdoor to those applications which "out of necessity makes use of human embryos for industrial and commercial purpose, but does not necessarily involve the destruction of embryos,"[74] for instance, the common applications about using stem cell lines as starting material. Thus the uncertainty of law had not been fully clarified.

8.6 Conclusion

The Biotechnology Directive had successively treated the patentability of gene-related inventions, the exceptions to patent and moral issues. In addition, the Directive generally achieved the goal of harmonization of patent laws among the member states. To some extent, the Directive simplified the uncertainty of the patent law which is benefit to increase the research investment and development funds.

[69] Opposition Decision re EP 0695351, Edinburgh University. Not published in OJEPO

[70] *University of Edinburgh/Stem Cell Isolation (Edinburgh),* T-1079/03 EP 949131742 unreported, July 21, 2003, Opposition Division, at 22

[71] Claudio Germinario, "The Value of Life" (2004) Patent World 16–18.

[72] Crespi R S, 'The human embryo and patent law-a major challenge ahead' (2006) European Intellectual Property Review

[73] G-02/06 of the Enlarged Board of Appeal of the European Patent Office

[74] Chapman P, 'Rejection of controversial stem cell patent fails to fully clarify law' M&C (London 28 November 2008)

First, the Directive ended the debate of discoveries and invention and affirmed that gene-related invention is eligible to the subject of patent. Second, the Directive expanded the standard of industrial applicability which must be disclosed in the application and required the indication of a function and technical information of DNA sequence. Third, the Directive clearly indicated the allowance of the genetic modified plant and animal. Also the Directive provided explicit answer to the degree of technical intervention to satisfy nonessential biological process. The Directive gave negative answer to the patentability of plant or animal varieties merely due to they are the result of microbiological or technical process. Last, considering the diversity of moral standard among member states, the Directive listed the immoral inventions and gave the space to the divergence.

It is not exaggerate to say that the Biotechnology Directive is a milestone of biotechnology patent law in Europe. However, it was challenged by various grounds.

Some objections were from the government of member state. The Government of Netherlands, which was supported by Italy and Norway, challenged the lawfulness of the Biotechnology Directive. They mainly argued that "it is incorrectly based on Article 100a (now Article 95) of the Treaty; that it is contrary to the principle of subsidiary; that it infringes the principle of legal certainty; that it is incompatible with international obligations; that it breaches fundamental rights, and that the procedure for its adoption was incorrect."[75]

The annulment was rejected by the ECJ. First, the court used the patentability of plant varieties and that of the human body as the evidence that Article 100a was the correct rational basis of enacting the Directive. Second, the ECJ indicated the guideline was given to the Article 6. Therefore the Directive had not exacerbated the legal uncertainty. Third, the court said the Directive handled the biotechnological inventions which could not be acted by member state alone. Thus the Directive did not violate the principle of subsidiarity. Fourth, the ECJ held that no evidence could be provided to prove the patent protection of biotechnological invention limited the purposes of the CBD. Last, Article 5 and Article 6 are the guarantee of human dignity. Besides, "only inventions that combined natural and technical elements in a way that allowed an industrial application to be isolated were patentable."[76]

Some objections were from the scholars. For instance, Amanda Odell-West, from the University of Sheffield, suggested that Recital 42 exclusion within the Biotechnology Directive could be invoked. According to Recital 42 of the Directive, the exclusion of Article 6(3) "in any case does not affect inventions for therapeutic or diagnostic purposes which are applied to the human embryo and are useful to it."[77]However, this provision seems not always suitable for Preimplantation Genetic

[75] Broensword R and Beyleveld D, 'Is patent law part of the EC legal order? A critical commentary on the interpretation of Article 6(1) of Directive 98/44/EC in Case C-377/98' (2002) Intellectual Property Quarterly

[76] Netherlands v European Parliament (c-377/98) [2002] All E.R. (EC)97; [2001] E.C.R. I-7079; [2001] 3 C.M.L.R. 49; [2002] F.S.R.36; (2002) 68 B.M.L.R.1

[77] West A O, 'Preimplantation Genetic Diagnosis, the medical exclusion and the biotechnology directive' (2007) Medical Law International Vol 8 pp. 239–250

Diagnosis (PGD), which can detect many different types of inherited diseases in embryos generated by in vitro fertilization. He provided the following reasons: First, PGD is not always beneficial and useful to the embryo. It can be used to genetically discriminate and select between embryos. Second, when PGD is used in private clinics, it essentially belongs to "the commercial use." Third, because of low percentage of success, a large number of embryos are made then discarded.[78] So, Recital 42 of the Directive seems to lack consideration in this situation and should be amended.

Among these challenges, the majority of disputes are related to moral issue. However, the scientific advancement could make these debates meaningless. Taking an instance of stem cell, the ethical controversies are mainly about human embryo stem cells. But the recent news reported that an ordinary adult stem cell can achieve the same goal of embryo stem cell.[79] This major breakthrough was gotten by a San Francisco research and development company. They claimed that those cells "can match tissues in the heart, lung, liver, pancreas, blood vessels, brain, muscle, bone, and fat." Moreover, adult stem cells can "avoid some of the complications found with embryonic stem cells, such as a specific type of cancer." This great news not only ended the dispute surrounding the Directive but also helped ignite the ethical contention of the use of embryos in stem cell research.

In sum, the Biotechnology Directive undeniably exist some insufficiency and uncertainty within itself.[80,81] However, we cannot neglect the contribution of the Directive because of its faults.

[78] West A O, 'Preimplantation Genetic Diagnosis, the medical exclusion and the biotechnology directive' (2007) Medical Law International Vol 8 pp. 239–250

[79] Unruh B, 'Scientific breakthrough! No embryonic stem cells needed' 22 August 2008 World net daily

[80] Singh HB, Keswani C, Singh SP (Eds.). Intellectual Property Issues in Microbiology (2019) Springer-Nature, Singapore. 425 pages, ISBN- 9789811374654

[81] Singh HB, Jha A, Keswani C (Eds.). Intellectual Property Issues in Biotechnology. (2016) CABI, Wallingford, UK. 304 pages, ISBN-13: 9781780646534

9 Assessing the Emergence of Bioeconomy in Transition Economies by a Future-Oriented Approach: The Case of Poland

Emmanuel Koukios, Lazaros Karaoglanoglou,
Danuta Ciechanska, Adam Mazurkiewicz,
Beata Poteralska, and Anna Sacio-Szymańska

Abstract

A seven-step approach is proposed for the assessment and management of bioeconomy-related developments affecting the pathway of transition economies and societies. The particular action steps of this approach include methodological elements from various fields, with emphasis on future-oriented methods, especially horizon scanning, identification of key drivers and barriers and scenario-building. These are combined with other relevant tools, such as mapping, techno-economic evaluation, technology assessment and strategy and policy analysis. This approach is applied in this paper to the case of Poland, a country currently in the process of preparing its bioeconomy strategy in the frame of regional and national smart specialisation efforts.

Keywords

Bioeconomy · Biomass · Mapping · Foresight · Transition economies · Poland · Policy · Strategy

E. Koukios (✉) · L. Karaoglanoglou
National Technical University of Athens, Athens, Greece
e-mail: koukios@chemeng.ntua.gr

D. Ciechanska · A. Mazurkiewicz · B. Poteralska · A. Sacio-Szymańska
Institute for Sustainable Technologies – National Research Institute, Radom, Poland

C. Keswani (ed.), *Bioeconomy for Sustainable Development*,
https://doi.org/10.1007/978-981-13-9431-7_9

9.1 Introduction and Methodology

9.1.1 An Emerging Socio-techno-economic Continent

According to the European Commission's definition, the use of the term "bioeconomy" includes all industrial and economic sectors that produce, manage and otherwise exploit biological resources and related services, supply or consumer industries, including a long list of activities (European Commission 2012; Koukios 2015): primary production sectors, such as agriculture, fisheries, forestry and aquaculture; the whole range of biological production- and conversion-based industries, such as agro-industries, as well as food, wood, fibre and other bioindustries; health-care and biomedical technologies, animal health applications and pharmaceuticals; biochemicals, biomaterials and "green" biochemistry; bioenergy and biofuels; waste and water management; and all kinds of biosystem design and operation tools and applications (Koukios 2014).

9.1.2 Expected Benefits for Transition Economies

The emergence of the whole spectrum of bioeconomic applications or some of its parts could open up a great number of strategic opportunities, especially for economies and societies that are currently in search of a new model and of the appropriate optimal change pathway leading to that. We will now summarise the main types of expected benefits (Koukios 2014, 2017).

Most of the bioeconomy sectors are of a high strategic interest, e.g. food, farming, plant and animal health and forest, thus raising the stakes for immediate action, especially at the research and policy levels. In addition, the strong momentum of bioeconomic "wave" of change could affect the modernisation efforts in almost all bioeconomy sectors, with beneficial effects on the competitivity of the emerging economies. This trend could be enhanced by the unlocking of the huge innovation potential of bioeconomy's "Pandora's box", thus leading to synergistic and accelerated development of novel processes, products, services, tools, methods and solutions.

In the area of environmental protection, the prudent use of the reservoir of biological knowledge, especially of the dimension of environmental biotechnologies, will facilitate the trends towards "greening" of the economy and society and catalyse a symbiotic development of environment-friendly applications in all emerging bioeconomy areas. A good example of this effect is that of the substitution of fossil-based carbon fuels and chemicals by "green" bio-based ones. These effects will particularly benefit the efforts for the mitigation of the climatic change vectors and contribute to the hopeful abatement of the climate change looming tsunami.

In the field of socio-economic and other "soft" types of effects, adopting a bioeconomy perspective could facilitate matching the rapidly changing societal needs and public concerns to an also rapidly changing production base, thus leading to increased business, employment, and local development opportunities. To take advantage of such benefits, decision- and policy-makers will have to fulfil the critical but extremely complex task of coordinating regional, national and other policies concerning all the involved bioeconomy sectors, as well as their interactions. The smooth deployment of bioeconomic applications could mobilize local and regional natural and human resources, while respecting cultures and traditions, on the way towards a more sustainable future (Bioeconomy Stakeholders Manifesto 2017; SCAR (Standing Committee for Agricultural Research) 2015).

9.1.3 Methodology

A seven-step approach is proposed for the assessment and management of bioeconomy-related developments affecting the pathway of transition economies and societies, such as those of Poland, which is specifically considered in this work. The particular action steps of this approach consist of the following:

(a) Overview of the state of the national economy and mapping of its relevant sectors
(b) Evaluation of the bioeconomy-related sectors
(c) Assessment of the relevant achievements and opportunities, as well as major risks and challenges
(d) Focus on the critical factors resulting from the preceding analysis
(e) Identification of the key factors affecting the bioeconomic developments
(f) Construction of scenarios of these developments based on appropriate combinations of the key factors
(g) Policy, strategy, research and innovation and other conclusions and recommendations resulting from the whole seven-step analysis and synthesis

The particular action steps of this approach include methodological elements from various fields, with emphasis on future-oriented methods, especially horizon scanning, identification of key drivers and barriers and scenario-building. These are combined with other tools, such as mapping, techno-economic evaluation, technology assessment and strategy and policy analysis (In 't Veld et al. 2007; Koukios et al. 2018; Sacio-Szymańska et al. 2015).

9.2 Results and Discussion

The six sections of this part and the Conclusions and Recommendations section following correspond to the seven steps (see the Methodology part) of the proposed approach for the assessment of the bioeconomy's emergence in a transition economy.

9.2.1 Mapping Bioeconomy

9.2.1.1 Overview

According to the latest report of the Bio-based Industries Consortium (BIC), titled "Mapping the Potential of Poland for the Bio-based Industry"(Bio-based Industries Consortium (BIC) 2019), Poland is a country with a great potential in the area of bioeconomy. In particular, Poland shows a great potential in the production of high-quality food and has large biological resources, so it could become an important point on the bioeconomic map of Europe.

The Polish bioeconomy focuses on traditional sectors: agriculture, forestry and food processing. This is an important branch of the national economy, responsible for almost 20% of employment and 10% of total production volume. Poland is seventh in Europe in terms of the value of its agricultural sector (France, Germany, Italy and Spain being at the forefront) and ranks fourth among EU countries in the production of rapeseed. Agriculture in Poland is the main source of biomass, where ~76% of biomass comes from this sector.

9.2.1.2 National Strategy

Poland has not published a strategy on bioeconomy until 2018. However, bioeconomic effects already play an increasingly important role in Poland and constitute an important element of the National Smart Specialization (NSS) document. In this document, agriculture and food-based, forest-based and environmental bioeconomy are defined as one of the main support areas alongside healthy society, sustainable energy industry, circular economy, water, fossil raw materials, waste and innovative technologies and industrial processes. They indicate preferences in providing support for the research, development and innovation (RDI) within the framework of the financial perspective for 2014–2020 (National Smart Specialization 2019).

Through these and other actions, bioeconomy has become one of the priority areas in Polish RDI that may contribute to the transformation of the national economy through modernisation, structural transformation, diversification of products and services and creation of innovative socio-economic solutions, also supporting the transition towards an efficient economy regarding natural and other resources.

9.2.2 Evaluation of Bioeconomy Sectors

9.2.2.1 Biomass and Bio-industry Prospects

In the report of the OECD Conference "Building a biomass innovation ecosystem in a circular bioeconomy in Poland" that took place in Krakow in 2017 (OECD 2017), several key conclusions of the discussion on biomass resources, their possibilities for utilization and recommendations for the development of the bioeconomy ecosystem in Poland were presented.

Poland is a country rich in biomass, with 18.8 Mha of agricultural land and 9.4 Mha of forest; together, they occupy about 90% of the national land area. Among EU countries, it is the third-largest producer of domestic biomass, fourth largest in biomass consumption and seventh in organic waste generation. Four countries account for slightly over half of the agricultural biomass supply potential of the EU, of which Poland's share is 11.7%. Similarly, five countries account for over half of the EU forest biomass supply potential, of which Poland's share is 8.4%.

However, Polish bio-based industries are still at a development stage, and thus there is much for Poland to gain from further development of its bioeconomy. The number of Polish biotechnology firms has been growing in recent years although this sector is still relatively small, and the number of those companies involved in industrial processing is only a limited percentage of the total number. Future bio-economic growth for Poland requires increased and sustained spending by both the public and private sectors to stimulate biotechnology RDI and deployment.

The same OECD Report (OECD 2017) concluded that better insights were needed on (i) the actual quantities and qualities of potential feedstock for the bio-based industry from the primary and the subsequent processing sectors; (ii) the possibilities to cope with varying and different quantities and qualities; and (iii) the readiness of actors in and across sectors to create new value chains, including primary sector actors as partners and beneficiaries.

9.2.2.2 The Bioeconomy Sectors

In Table 9.1, we present a rather exhaustively complete list of all the bioeconomy-related sectors in any EU economy, national and regional. This list can be used as a template for the qualitative and quantitative assessment of the national and regional bioeconomies.

In Fig. 9.1, the major fields and sectors of Polish bioeconomy are being evaluated as far as their contribution to the national economy. The already significant effects of bioeconomy, along with the key role of the food sector, and at the same time the significant presence of biomass-based activities should be noted.

Table 9.1 The bioeconomy-related fields and sectors of an EU economy

Economy – the big picture
Agriculture
Agro- and food industries
Forest and wood industries
Traditional/conventional bioenergy
New bioenergy – biofuels
Pharmaceuticals – cosmetics
Aquaculture – fishing
Old and new service sectors (health, environment, tourism, etc.)
New bioindustries – biorefineries
Inland waters
Education, research – strategy and policy aspects

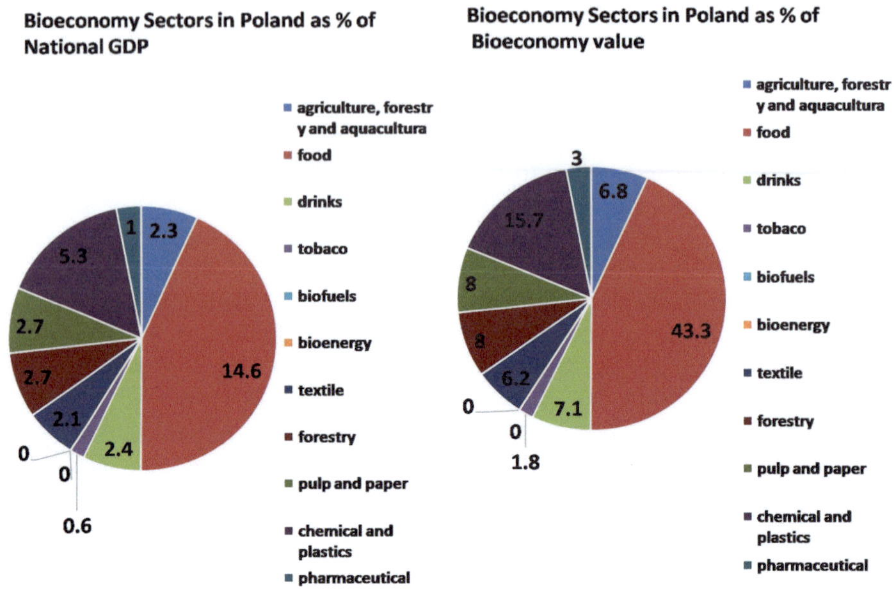

Fig. 9.1 The economic contribution of the major bioeconomic sectors of Poland. (Source: Authors' Estimates – National Statistical Yearbook, 2017, Polish Bioeconomy Platform (Bielecki 2017))

Fig. 9.2 Results of an assessment of the sustainably available yearly biomass potential of Poland, expressed in energy terms. (Source: Authors' Estimates, based on source Koukios 2017)

PL Bioresource Potential (Sustainable, 2030)

❖ BIOMASS SOURCE	Mtoe/Year
- Straw & Other Agro-Residues	5.0
- Animal Manure	11.7
- Landscape Management	1.1
- Sawmill & Other Wood Residues	2.5
- Wood Used for Energy	4.6
- Co-firing with Coal for Electricity	4.9
- Transport Biofuels	0.6
- Food, Landfill, MSW, Sludges, Fats	2.3
- Paper Wastes	2.3
TOTAL BIOMASS	35.0

In Fig. 9.2, we summarise the results of the authors' assessment of the national biomass-based energy potential. From this analysis, we can assess the very promising potential growth of the contribution of bioenergy and biofuels.

9.2.3 Assessment of the Opportunities and Challenges

According to the horizon scanning (In 't Veld et al. 2007) foresight approach adopted in the research reported here, two types of observed findings are being identified: on the one hand, achievements and opportunities, as summarised in Table 9.2, and, on the other hand, risks and challenges, as summarised in Table 9.3. These findings are based on a detailed desk research, the results of which are presented in the source documentation cited (Koukios, 2017).

9.2.4 Key Factors Determining Bioeconomy Dynamics (2015–2030)

Based on the information provided in Tables 9.2 and 9.3, and according to the approach for the assessment of the emerging bioeconomy as followed in this work (Koukios 2017; In 't Veld et al. 2007), we have distinguished the following as key drivers, key barriers and other key pathway-determining actions, with a time horizon of the year 2030.

Table 9.2 Major achievements and opportunities of polish bioeconomy

Fields and sectors	Attributes
The big picture of the economy	Smooth transition – fast convergence to EU – not affected by financial crisis. GDP growth: from 1990 = 100 to 2015 = 250
Agriculture	Large sector (in GDP and employment terms) – good performance (potato, grains, meat, apples, rape)
Food industry	Competitive industries – innovation inputs high exports and direct foreign investments
Forest and wood industries	Three healthy sub-sectors (forest, wood, paper) – keeping up with technological innovation
Traditional/conventional bioenergy	Co-firing of biomass with coal for electricity, biomass-based heat and electricity generation planned
New bioenergy – biofuels	Liquid biofuel (bioethanol, biodiesel) production – biogas generation from bio/waste-sources
Pharmaceutical – cosmetics	Robust, chemistry-based pharma sector – the "success story" of national cosmetics sector
Aquaculture – fishing	Ca. 500 freshwater farms (carp, trout, stocking) – rural development impact – modernisation efforts
Old and new service sectors	Modernising national health-care system – greening options considered, e.g. for tourism
New bioindustries – biorefineries	Biotechnology in a product-oriented frame – national platform – reports RRB conf (Renewable Resources & Biorefineries), Wrocław, June 2017
Inland waters	Megaproject of inland navigation – touristic interests in water activities
Education, research – strategy and policy	Education support of some bioeconomy skills, e.g. cosmetics – 2 regions (Łodż, Lublin) with bioeconomy strategies

Source: Koukios (2017)

Table 9.3 Major risk and challenges of polish bioeconomy

Fields and sectors	Attributes
The big picture of the economy	Need to adopt national/regional bioeconomy strategies – negative effects of ageing and demographics
Agriculture	Modernised ca. 50% of sector (small family farms) – soil and water supply (quantity/quality) problems
Food industry	Need to modernise the less competitive units of sector – branding of polish food products needed
Forest and wood industries	Accelerate sector's technological modernisation – develop integrated resource-based strategies
Traditional/conventional bioenergy	Coal replacement in electricity by renewables – air pollution control in biomass-based systems
New bioenergy – biofuels	Introduction of energy crops and 3G biofuels – innovative utilisation of lignocellulosic biomass
Pharmaceutical-cosmetics	Coordinate pharma sector with health-care trends – combine cosmetics with anti-ageing trend
Aquaculture – fishing	Modernise sector – enhance societal aspects – extend aquafarming to new outlets (micro-algae)
Old and new service sectors	Branding and strategy for tourism (agro, eco, theme) – greening targeted – esp. for clean air and water
New bioindustries – biorefineries	Biotech molecular roles in a bioeconomy – biorefineries for resource and market efficiencies
Inland waters	Environmental concerns on megaproject(s) – water quality targets – synergies with biomass
Education, research – strategy and policy	Research and innovation as key parts of polish bioeconomy – climate change factors and resilience needs

Source: Koukios (2017)

9.2.4.1 Key Drivers

(A1) Continuing Growth of GDP – Smooth Transition to a Market Economy
(A2) Agricultural Production and Food Industry – Exports and Foreign Investments
(A3) Momentum in Relevant Industrial Sectors – Wood, Bio, Pharma, Cosmetics

9.2.4.2 Key Barriers

(B1) Ageing of Population – Unfavourable Demographic Trends at All Levels
(B2) Coal-based Electricity – High Energy Consumption – Climate Change Effects
(B3) Environmental Pollution – Air, Water, Soil, Solid Wastes – Low Greening Pressure

9.2.4.3 Key Path-Determining Actions Towards 2030

(C1) Modernisation – Re-organisation – Management and Leadership Options
(C2) Innovation Eco-system – Research and Development – Education and Training
(C3) Policy Frame – Smart Regional Specialisation – Strategies and Plans

9.2.5 Critical Factors for the Emergence of Bioeconomy in Poland

9.2.5.1 Strategic Consultation Workshop

The key factors identified in the previous step of the whole process were evaluated at this step with the help of a *Consultation and Discussion Workshop*, which took place at the Institute for Sustainable Technologies – National Research Institute (ITeE-NRI) in Radom, Poland, on June 26, 2017. The aim of the Workshop was to assess the research findings on the emergence of bioeconomy in Poland by a multidisciplinary audience of ca. 20 stakeholders from a large spectrum of relevant fields, including research, innovation, agriculture, industry, technology, economics, humanistic studies, materials, bioengineering, environmental studies, foresight and policy-making. For more details on the Workshop proceedings, see ref. Koukios (2017).

The landscape of the dynamics of change according to this assessment is *a multi-parameter* one, as shaped by three high and very high strength factors, specifically:

- (C2) *Innovation Eco-system – Research and Development – Education and Training*; being the only very high overall strength factor, and the strongest action factor;
- (A3) *Momentum in Relevant Industrial Sectors – Wood, Bio, Pharma, Cosmetics*; and
- (B2) *Coal-based Electricity – High Energy Consumption – Climate Change Effects*; being the strongest driver and barrier factors, respectively, and the high strength ones overall.

The rest of the bioeconomy emergence dynamics consists of

- three moderate strength factors (A2, B3, C3), one driver, one barrier, one action; whereas
- the remaining three factors (A1, B1, C1) are expected to have rather weak but still not negligible effects, which will possibly affect the fine tuning of the major developments.

A small number of participants proposed additional key factors, which were found to have either weaker or complementary effects compared to those of the already identified factors, thus indirectly validating/confirming the "saturation" of the workshop issue by the nine tabled factors.

9.2.5.2 The Factors of High Importance

(C2) *Innovation Eco-system – Research and Development – Education and Training* is the single key factor with the highest comparative strength of effects, being one of the three action types considered. The great majority of participants indicated the particular importance of this factor for the emergence and development of bioeconomy in Poland. Its strategic actions are found to have triggering effects on the whole dynamics of change, including amplification effects on the key drivers, disrupting effects on the key barriers and shaping effects on the other key actions. The participants were specifically asked for their opinion on the presentation of R&D in (C2) together with education and training, in an innovation ecosystem format, which was welcomed by all of them, while stressing the key role of education.

(A3) *Momentum in Relevant Industrial Sectors – Wood, Bio, Pharma, Cosmetics* is the key driver of the highest comparative strength, which was also mentioned by the majority of participants as the most important driver. The importance of the particular driver results, among other reasons, from the fact that all the sectors covered by this driver are important for the Polish economy and that they complement each other. According to many participants, relevant industrial sectors are not only the few mentioned by this driver, but any sectors in which bioresources and bioprocesses are used, i.e. all bio-based industries.

(B2) *Coal-based Electricity – High Energy Consumption – Climate Change Effects* was mentioned by the majority of participants as the strongest barrier to the emergency of bioeconomy in Poland in the period until 2030. As coal is a fossil energy source, the shift towards renewable resources is inevitable, through a systemic approach at the national level. At the same time, an inter-sectoral approach is needed for the Polish economy to overcome the additional serious barriers of a fragmented industry. Another obstacle for such a major re-orientation of the Polish economy regarding energy sources is the number of jobs linked with the coal industry and other vested interests in the energy sector.

9.2.6 Synthesis of Change Dynamics: From Clusters to Scenarios

9.2.6.1 Identification of Clusters of Factors

The results of the assessment presented above make possible the final step of the process of mapping the landscape of dynamics affecting the emergence of the bioeconomy in Poland by the year 2030. According to the so far results of mapping, there appear to be in operation two major clusters of forces shaping this landscape, i.e.:

CLUSTER I: The development of the appropriate National Innovation Ecosystem, including interdisciplinary research, technology, education and training – as described in action (C2) *Innovation Eco-system – Research and Development – Education and Training*. This is directly linked with the strategic target of upgrading the position of the country on the EU innovation score board but should also cover obtaining excellence on specific bioeconomic lines. Developing appropriate policies and strategies – see action (C3) – is an essential part of this cluster, which guarantees the feasibility of the emerging bioeconomy developments.

CLUSTER II: The establishment of an integrated and coordinated strategy for a national *Sustainable Bioeconomy*, as part of the Responsible Development Plan. This should include primarily bio-based industries (see driver (A3)) and clean energy (see barrier (B2)) but also agriculture and food (see driver (A2)) as well as bio-greening solutions for clean environment (see barrier (B3)). A main role of this cluster will be to guarantee the socio-economic desirability of the emerging bioeconomy developments.

9.2.6.2 Emergence Scenarios

The following scenarios result from the above assessment of the bioeconomy dynamics:

The key dialogue of the clusters of forces that generate the five scenarios shown in Fig. 9.3 is found to be the one of *Socioeconomic Acceptance/Partnership* vs. *Scientific/Technological (S/T) Excellence*, leading to the following types of emergence outcomes:

- PROUD EAGLE: High S/T excellence and high socio-economic support, leading to great innovation-based success stories and highly competitive, modernised applications and multi-sector, "Made in Poland" branding
- GALLOPING HORSE: High S/T excellence bur low socio-economic support, leading to limited technology-based success stories and selective cases of branding based on sector-specific competitivity and modernisation

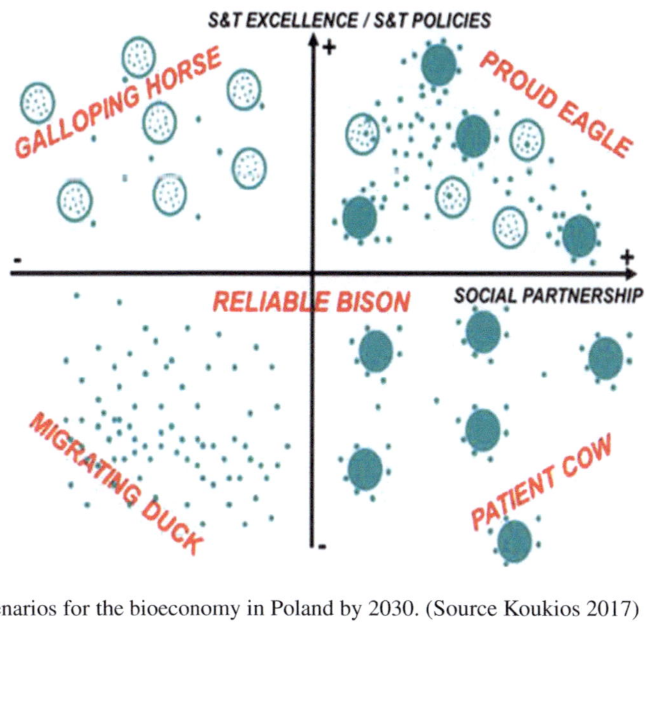

Fig. 9.3 Scenarios for the bioeconomy in Poland by 2030. (Source Koukios 2017)

- PATIENT COW: Low S/T excellence but high socio-economic support, leading to few success stories and traditional/low-to-medium tech cases of branding, spread across several low modernisation sectors
- MIGRATING DUCK: Low S/T excellence and low socio-economic support, leading to very few innovation-based success stories and even less cases of branding "islands", catalysing brain-drain and other types of losses of capital and resources
- RELIABLE BISON: Moderate S/T excellence and moderate socio-economic support, leading to the continuation of the present dynamics of "business as usual" and accompanied by a mixture of limited elements from the other scenarios

9.3 Conclusions and Recommendations

Of the highest priority for the future of Polish bioeconomy will be a dual-target scheme, consisting of the development of the appropriate Innovation Ecosystem, especially focusing on the bio-industrial sectors, along with sustainability friendly policies, especially focusing on the energy, food and resources sectors (Polish Government 2017).

While assessing action types, a number of additional policy and/or innovation priorities were identified, in an effort to complement and elaborate the ones already considered: networking and international cooperation; human resources, especially in education; logistics; modelling bioeconomy systems; opinion leadership; legal framework of policies; and updating the smart regional specialisation option by formulating regional platforms.

Regarding strategic priorities, it is recommended that bioeconomy-oriented policies should develop symbiotic relationships with the Polish Government Strategy for Responsible Development (Polish Government 2017), where many of the above-discussed issues are tackled, and many strategic actions are proposed, which should integrate selected bioeconomic elements and help re-orient the Polish economy towards more competitive, more innovative and more sustainable, socially and environmentally, targets.

Acknowledgements This paper was based on the original research prepared in the frame of the project titled "The New Tsunami of Socio-Technical Change: Foresight Agro/Bio/Chemo/Eco/Cogno … Convergence (BIO-TSUNAMI)", supported by the POLONEZ 1 Programme of the National Science Centre (NCN) with contract no. UMO-2015/19/P/HS4/04103. This project has received funding from the European Union's Horizon 2020 Research and Innovation Programme under the Marie Skłodowska-Curie grant agreement No. 665778.

The authors are particularly grateful to the Institute for Sustainable Technologies – NRI, Radom, Poland, for their permission of the use of material included in the ITeE monograph: Koukios E., Mazurkiewicz A., Sacio-Szymańska A., Poteralska B. (eds.), *The Emergence of Bio-Economy, Opportunities and Risks – A Forward-Looking Study*, ISBN 978-83-7789-484-2. ITeE-PIB, Radom, 2017, which was based on the above-mentioned project, and of which monograph the four co-editors are co-authors of the present work.

References

Bielecki S (2017) Bioeconomy in Poland – the road to the future. IFIB, Rome

Bio-based Industries Consortium (BIC) (2019) Report. Mapping the potential of Poland for the bio-based industry. https://biconsortium.eu/news/bioeconomy-potential-portugal-romania-and-poland-profiled-new-bic-country-reports. Accessed 1 Mar 2019

Bioeconomy Stakeholders Manifesto (2017). https://ec.europa.eu/research/bioeconomy/pdf/european_bioeconomy_stakeholders_manifesto.pdf. Accessed 1 Mar 2019

European Commission (2012) Innovating for sustainable growth: a bioeconomy for Europe, 60 final. COM, Brussels. 13.2.2012

Koukios E (2014) Technology management for the bio-based economy: mapping, dynamics and policies – the case of Greece. In: Trzmielak DM, Gibson DV (eds) International cases on innovation, knowledge and technology transfer. Centre of Technology Transfer, University of Łodz, Łodz

Koukios E (2015) From biotechnologies to bioeconomy and biosociety. In: Agrafiotis D (ed) Technology foresight. Ellinika Grammata, Athens. (in Greek)

Koukios E (2017) The bio-tsunami project. In: Koukios E et al (eds) The emergence of bioeconomy – opportunities and risks – a forward-looking study. ITeE, Radom

Koukios E et al (2018) Targeting sustainable bioeconomy. J Clean Prod 172:3931–3941

In 't Veld R et al (eds) (2007) Horizon Scan Report: Towards a future oriented policy and knowledge agenda. Horizon Scan Publication, The Hague

National Smart Specialization (2019). https://www.smart.gov.pl/en/. Accessed 4 Mar 2019

OECD (2017) Conference report. Building a biomass innovation ecosystem in a circular bioeconomy in Poland. Krakow

Polish Government (2017) National responsible development plan

Sacio-Szymańska A, Mazurkiewicz A, Poteralska B (2015) Corporate foresight at the strategic research institutes. Bus: theory Pract/Verslas: Teorija Prakt 16(3):316–325

SCAR (Standing Committee for Agricultural Research) (2015) Sustainable agriculture, forestry and fisheries in the bioeconomy – a challenge for Europe. Report, 4th Foresight Exercise, 2015, http://ec.europa.eu/research/scar/pdf/feg4-draft-15_mawy_2015.pdf. Accessed 1 Mar 2019

Enabling Bioeconomy with Offshore Macroalgae Biorefineries

10

Alexander Golberg, Meiron Zollmann, Meghanath Prabhu, and Ruslana Rachel Palatnik

Abstract

The bioeconomy provides a possible solution for the increasing demand on natural resources by substitution of the nonrenewable resources with resources derived from biomass, thus reducing the environmental impact of fossil fuels. A fundamental unit that will enable the bioeconomy implementation is biorefinery. The bioeconomy is a collective term for the complex system that includes biomass production, transportation, conversion into products, and product distribution. In this chapter, we introduce the concept of offshore marine biorefineries as potential drivers for the bioeconomy of the future. We discuss fundamental thermodynamics principles that determine the optimum scale of biorefineries and put the limit for the services area for a single-processing unit. We provide a review of the current methods to produce biomass offshore. Next, we exemplify the marine biorefineries, which show co-production of several products from the same biomass, thus reducing the waste and maximizing economic benefit from the unit. In addition, we discuss the economic and environmental challenges of marine biorefineries as an emerging platform for society transition to low-carbon economy.

Keywords

Biorefineries · Bioeconomy · Green technology · Renewable energy · Biofuel · Biomass

A. Golberg (✉) · M. Zollmann · M. Prabhu
Porter School of Environment and Earth Sciences, Tel Aviv University, Tel Aviv, Israel

R. R. Palatnik
Department of Economics and Management, and SEED – the Sustainable Economic and Environmental Development Research Center, The Max Stern Yezreel Valley College, Afula and Nazareth, Israel

NRERC- Natural Resource and Environmental Research Center, University of Haifa, Haifa, Israel

C. Keswani (ed.), *Bioeconomy for Sustainable Development*,
https://doi.org/10.1007/978-981-13-9431-7_10

10.1 Transition to Low-Carbon Societies with Bioeconomy

Growing population, increasing quality of life, and longevity impose new pressures on all industrial sectors involved in the production of food, chemicals, and fuels. This increasing pressure is expected to increase the use of land, potable water, fossil fuels, and other natural resources. This increased use of natural resources could lead to unpredictable changes in climate, loss of biodiversity, and reduction of the ability to maintain ecosystems sustainably. The bioeconomy provides a possible solution for this increasing demand to natural resources by substitution of the depletable resources with biomass-based commodities, thus reducing the environmental impact of fossil fuels. The bioeconomy describes the global industrial transition of sustainably utilizing renewable aquatic and terrestrial resources in energy, intermediates, and final products for economic, environmental, social, and national security benefits.

For bioeconomy implementation, the optimal supply chain should be designed in terms of procurement of feedstock (intermediate inputs), production and processing at the biorefinery, transportation, and marketing. The entrepreneurs must decide how much to produce, what segments of the supply chain to undertake in-house versus sourcing externally, and what institutions such as contracts and standards they will use to coordinate the suppliers assuring its external sourcing (Zilberman et al. 2019). These decisions are affected by the investor's financial situation, the political and social system, the technology available, etc. (Du et al. 2016).

As supply chains increasingly encompass far-flung markets and supply sources, manufacturers and retailers are susceptible to various types of supply chain risks. There are diverse supply chain risks associated with disruptions or delays that could be categorized into supply risks, process risks, demand risks, intellectual property risks, behavioral risks, and political/social risks. Supply chain contract uncertainties may occur due to asymmetric information (Du et al. 2016). That is, the innovator may not observe the ability of an effort being devoted by the contracted supplier, or the quality of his product. Entrepreneurs may invest in protective measures to increase the resilience of their supply chains to extreme weather risks. They may geographically diversify their external sources of feedstock to reduce exposure to weather shocks. Therefore, incorporating risk considerations may actually increase the cost of investment in implementing an innovation, especially if the enterprise is constrained by credit.

The supply chain design of industries in the bioeconomy may require determining strategies for the production and processing of the feedstock to produce multiple products (Zilberman et al. 2019). There are established supply chains for seaweed-based food production (Valderrama et al. 2015) and for bioethanol. Evidently, corn is used to produce ethanol as well as the residue product, Distillers Dried Grains (DDGs), which is being sold as animal feedstock (Taheripour et al. 2010). Many of the agrifood innovations increased the value added of agricultural resources either by identifying non-food uses of agricultural products and residues as part of the bioeconomy or producing differentiated products by increasing their convenience and quality (Zilberman et al. 2019).

10.2 Biorefineries as Essential Technological Platforms for Bioeconomy

A fundamental unit that will enable the bioeconomy implementation is biorefinery. The bioeconomy is a collective term for the complex system that includes biomass production, transportation, conversion into products at the biorefinery, and product distribution and marketing. Biorefineries convert renewable biomass into biofuels, food, chemicals, and other bio-based products. Some biorefinery technologies include power generation. Potentially, biorefineries create products with higher added value to the benefit of the economy and the environment.

The use of versatile, robust technologies is one of the key factors in biorefineries. The synergetic combination of process technologies can lead to the development of advanced biorefineries where non-food biomass is converted by a combination of mechanical, thermochemical, chemical, and biochemical processes into a range of bio-based chemicals, e.g., materials, chemicals, and energy. Hence, the maximum value is achieved from each feedstock (De Jong and Jungmeier 2015).

The main output of the biorefineries is bio-based chemicals. Bio-based chemicals can be defined as those classes of chemicals, which are produced by using natural feedstock and have minimal impact on the environment. Examples for bio-based chemicals include (but not limited to) carboxylic acids, polylactic acid, fatty acids, isoprene, biosolvents (e.g., bioethanol), amino acids, vitamins, bio-pesticides, bio-fertilizers, antioxidants, sterols, and even industrial enzymes (Golden et al. 2015; De Jong et al. 2012). The major market demand-driving factors that are expected to boost the demand include the availability of raw materials at a reduced cost, increasing consumer awareness toward and subsequent demand for bio-based products and government initiatives to promote green products among others.

The major drivers for the deployment of biorefineries are:

(i) Sustainable and renewable energy supply – as biorefineries utilize renewable feedstock
(ii) Saving foreign exchange reserves – required alternatively for importing fossil fuels and other chemicals
(iii) Reduced dependency on imported crude petroleum and other chemicals – due to locally grown feedstock for biorefineries
(iv) Establishment of carbon-neutral and circular economy – allowed by low-carbon footprint and net positive environmental impact of biorefineries

A comprehensive review of optimization-oriented biomass supply-chain designs shows numerous prior works that addressed various important conditions for a profitable supply chain (Ghaderi et al. 2016). Surprisingly, this review of 146 studies concluded that researchers have been mostly orientated toward single-feedstock, single-product, single-period, single-objective, and deterministic models without considering all the dimensions of sustainability. An alternative to this is a co-production of multiple products from the same biomass. Such processes are very common in the petrochemical industry and lead to almost complete use of the raw material with close to zero waste and maximum valorization. However, for biomass feedstock-based biorefineries, integrated production of food, energy, and other valuable products with zero waste is a relatively new and novel idea and hence, limited literature is available. We discuss the co-production option for biorefineries in the following sections.

10.3 Offshore Marine Biorefineries

The choice of raw biomass material is critical to ensuring the efficient production of biofuels (Bentsen and Felby 2012). The currently used crops and cultivation methods supply raw biomass for the food and feed sectors for hundreds of years; however, most recently they also started to supply biomass for the transportation energy production. The first-generation liquid biofuel feedstock includes traditional agriculture crops (cereals, potatoes, sugar beet, and rapeseed), wood, and dedicated energy crops, while first-generation fuel products include ethanol and biodiesel (International Energy Agency 2011). The second-generation biomass feedstock includes animal fat and dedicated lingo-cellulosic crops and produces hydro-treated vegetable oil, cellulosic-ethanol, biomass-to-liquids (BtL)-diesel, bio-butanol, and advanced drop-in replacement fuels such as fatty acid ethyl esters, alkanes, alkenes, terpenes, and methyl ketones (Keasling and Chou 2008; Dunlop 2011; Lee et al. 2008; Bokinsky et al. 2011; Steen et al. 2010; Peralta-Yahya et al. 2011; Zhang et al. 2014). However, recent studies indicate that the future of biomass sector development is under a high degree of uncertainty mainly due to the limited crop yields and land availability (Bentsen and Felby 2012; Star-coliBRi 2011).

Alternative sources for biomass are offshore grown macroalgae. Macroalgae (or seaweeds) have been harvested throughout the world as a food source and as a commodity for the production of hydrocolloids for centuries. To date, macroalgae still present only a tiny percent of the global biomass supply ($\sim 17 \cdot 10^6$ wet weight macroalgae in comparison to $16 \cdot 10^{11}$ ton of terrestrial crops, grasses, and forests) (Roesijadi et al. 2010; Pimentel and Pimentel 2008; Pimentel 2012). However, world macroalgae biomass cultivation has continuously increased over the last 10 years at an average of 10% and is considered as new promising biomass for low-carbon economy (Jung et al. 2013; Balina et al. 2017).

Macroalgae are photosynthetic organisms living in damp places. As per classification, macroalgae are of three different kinds, green (*Chlorophyta*), brown (phaeophyta), or red (*Rhodophyta*) macroalgae, based on the composition of their photosynthetic pigments (Jung et al. 2013). In addition to photons, the algal plant needs nutrients (mostly nitrogen and phosphorus) and a carbon source to grow. These features are of major interest regarding two points. First, since algae fix carbon (annual cultivation and processing of 1 ton of dry weight of seaweed evaluated over a time horizon of 100 years result in a net reduction of 9.3 tons of atmospheric carbon, equivalent to 34 ton CO_2) (Seghetta et al. 2016a), it can be used as carbon storage and then as fuel. Second, waters polluted with excessive nutrient levels can be cleaned through growing and harvesting of algal biomass. Moreover, an expanding body of evidence has demonstrated that marine macroalgae, which have unique chemical composition (Nunes et al. 2017; Patarra et al. 2011), have high growth rate (Jung et al. 2013; Chemodanov et al. 2017a), contain very little lignin, and do not compete with food crops for arable land or potable water, can provide a sustainable alternative source of biomass for sustainable food, fuel, and chemical generation (Roesijadi et al. 2010; Van Hal et al. 2014; Wei et al. 2013; Enquist-Newman et al. 2014; Potts et al. 2012; Hannon et al. 2010; Kraan 2013; Wargacki et al. 2012; van der Wal et al. 2013). The conceptual framework of offshore marine biorefineries is shown in Fig. 10.1.

10.4 Offshore Biomass Cultivation

Macroalgae feedstock for biorefineries cannot be based on the harvesting of wild stocks or on cultivation in onshore or nearshore farms. Wild-stock harvesting leads inevitably to over-exploitation, while on- or nearshore farming competes with food crops or coastal uses (Buschmann et al. 2017) and is limited by decreasing available areas (Möller et al. 2012). Two main solutions withstand the conditions above. One is envisioning of construction of very large seaweed farms in coastal unfertile deserts (Buschmann et al. 2017). The second, with a wider potential for global implementation, is the offshore cultivation.

Early reports of the offshore algae cultivation concept proposed to release juvenile *Sargassum* sp. 500 miles offshore the US-Canada border and harvest them

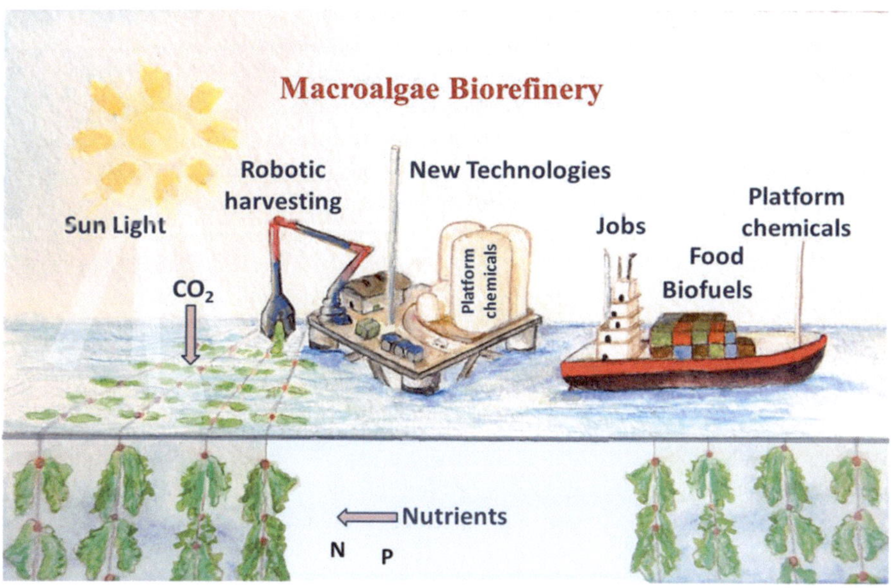

Fig. 10.1 The concept of offshore biorefineries for the production of food, platform chemicals, and biofuels in the ocean (Figure adapted from Lehahn et al. 2016 with permit)

offshore the USA-Mexico border, for the production of methane in onshore anaerobic digesters (Szetela et al. 1976). This proposal inspired in the late 1960s by Howard Wilcox from the San Diego Naval Undersea Center to envision and develop the first multi-product floating seaweed farm, called the "Ocean Food and Energy Farm Project" (Roesijadi et al. 2008). Due to the energy crisis of the 1970s, this project was stopped in favor of prioritized biofuel production programs. The "Marine Biomass Program" which operated during the 1970s and the early 1980s in California (Roesijadi et al. 2008) has made significant advances in understanding the complexity of the marine biological system and in enhancing growth data but failed to overcome the difficulties of working in the open ocean, especially the stability of the cultivation systems and of the attachment of the algae to the systems (Roesijadi et al. 2008).

Following the beginning of the new millennium, with increasing awareness of the environmental effects of the industrial era (Suutari et al. 2015), scientific engagement with offshore biomass cultivation has become significant again (Roesijadi et al. 2008, 2010; Suutari et al. 2015; Reith et al. 2005; Buck and Buchholz 2004, 2005; Buck et al. 2004; van den Burg et al. 2013; Hughes et al. 2012; Korzen et al. 2015a). Although previous techno-economic assessments were not favorable of offshore algae cultivation, four decades of technological evolution, casted into the current political-environmental context, has led to a reexamination of this idea (Feinberg and Hock 1985). This technological evolution includes experience gained through oil and gas exploration, advancements in the oceanographic and atmospheric sciences, and major improvements in both tensile strength and weight of materials that can be used at sea (Roesijadi et al. 2008). These new technologies include also the development of flexible and submersible offshore aquaculture structures, such as the SUBflex which is being operated offshore Israel since 2006 (Drimer 2019). Simultaneously, the establishment of offshore wind farms (Reith et al. 2005) and the inevitable distancing of aquaculture facilities from the coast (Troell et al. 2009) facilitated an additional potential reduction in cultivation costs via integration of infrastructure and operations (Reith et al. 2005; Buck and Buchholz 2004).

Traditional offshore algae cultivation systems include ropes, lines, nets, rafts, and cages, which are all popular due to inexpensive installation and maintenance (Table 10.1) (Fernand et al. 2017).

For example for the production of green macroalgae biomass, Liu et al. (2010) cultivated *Ulva prolifera* and *Ulva intestinalis* on rafts in the Yellow Sea offshore Jiangsu coastline, China, and measured yields of 198.6 and 89.2 kg ww ha^{-1} 5 $months^{-1}$, respectively (Liu et al. 2010). The goal of this cultivation experiment was to examine the potential of these two species to exploit aquaculture rafts and cause green-tide events. Smaller-scale offshore experiments have demonstrated the cultivation of *Ulva rigida* in the Eastern Mediterranean Sea. Korzen et al. (2015b) used nylon net cages integrated with fish cages offshore Mikhmoret, Israel, and achieved maximal specific growth rates of 16.8% per day along 2 weeks cultivation periods (Korzen et al. 2015b). This maximal growth rate was measured 22 m downstream the fish cages where nutrients were sufficient. Chemodanov et al. (2017a) used flat double-layer net reactors at the nearshore location at the Reading Power

Table 10.1 Offshore-cultivated macroalgae biomass productivity yields (reports from 1980 to 2015)

Cultivation system	Species	Location	Yield	Units	Reference
Rope, vertical	*Undaria pinnatifida*	North western coastal bay of Spain	8.3	kg ww/m/139 days	Peteiro and Freire (2012)
			21[a]	t ww/ha/year	Peteiro and Freire (2012)
Rope, horizontal	*Undaria pinnatifida*	North Western coastal bay of Spain	5.9	kg ww/m/147 days	Peteiro and Freire (2012)
Rope (concentrical)	*Laminaria saccharina*	German North Sea	4	kg ww/m/6 months	Buck and Buchholz (2004)
Ropes farm, horizontal	*Laminaria japonica*	Hokkaido, Japan	10^6	t ww/41.2km^2/year	Yokoyama et al. (2007)
Rope, horizontal, transplanted[b]	*Saccharina latissima*	Northern Spain, Bay of Biscay	7.8	kg ww/m/106 days	Peteiro et al. (2014)
			45.6	t ww/ha/106 days	Peteiro et al. (2014)
Rope, horizontal	*Laminaria saccharina*	British Columbia, Canada	3–8	kg ww/m/8 months	Druehl et al. (1988)
Rope, horizontal	*Laminaria groenlandica*	British Columbia, Canada	2.6–20.5	kg ww/m/18 months	Druehl et al. (1988)
Rope, horizontal	*Cymathere triplicata*	British Columbia, Canada	1.1–2.7	kg ww/m/7 months	Druehl et al. (1988)
Rope, vertical[c]	*Palmaria palmata*	Northwest Scotland	1	kg ww/horizontal meter of top rope/year	Sanderson et al. (2012)
Rope, vertical[c]	*Saccharina latissima*	Northwest Scotland	28[d]	kg ww/horizontal meter of top rope/year	Sanderson et al. (2012)
Rope, horizontal	*Alaria esculenta*[e]	Ard Bay, Carna, Co. Galway, Ireland	45.7	kg ww/m/year	(Kraan and Guiry 2001)
Rope, horizontal	*Saccharina latissima*	Isle of Man, Irish Sea	2.8	kg dw/m/year	Holt (1984)
Cage[f]	*Gracilaria tikvahiae*	Indian river lagoon, Florida	9.7[g]	g dw/m^2/day	Hanisak (1987)

(continued)

Table 10.1 (continued)

Cultivation system	Species	Location	Yield	Units	Reference
Cage[f]	*Gracilaria tikvahiae*	Hutchinson Island, Florida	22.4	g dw/m²/day	Hanisak (1987)
Nylon line attached to stakes fixed in sea bottom	*Eucheuma spinosum* (Bohol)	Zanzibar Island, Tanzania	5.4–7%[h]	Daily growth rate	Lirasan and Twide (1993)
Floating raft with rope	*Sargassum naozhouense* (Tseng et Lu)	Liusha Bay, Xuwen, Guangdong, China	1750	kg ww/km/95 days	Xie et al. (2013)
Rope net with two bamboo poles	*Ulva prolifera*	Jiangsu coastline, China	198.6[i]	kg ww/ha/5 months	Liu et al. (2010)
Rope net with two bamboo poles	*Ulva intestinalis*	Jiangsu coastline, China	89.2[i]	kg ww/ha/5 months	Liu et al. (2010)

Table adapted from Fernand et al. (2017) with permit
ww: wet weight; dw: dry weight.
[a]Estimated value
[b]Transplants were 2.1 kg fresh wt m-1 rope
[c]Droppers 1 m apart, with one 10 cm section of seeded string for every 1 m of dropper to 7 m depth
[d]Highest mean yield obtained for a longline
[e]High-yielding strain
[f]2 cm plastic mesh on 2.5–5.0 cm diameter PVC pipe frames measuring 1 × 1 × 0.25 m deep (0.6 m² cage)
[g]Average between two stations (7.8 and 11.6 g dw/m²/day)
[h]Minimum and maximum of five test plants at different locations
[i]Average of six stations

Station in Tel Aviv, Israel, and measured a mean daily growth rate of 4.5 ± 1.1%, corresponding to an annual average productivity of 5.8 ± 1.5 g DW m^{-2} day^{-1} (Fig. 10.2) (Chemodanov et al. 2017a). More advanced systems are the offshore-ring that was developed by Buck et al. (2004, 2005) (Buck and Buchholz 2005; Buck et al. 2004) and the moored multi-body seaweed farm that was developed by Olanrewaju et al. (2017), both supposed to withstand rough offshore conditions.

Different approaches have been suggested regarding future design of offshore cultivation systems. The commonly used extensive approach allows the algae to grow without adding nutrients or applying external mixing. Using a combination of climate models and seaweed metabolic models, global biomass potential for offshore cultivation was established if extensive approaches are used (Fig. 10.3). The main advantage of this approach is decreased labor, technology, and energy inputs, thus improving energy balance, while the main disadvantage is decreased biomass yields, leading to a large area-demand (Buck et al. 2008). Extensive cultivation can be performed on anchored platforms or on free-floating enclosures (Roesijadi et al. 2008).

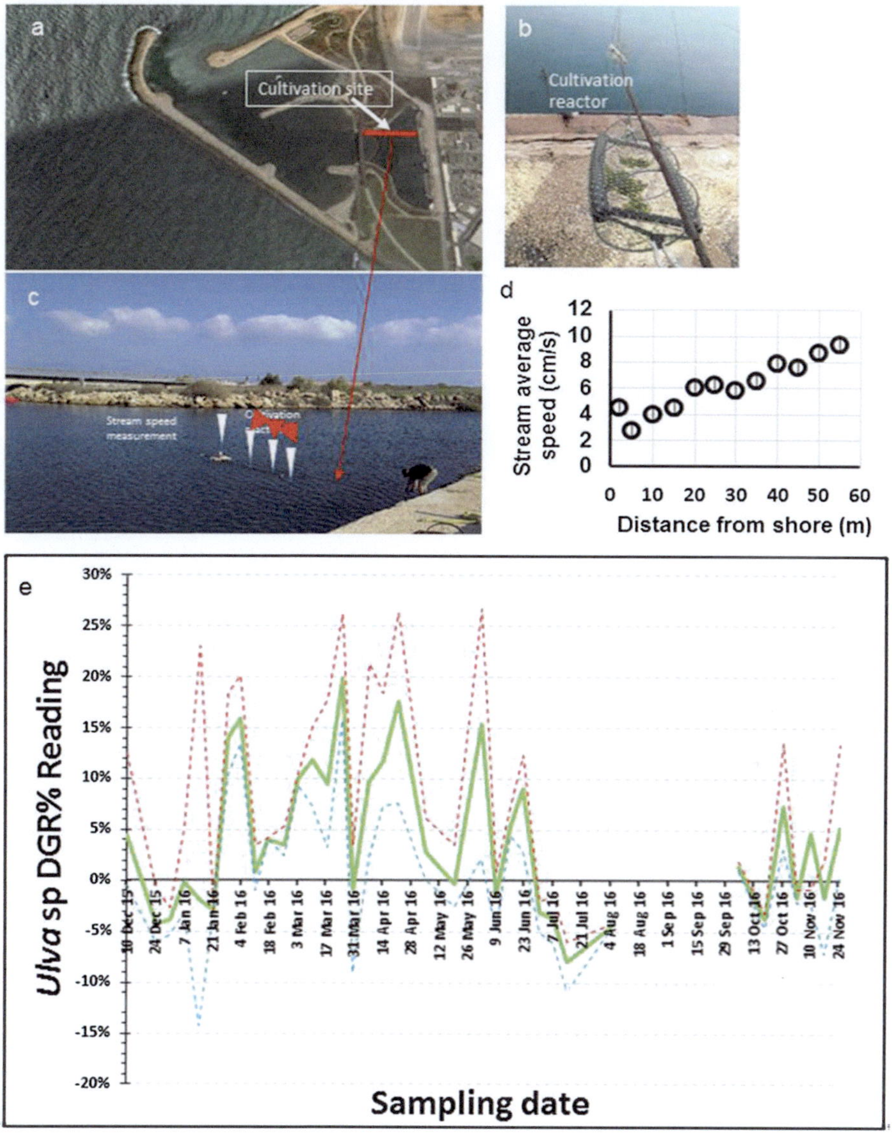

Fig. 10.2 Experimental setup for daily growth rate (DGR) measurements. (**a**) Macroalgae cultivation site at Reading Power Station in Tel Aviv, Israel. (**b**) Flat thin cultivation reactor with a signal cultivation depth and double net design. (**c**) Positions of cultivation reactors during 1 year measurements. (**d**) Water current speed profile at the cultivated area, measured at the same depths as the flat thin cultivation reactor ($N = 10$ for each point). (**e**) Measured annual daily growth rate (%DGR) of *Ulva* biomass at reading ($N = 3$ for each point). Green line shows an average value, red line shows a maximum value, and blue line shows the minimum value of % DGR. (Images adapted from Chemodanov et al. 2017a with permit)

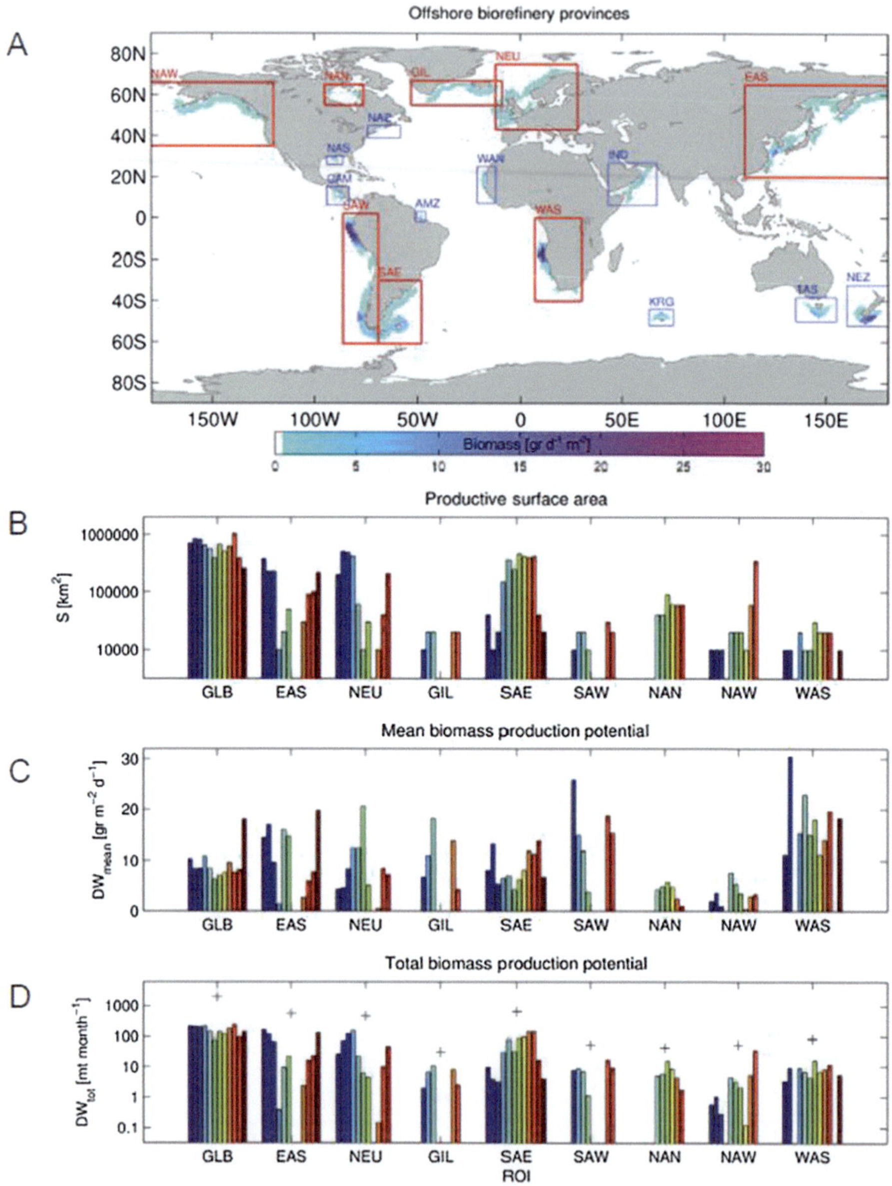

Fig. 10.3 Regional potential for offshore biomass production with extensive methods. (**a**) Potential for biomass production at a distance of less than 400 km from land. Boxes delineate major offshore biorefinery provinces, with that permitting biomass production at water depth of up to 100 m (defined as near-future deployable biorefinery provinces – NDBP) marked in red and that permitting biomass production only at deeper waters marked in blue. (**b–d**) Monthly estimates of (**b**) productive surface area, (**c**) mean biomass production potential, and (**d**) total production potential within the 8 NDBP (red boxes and associated abbreviations in panel A) and integrated globally (denoted GLB). Colors denote different months of the year. The analysis is

Free-floating enclosures can be released in areas with predicted currents, or alternatively followed with tracking devices, and finally collected and harvested when time and location are suitable and biomass weight is satisfactory. Such a system was suggested by Notoya (2010) that proposed to grow seaweed beds on 100 km^2 rafts, floating away from shipping lanes until ready to harvest. However, this concept is yet to be demonstrated. Anchored platforms can be sited in areas that are favorable for cultivation, aiming for optimal temperatures and sunlight, water motion which is sufficient to break down diffusion barriers, and natural supply of nutrients, for example, in natural upwelling zones (Roesijadi et al. 2008).

When the environmental concentration of nutrients are low, nutrients may be provided by artificial upwelling of deep nutrient-rich water as suggested already in the 1970s in the "Marine Biomass Program" (Roesijadi et al. 2008). The artificial upwelling solution can be potentially combined with Ocean Thermal Energy Conversion (OTEC) ventures, utilizing deep seawater for both nutrient enrichment and temperature difference-based power generation, thus reducing costs and environmental footprints while harnessing energy and nutrients (Roels et al. 1979). Another solution for supplying nutrients offshore is the polytrophic aquaculture, also known as integrated multi-trophic aquaculture (IMTA) (Ashkenazi et al. 2019; Neori et al. 2004). This approach can significantly increase system sustainability, inter alia by recycling waste nutrients from higher trophic-level species into the production of lower trophic-level crops of commercial value, such as macroalgae (Troell et al. 2009). Furthermore, co-cultivation of different species can increase productivity by increasing the light-harvesting efficiency. This can be done, for example, in a layered seaweed cultivation system, employing typical light absorption characteristics of green, brown, and red macroalgae, respectively, thus improving light use (Reith et al. 2005).

In contrast to the extensive approach, the intensive approach emphasizes the importance of achieving maximal biomass yields, even at the expense of energy costs. Following this approach, Golberg and Liberzon (2015), for example, have modeled smart mixing regimes to improve biomass productivity by enhancing light-harvesting and carbon fixation (Golberg and Liberzon 2015). Mixed water cultivation is commonly applied to onshore reactor cultivation of free-floating green algae (Chemodanov et al. 2017b). However, applying free-floating algae cultivation offshore, mixed or non-mixed, is challenging due to forceful ocean currents and increased loss risks, which may lead to uncontrolled macroalgal blooms (Liu et al. 2009).

Fig. 10.3 (continued) performed over locations associated with water depth of 100 m or shallower. The + signs mark annually integrated biomass production potential at each region. East Asia offshore waters (EAS); Northern Europe offshore waters (NEU); Greenland and Iceland offshore waters (GIL); North America offshore waters, north (NAN); North America offshore waters, west (NAW); South America offshore waters, east (SAE); South America offshore waters, west (NAW); West Africa offshore waters, south (WAS). Amazon River estuary (AMZ); Central America offshore waters (CAM); Indian Ocean (IND); Kerguelen (KRG); New Zealand (NEZ); Tasmania (TAS); North America offshore waters, south (NAS); North America offshore waters, east (NAE); West Africa offshore waters, north (WAN). Images adapted from ref (Lehahn et al., 2016)with permit.

10.5 Co-production of Multiple Products from Macroalgae

The global potential of the near-future achievable deployment offshore biomass production (i.e., in regions extending up to 400 km distance from the shore and with a water depth of up to 100 m) can provide $1.96 \bullet 10^9$ ton DW $year^{-1}$. This is equivalent to the 37 EJ $year^{-1}$ of primary energy potential (calculated as LHV). In comparison, the predicted bioenergy potential from agricultural land in 2050 is expected to be 64–161 EJ $year^{-1}$ (Haberl et al. 2011). Based on already available protocols in literature, this biomass can be converted to multiple products such as ethanol, butanol, acetone, methane, proteins, and others (Table 10.2).

A recent study developed a methodology to assess the performance of the integrated two-stage supply chain – feedstock farming and processing into multiple outputs. The results from the nonlinear dynamic model clarify that learning (by doing/researching) in a multistage supply chain creates a positive externality of co-outputs. Moreover, if the learning rate is faster than cost increase, then output grows faster than prices. Next, they demonstrated the application of the modeling framework on macroalgae (seaweed) farming and processing in the biorefinery into crude proteins and polysaccharide (carrageenan). The results indicated that for average prices of proteins and carrageenan, and for average costs of investment in cultivation farm and the biorefinery, macroalgae utilization is cost-efficient. The study indicated that using near-future aquaculture technologies, offshore cultivation of macroalgae has the potential to provide some of the basic products required for human society in the coming decades. However, the profitability of this supply chain is fragile due to the high volatility of outputs' prices, as well as a wide range of feedstock growth rate and chemical composition. Notably, the researchers identified the first stage of the supply chain, namely, macroalgae marine cultivation, as the main constraint for commercialization.

Studies have shown that the remaining pulp after extraction of high-value polysaccharides such as agar, alginates, and carrageenan still contain high amount of carbohydrates and other nutrients including protein, lipids, and ash, which may be used as a source of raw material for extraction of various other materials rather than treating as a waste (Kumar and Sahoo 2017; Alvarado-Morales et al. 2015). Utilization of all the organic content to useful, high-value products would make the biorefinery process most profitable and sustainable by maximizing the biorefinery's overall economic performance (Laurens et al. 2017). Notably, the outputs from macroalgae-based biorefinery vary with the species of the seaweed, as presented below.

Considering this, co-production of two or more products from green macroalgae in an integrated, cascading, biorefinery approach has been followed, thus maximizing the benefits of seaweed biomass (Postma et al. 2017; Trivedi et al. 2016; Ben Yahmed et al. 2016; Bikker et al. 2016a). Experimentally, sequential recovery of four economically important fractions, a mineral-rich liquid extract, lipid, ulvan (a sulfated polysaccharide, S-PS, of the genus *Ulva*), and cellulose from *Ulva fasciata*, was reported by Trivedi et al. (2016). The mineral-rich liquid extract was extracted by mechanical grinding and ulvan by a hydrothermal process. Bikker et al. (2016b)

Table 10.2 Potential for offshore production cf biomass and derived products for various cultivation stocking densities. The notion "All waters" refers to all locations regardless of water depth and distance from the coast, while "shallow nearshore waters" refers to areas associated with water depths smaller than 100 m and located less than 400 km from the ccast

Biomass stocking density	1 kg m^{-2}		2 kg m^{-2}		4 kg m^{-2}		6 kg m^{-2}		8 kg m^{-2}	
	All waters	Shallow near shore waters	All waters	Shallow near shore waters	All waters	Shallow near shore waters	All waters	Shallow near shore waters	All waters	Shallow near shore waters
Biomass (10^6 t $year^{-1}$) (DW)	67,500	591	81,000	710	108,000	946	54,000	473	40,500	355
Ethanol (10^6 t $year^{-1}$)	2025–15,525	18–136	2430–18,630	21–163	3240–24,840	28–218	1620–12,420	14–109	1215–9315	11–82
Butanol (10^6 t $year^{-1}$)	2025–4050	18–35	2430–4860	21–43	3240–6480	28–57	1620–3240	14–28	1215–2430	11–21
Acetone (10^6 t $year^{-1}$)	675–1350	6–12	810–1620	7–14	1080–2160	9–19	540–1080	5–9	405–810	4–7
Methane (10^6 m^3 $year^{-1}$)	675–6480	6–57	810–7776	7–68	1080–10,368	9–91	540–5184	5–45	405–3888	4–34
Protein (10^6 t $year^{-1}$)	3375–16,200	30–142	4050–19 440	35–170	5400–25,920	47–227	2700–12,960	24–114	2025–9720	18–85
Energy (10^{12} kJ $year^{-1}$)	1,282,500	11,234	1,539,000	13,481	2,052,000	17,974	1,026,000	8987	769,500	6740

Table adapted from Lehahn et al. (2016) with permit

demonstrated the co-production of a sugar-rich hydrolysate (38.8 g l^{-1}), used for the production of acetone, butanol, ethanol, and 1,2-propanediol by clostridial fermentation, and a protein-enriched (343 g kg^{-1} in dry matter) extracted fraction, used as animal feed, out of *Ulva lactuca* biomass. The extraction procedure included solubilizing the sugars by hot water treatment followed by enzymatic hydrolysis and separating solid and liquid fractions by centrifugation. Ben Yahmed et al. (2016) demonstrated the integrated biorefinery approach for co-production of bioethanol and biogas using fermentation and anaerobic digestion of *Chaetomorpha linum* hydrolysate obtained by thermochemical and enzymatic hydrolysis (Ben Yahmed et al. 2016). Postma et al. (2017) have co-extracted water-soluble proteins and carbohydrates from fresh *U. lactuca* biomass using osmotic shock, enzymatic hydrolysis, the pulsed electric field (PEF), and high shear homogenization (Postma et al. 2017). Glasson et al. (2017) reported the extraction of salts, pigments, ulvan, and monosaccharides from *Ulva ohnoi*, and more recently Gajaria et al. (2017) reported the extraction of five different chemical products: minerals, lipids, ulvan, protein, and cellulose, from *U. lactuca*, both in a cascading biorefinery process (Gajaria et al. 2017). Very recently, an additional work showed the liquid fraction obtained after homogenization of fresh *Ulva* biomass, and filtration can be processed for the effective extraction of starch in its native form (Prabhu et al. 2019). We also proposed that the starch extraction can be effectively integrated into the biorefinery, and the leftover biomass can be processed for the extraction of other various products. Based on various integrated biorefinery concepts mentioned above, a process design was developed for co-production of six different products and applications (Fig. 10.4).

Brown seaweeds are interesting feedstock for biorefineries as they contain a diverse array of metabolites including extracellular matrix polysaccharides such as alginates and fucoidans, storage polysaccharides such as laminarin and mannitol, and bioactive polyphenolic compounds and pigments such as fucoxanthin with potential applications in pharmaceutical, food, cosmetic, and biotechnology industries (Kostas et al. 2017). Using brown macroalgae *Laminaria digitata*, fucoidan, alginate, and bioethanol were extracted in a cascading biorefinery by Kostas et al. (2017). They also showed that the methanol fraction extracted using the waste stream after fucoidan extraction had antioxidant and antimicrobial activity. Yuan and Macquarrie (2015) indicated that *Ascophyllum nodosum* could be potentially used as feedstock for a cascading biorefinery process to produce valuable chemicals and fuels (fucoidan, alginates, sugars, and biochar). Van Hal et al. (2014) demonstrated extraction scheme for mannitol, alginate, laminarin, and sugars from *Saccharina latissima* in a cascading biorefinery. The mannitol was converted to isomannide; the laminarin was fermented to acetone, butanol, ethanol (ABE); and the alginate fraction was converted to furan dicarboxylic acid (FDCA) (Van Hal et al. 2014). Co-producing succinic acid, phenolic antioxidants, and fertilizer from *Saccharina latissima* indicate that the economic profit of the biorefinery is positive (Marinho et al. 2016).

Using red macroalgae, *Gracilaria corticata*, Baghel et al. (2016) demonstrated the simple process for recovery of mineral-rich liquid extract (MRLE), pigments,

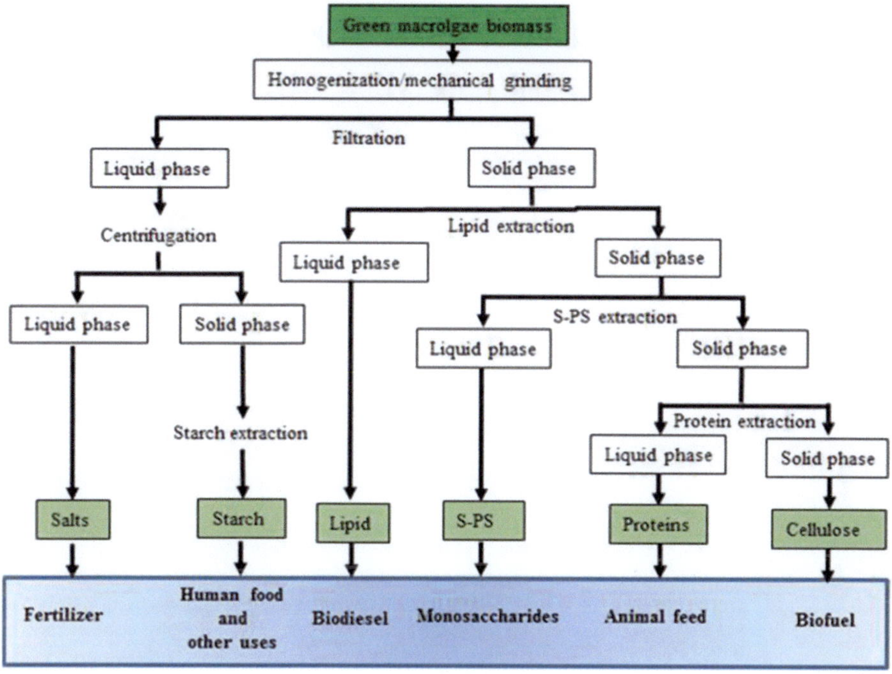

Fig. 10.4 Green macroalgae biorefinery process for co-production of a wide range of valuable products. (Figure adapted from Prabhu et al. 2019 with permits)

crude lipid, agar, soil conditioner, and bioethanol using various techniques such as homogenization, ultra-membrane filtration solvent extraction, enzymatic and thermal hydrolysis, and fermentation. An example of the biorefinery based on the currently widely cultivated red macroalgae *Kappaphycus alvarezii* being biorefined for the production of bioethanol, carrageenan, fertilizer, and biogas is shown in Fig. 10.5. Such studies enable to realize the full potential of marine macroalgal feedstock for production of fuel and chemicals. Such a production of high-value multiple products at a time, in the integrated process, is necessary in order to meet current bioeconomy challenges (Chandra et al. 2019). Various integrated biorefinery studies producing various streams of products in cascading fashion involving green, brown, and red macroalgae species are shown in Table 10.3.Various products and by-products can be derived from integrated alga (Chandra et al. 2019; Sahoo et al. 2012; Milledge et al. 2016). The most important biomass products in algal biorefineries are:

- Biomass – health food, functional food, feed additive, aquaculture, biofertilizer
- Phycocolloids – agar, carrageenan, alginates
- Pigments/carotenoids – astaxanthin, phycocyanin, phycoerythrin, fucoxanthin
- Vitamin – A, B1, B6, B12, C, E, biotin, riboflavin, nicotinic acid, pantothenate, and folic acid

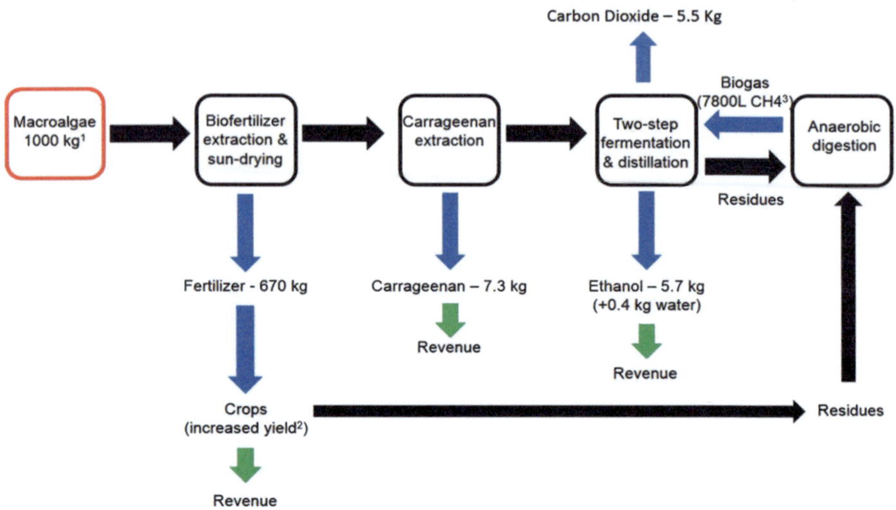

Fig. 10.5 Red seaweed *Kappaphycus*-based biorefinery for the co-production of fertilizers, carrageenan, ethanol, and biogas. (1) 1kg of freshly harvested algae (fresh weight), (2) (Eswaran et al. 2005), (3) yield from digestion of algal biomass only (Park et al. 2012). Calculations were done using the following yield and assumption: fertilizer yield (67%), residue moisture content (25%), carrageenan yield (12% g/kg dry algae), ethanol yield (minimum scenario of 77.6 g/kg dry algae), ethanol purity after distillation (95:5 v:v ethanol-water mixture), 1 mol of produced ethanol = 1 mol of produced CO_2, 141 L CH_4/kg of algal dry matter before ethanol production. (Figure adapted from Ingle et al. 2017 with permits)

- Antioxidants – β-carotene, tocopherol
- Antioxidant extracts – PUFA extracts (polyunsaturated fatty acids), arachidonic acid (ARA polyunsaturated omega-6 fatty acid, docosahexaenoic acid) (DHA omega-3 fatty acid)
- Other/pharmaceuticals – antifungal, antimicrobial, antiviral, toxins, amino acids, proteins, and sterols
- Biofuels – bioethanol, biodiesel, bio-butanol, biomethane, and biochar

10.6 Economic Challenge of Offshore Marine Biorefineries

There are numerous challenges associated with the successful deployment of marine biorefinery operations as summarized in Fig. 10.6. The profitability of marine biorefinery is subject to various sources of uncertainty such as that of feedstock supply, processing technology, investment, contracting, and demand (Palatnik and Zilberman 2017).

Selection of right biomass is critical as there are numerous species of macroalgae. As presented in Table 10.3, each of them differs for chemical structure and therefore differs for bio-based chemicals to be produced (Jung et al. 2013). The

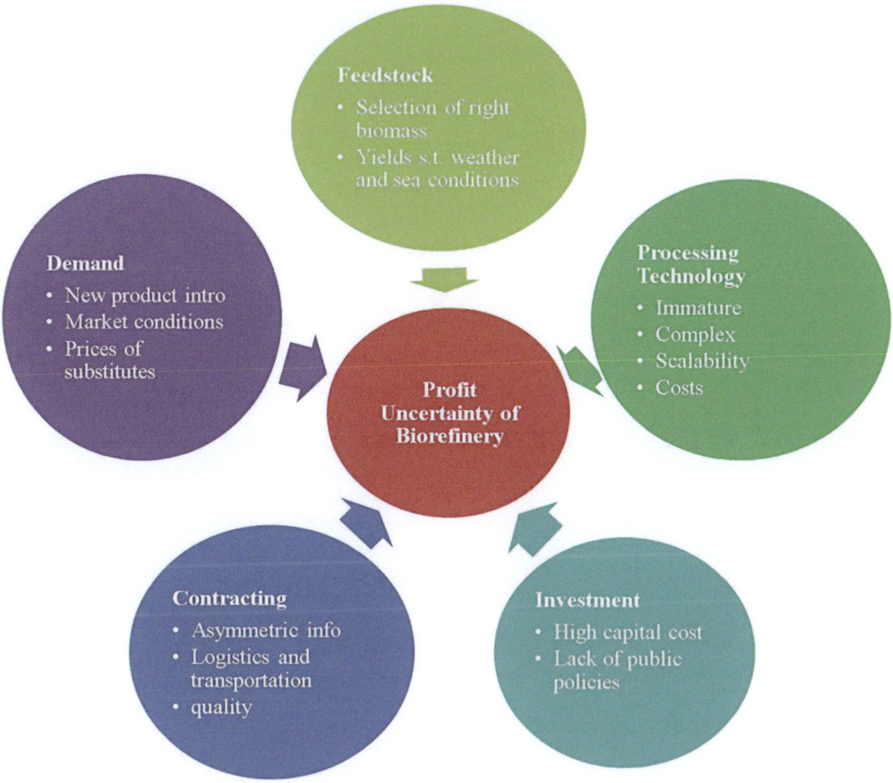

Fig. 10.6 Major sources of uncertainty for marine biorefinery deployment

entrepreneur should determine which algae-based activities are profitable under multidimensional uncertainty outlined below.

The rate of feedstock growth shows a wide range of values. Studies report the range of 6–108 tonnes/ha per year (Valderrama et al. 2015). This uncertainty in feedstock yield has a major impact on the cost-effectiveness of the technology. Feedstock growth depends on saturation kinetics by light intensity, ambient dissolved inorganic nutrient concentrations, and temperature (Lehahn et al. 2016). Cultivation uncertainty is exacerbated by stochastic weather, seasonal variability between regions, within years, and between years. Studies point at the biomass productivity as the main constraint against being competitive with other energy- and protein-producing technologies (Seghetta et al. 2016b).

Anaerobic digestion, fermentation, transesterification, liquefaction, and pyrolysis can convert algal biomass into proteins and sugars that can result in food, chemicals, and biofuels. At each stage of the production process, the entrepreneur should decide between various options that ultimately affect the irreversible (sunk) and variable costs of the production, the productivity, and the output, therefore affecting the total profitability. Yet, the biorefinery yields are highly uncertain (Lehahn et al.

Table 10.3 Green, brown, and red macroalgae species-based biorefinery studies carried out for production of various products

Species	Biorefinery products	Technologies/methods	Reference
Green macroalgae			
U. lactuca	Proteins and carbohydrates	Osmotic shock, enzymatic hydrolysis, PEF or high shear homogenization	Postma et al. (2017)
U. lactuca	Animal feed (343 g protein kg^{-1} dry matter), acetone, butanol, ethanol, and 1,2-propanediol	Thermal and enzymatic hydrolysis and fermentation	Bikker et al. (2016c)
Chaetomorpha linum	Bioethanol and biogas	Thermochemical and enzymatic hydrolysis, fermentation, and anaerobic digestion	Ben Yahmed et al. (2016)
Ulva fasciata	MRLE, lipid, ulvan, and cellulose	Mechanical grinding, thermal and chemical extraction, and water extraction	Trivedi et al. (2016)
Ulva ohnoi and *Ulva tepida*	Mainly salt (demonstrating the use of leftover biomass for protein, fertilizer, animal feed and fuel)	Aqueous washing and drying	Magnusson et al. (2016)
U. lactuca	MRLE, lipid, ulvan, protein and cellulose	Mechanical pressing and crushing, heat treatment and organic solvent, and alkali extraction	Gajaria et al. (2017)
U. lactuca	Acetone, butanol and ethanol (ABE)	Pretreatment, enzymatic saccharification and fermentation	van der Wal et al. (2013)
Ulva rigida	Liquid stream with carbohydrate and salt; remaining stream with concentrated protein	Ionic liquid deconstruction	Pezoa-Conte et al. (2015)
U. ohnoi	Salt, pigment, ulvan, and protein	Aqueous pretreatment, thermal and chemical extraction	Glasson et al. (2017)
U. lactuca	MRLE, ulvan, protein, and methane	Aqueous, thermal and chemical extraction, and anaerobic digestion	Mhatre et al. (2018)
Brown macroalgae			
Laminaria digitata	Succinic acid, feed and energy	Enzymatic hydrolysis, fermentation, anaerobic digestion	Alvarado-Morales et al. (2015)
Laminaria digitata	Alginate, fucoidan, alginate, bioethanol	Acid hydrothermal, enzymatic scarification, fermentation	Kostas et al. (2017)
Ascophyllum nodosum	Fucoidan, alginates, sugars, and biochar	Thermal, acid hydrolysis	Yuan and Macquarrie (2015)

(continued)

Table 10.3 (continued)

Species	Biorefinery products	Technologies/methods	Reference
Saccharina latissima	Isomannide, butanol, furan dicarboxylic acid, biogas	Shredding and pressing, fermentation	Van Hal et al. (2014)
Saccharina latissima	Succinic acid, fertilizers, and antioxidants	Enzymatic hydrolysis and fermentation, solvent extraction	(Marinho et al. 2016)
Durvillaea potatorum	Alginate, fucoidan, and laminarin	Mechanical grinding, acid and alkali extraction	Abraham et al. (2019)
Red macroalgae			
Gracilaria verrucosa	Agar, bioethanol, and biofertilizer	Thermochemical, enzymatic hydrolysis, and fermentation	Kumar et al. (2013)
Gracilaria corticata	MRLE, pigments, lipid, agar, soil conditioner, and bioethanol	Homogenization, ultra-membrane filtration solvent extraction, enzymatic and thermal hydrolysis, fermentation	Baghel et al. (2016))
Porphyra umbilicalis	Proteins, carrageenan, pectin, and cellulose	Cold alkali extraction, thermochemical, solvent extraction	Wahlström et al. (2018)

2016) signaling the immaturity of the technology. The upper value of yields can be ten times larger than the lower one (Table 10.4), significantly affecting the potential profitability of the process.

Numerous studies are focusing on the effort to evaluate future costs of the process that is currently available mostly in small (lab) scale (e.g., Seghetta et al. 2016b; Korzen et al. 2015c). These studies, however, do not report a structured production function that leads to a cost function. The common assumption is a linear approximation. This assumption should be treated cautiously and verified against actual data when production is scaled up.

The development of a new biorefinery, its design, and construction requires huge investments (Stichnothe et al. 2016). The strategy about the capacity of the biorefinery may change over time; the innovator may experiment by starting at a small scale. Once the production system is established, the innovator may either expand operations or reach out to cooperatives to provide it with inputs.

Moreover, introducing and perfecting innovations is a random process, and the economic conditions that face technology vary over time. Learning takes time, and the dynamics of knowledge accumulation affect the timing of introduction of innovations, their refinement, and their commercialization. Timing can also affect the decision regarding both the capacity of innovation and the extent of reliance on external sources.

Lack of public policies supporting biorefinery sector limits the long-term investment decision required. There are various strategies, but there are no distinct policy drivers for the utilization of bio-based chemicals, in direct contrast to the biofuels industry where various national regulations are driving rapid growth.

Table 10.4 Examples for costs of macroalgae production and commodities' prices

Description	Average value	Range	Source
First unit cost of *carrageenan*	USD 4500/ ton 2014	USD 4000–6500/ton 2010	Brown (2015)
First unit cost of feedstock seaweed *Kappaphycus*	USD 1600/ ton 2016	USD 600–7000/ ton 2010	Calculated based on Ricardo et al. (2015)
Price of protein	USD 5000/ Ton	USD 1000–15,000/ton 2016	Price calculated from value and quantity world *2016 Source:* UN COMTRADE; commodity 210610 protein; concentrates and textured protein substances
Annual growth of the price of protein	7%	−43 to 92% in 1991–2016	Price calculated from value and quantity of Philippines and US export 1991–2016
			Source: UN COMTRADE; commodity 210610 protein; concentrates and textured protein substances
Price of *carrageenan*	USD 5500/ ton	USD 3000–6000/ton 2016	https://www.alibaba.com/trade/search?fsb=y&IndexArea=product_en&CatId=&SearchText=carrageenan
Annual growth of the price of *carrageenan*	4%	−11 to 53% in 1991–2016	Price calculated from value and quantity of Philippines export
		S. D. 16%	*Source:* UN COMTRADE; commodity HS130239 (mucilages and thickeners ones)

The impact of price variation should be analyzed in several aspects: price uncertainties that face the aqua-farmer, price uncertainties of feedstock for the biorefinery, and the price uncertainty of competitive outputs (backstop technology). A seaweed industry that contains many small-scale price-takers is especially prone to boom-bust cycles. For example, the strong demand from China drove the price of dry cottonii in the Philippines from USD 900/ton in 2007 to almost USD 3000/ton in 2008 causing the Philippines production to double from 1.5 million tons (wet weight) in 2007 to 3.3 million tons in 2008. The "seaweed rush" lasted only 1 year – the price dropped to USD 300/ton in 2009 (Ricardo et al. 2015). Table 10.4 exemplifies the range of prices as well as annual growth rates for one of the macroalgae species – *Kappaphycus* – and for two possible biorefinery outputs: *carrageenan* and proteins in the years 1991–2016. Generally, when strong demand for dry seaweeds drives up the price, seaweed farmers tend to increase their planting efforts and/or harvest immature crops. However, if the price is low, seaweed farmers tend to reduce production, which creates sourcing difficulties for the biorefineries. On the other hand, biorefineries would tend to reduce demand as prices of feedstock rise by substituting cheaper alternatives. A likely result would then be that feedstock supply exceeding demand and consequently a collapse in price.

The economics of biorefinery based product depends heavily on drop-in versus non-drop-in (existing demand and infrastructure). Therefore, demand may be very strong or very weak, leading to general uncertainty. It is difficult to know, for example, if an investment in the bio-based supply chain will make economic sense. It might not be possible to sell the produced bio-based chemicals at a price necessary to make the investment profitable. Of course, these are the kinds of decisions that all businesses face, but the reliance of biomass markets on policy measures and the lack of long-term signals in, for example, EU policy regarding biomass means that uncertainties are unusually high. In addition to the production cost, the value of biorefinery products when reaching end users may also reflect the expenses on research and development (R&D), formulation, marketing, etc. (Ricardo et al. 2015). Specific information on these aspects is generally lacking.

10.7 Sustainability and Environmental Impacts

The sustainability of seaweed biorefineries was assessed in various life cycle assessment (LCA) studies (Seghetta et al. 2016b, Seghetta et al. 2017; Alvarado-Morales et al. 2013; Czyrnek-Delêtre et al. 2017; Langlois et al. 2012; van Oirschot et al. 2017; Aitken et al. 2014). Overall, seaweed cultivation was found to contribute to the environmental restoration and climate mitigation. However, several parameters have been pointed out to have significant effects on the environmental performance of the complete biorefinery process and should be optimized.

Large-scale macroalgae cultivation can be responsible for positive and negative impact on coastal and marine ecosystems (Hughes et al. 2012). Therefore, the balance is necessary to attain in between food, chemicals, and fuel production and its environmental cost (Wei et al. 2013). Although scale-up reduces production costs of macroalgae, the offshore cultivation is challenging because of the harsh environment and also could possess risks to the environment. Risk management framework should be developed for each individual case to address these factors.

The overall risk management framework is generally used for the decision-making process and provides a more clear idea to make a decision about any technological term and includes and comprises defining the challenge or problem; stakeholders involved; consideration of almost possible concerns; identification of actual risk, review, and judgment; and finally the decision (Singh et al. 2016; Keswani and Singh 2019). The proposed framework of risk management for offshore macroalgae cultivation is shown in Fig. 10.7. The framework is divided into three sections as follows. Section 10.1 shows the possible risks, which can be prevented before the cultivation or during the cultivation and are related to the requirement of macroalgae cultivation. Section 10.2 shows the risks that might be controlled in the process production of biomass and harvesting. Section 10.3 shows the risks that will need to be mitigated, as these are the potential impacts of cultivation of macroalgae on the marine environment.

To summarize, we show that offshore macroalgae biorefinery concepts are emerging for co-production of multiple products, which could reduce the environmental burden of fossil fuels and agriculture. Technologies for offshore biomass

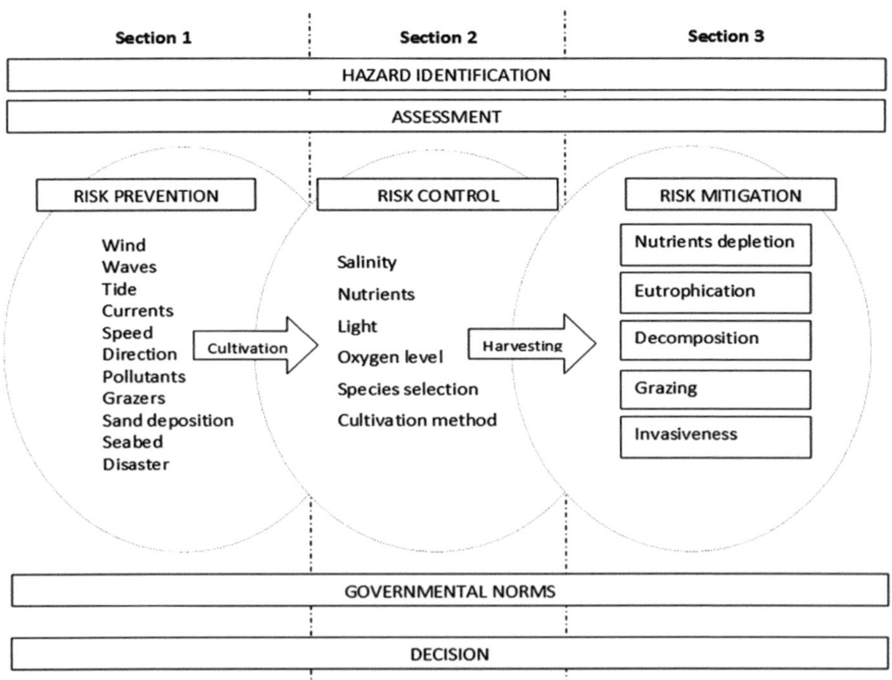

Fig. 10.7 The entire framework of the risk management for offshore macroalgal cultivation. (Figure adapted from Lehahn et al. 2016 with a permit)

cultivation are being developed worldwide. Biomass fractionation technologies are emerging and provide a broad spectrum of products. Yet the challenging, high-energy sea environment and unusual composition of the biomass still result in high levels of uncertainty of technological and economic feasibility of these projects. To decrease this uncertainty, demonstration units with different scales, technologies, species, and products are needed.

Acknowledgments The authors thank the Israel Ministry of Energy, Israel Ministry of Science and Technology, and Israel Innovation Authority for the support.

References

Abraham RE, Su P, Puri M, Raston CL, Zhang W (2019) Optimisation of biorefinery production of alginate, fucoidan and laminarin from brown seaweed Durvillaea Potatorum. Algal Res 38:101389

Aitken D, Bulboa C, Godoy-Faundez A, Turrion-Gomez JL, Antizar-Ladislao B (2014) Life cycle assessment of macroalgae cultivation and processing for biofuel production. J Clean Prod 75:45–56

Alvarado-Morales M, Boldrin A, Karakashev DB, Holdt SL, Angelidaki I, Astrup T (2013) Life cycle assessment of biofuel production from brown seaweed in nordic conditions. Bioresour Technol 129:92–99

Alvarado-Morales M, Gunnarsson IB, Fotidis IA, Vasilakou E, Lyberatos G, Angelidaki I (2015) Laminaria digitata as a potential carbon source for succinic acid and bioenergy production in a biorefinery perspective. Algal Res 9:126–132

Ashkenazi DY, Israel A, Abelson A (2019) A novel two-stage seaweed integrated multi-trophic aquaculture. Rev Aquac 11:246–262

Baghel RS, Trivedi N, Reddy CRK (2016) A simple process for recovery of a stream of products from marine macroalgal biomass. Bioresour Technol 2016(203):160–165

Balina K, Romagnoli F, Blumberga D (2017) Seaweed biorefinery concept for sustainable use of marine resources. Energy Procedia 128:504–511

Ben Yahmed N, Jmel MA, Ben Alaya M, Bouallagui H, Marzouki MN, Smaali I (2016) A biorefinery concept using the green macroalgae Chaetomorpha linum for the coproduction of bioethanol and biogas. Energy Convers Manag 119:257–265

Bentsen NS, Felby C (2012) Biomass for energy in the European Union – a review of bioenergy resource assessments. Biotechnol Biofuels 5(1):25

Bikker P, Krimpen MM, Wikselaar P, Houweling-Tan B, Scaccia N, Hal JW, Huijgen WJJ, Cone JW, López-Contreras AM, van Krimpen MM et al (2016a) Biorefinery of the green seaweed Ulva Lactuca to produce animal feed, chemicals and biofuels. J Appl Phycol 28:3511–3525

Bikker P, van Krimpen MM, van Wikselaar P, Houweling-Tan B, Scaccia N, van Hal JW, Huijgen WJJ, Cone JW, Lopez-Contreras AM (2016b) Biorefinery of the green seaweed Ulva lactuca to produce animal feed, chemicals and biofuels. J Appl Phycol 28:1–15

Bikker P, van Krimpen MMM, van Wikselaar P, Houweling-Tan B, Scaccia N, van Hal JWW, Huijgen WJ, Cone JWW, López-Contreras AM, Scaccia NazarenoScaccia N et al (2016c) Biorefinery of the green seaweed *Ulva Lactuca* to produce animal feed. Chem Biofuels 28:1–15

Bokinsky G, Peralta-Yahya PP, George A, Holmes BM, Steen EJ, Dietrich J, Soon Lee T, Tullman-Ercek D, Voigt CA, Simmons BA et al (2011) Synthesis of three advanced biofuels from ionic liquid-pretreated switchgrass using engineered Escherichia Coli. Proc Natl Acad Sci 108:19949–19954

Brown TR (2015) A techno-economic review of thermochemical cellulosic biofuel pathways. Bioresour Technol 178:166–176

Buck BH, Buchholz CM (2004) The offshore-ring: a new system design for the open ocean aquaculture of macroalgae. J Appl Phycol 16(5):355–368
Buck BH, Buchholz CM (2005) Response of offshore cultivated Laminaria saccharina to hydrodynamic forcing in the North Sea. Aquaculture 250(3–4):674–691
Buck BH, Krause G, Rosenthal H (2004) Extensive open ocean aquaculture development within wind farms in Germany: the prospect of offshore co-management and legal constraints. Ocean Coast Manag 47(3–4):95–122
Buck BH, Krause G, Michler-Cieluch T, Brenner M, Buchholz CM, Busch JA, Fisch R, Geisen M, Zielinski O (2008) Meeting the quest for spatial efficiency: progress and prospects of extensive aquaculture within offshore wind farms. Helgol Mar Res 62(3):269–281
Buschmann AH, Camus C, Infante J, Neori A, Israel Á, Hernández-González MC, Pereda SV, Gomez-Pinchetti JL, Golberg A, Tadmor-Shalev N et al (2017) Seaweed production: overview of the global state of exploitation, farming and emerging research activity seaweed production. Eur J Phycol 52:391
Chandra R, Iqbal HMN, Vishal G, Lee H-S, Nagra S (2019) Algal biorefinery: a sustainable approach to valorize algal-based biomass towards multiple product recovery. Bioresour Technol 278:346–359. No. November 2018
Chemodanov A, Robin A, Golberg A (2017a) Design of marine macroalgae photobioreactor integrated into building to support seagriculture for biorefinery and bioeconomy. Bioresour Technol 241:1084–1093
Chemodanov A, Jinjikhashvily G, Habiby O, Liberzon A, Israel A, Yakhini Z, Golberg A (2017b) Net primary productivity, biofuel production and CO_2 emissions reduction potential of Ulva Sp. (Chlorophyta) biomass in a coastal area of the Eastern Mediterranean. Energy Convers Manag 148:1497–1507
Czyrnek-Delêtre MM, Rocca S, Agostini A, Giuntoli J, Murphy JD (2017) Life cycle assessment of seaweed biomethane, generated from seaweed sourced from integrated multi-trophic aquaculture in temperate oceanic climates. Appl Energy 196:34–50
De Jong E, Jungmeier G (2015) Bioreenery concepts in comparison to petrochemical Reeneries. In: Industrial Biorefineries White Biotechnol, pp 3–33
De Jong E, Higson A, Walsh P, Wellisch M (2012) Product developments in the bio-based chemicals arena. Biofuels Bioprod Biorefin 6(6):606–624
Drimer N (2019) First principle approach to the design of an open sea aquaculture system. Ships Offshore Struc. https://doi.org/10.1080/17445302.2016.1213491
Druehl LD, Baird R, Lindwall A, Lloyd KE, Pakula S (1988) Longline cultivation of some laminariaceae in British Columbia, Canada. Aquac Fish Manag 19:253–263
Du X, Lu L, Reardon T, Zilberman D (2016) Economics of agricultural supply chain design: a portfolio selection approach. Am J Agric Econ 98:1377–1388
Dunlop MJ (2011) Engineering microbes for tolerance to next-generation biofuels. Biotechnol Biofuels 4:32
Enquist-Newman M, Faust AME, Bravo DD, Santos CNS, Raisner RM, Hanel A, Sarvabhowman P, Le C, Regitsky DD, Cooper SR et al (2014) Efficient ethanol production from brown macroalgae sugars by a synthetic yeast platform. Nature 505(7482):239–243
Eswaran K, Ghosh PK, Siddhanta AK, Patolia JS, Periyasamy C, Mehta AS, Mody KH, Ramavat BK, Prasad K, Rajyaguru MR (2005) Integrated method for production of carrageenan and liquid fertilizer from fresh seaweeds. US Patent 6,893,479
Feinberg D, Hock S (1985) Technical and economic evaluation of macroalgae cultivation for fuel production (draft). NREL Report. https://www.nrel.gov/docs/legosti/old/2685.pdf
Fernand F, Israel A, Skjermo J, Wichard T, Timmermans KR, Golberg A (2017) Offshore macroalgae biomass for bioenergy production: environmental aspects, technological achievements and challenges. Renew Sust Energ Rev 75:35–45
Gajaria TK, Suthar P, Baghel RS, Balar NB, Sharnagat P, Mantri VA, Reddy CRK (2017) Integration of protein extraction with a stream of byproducts from marine macroalgae: a model forms the basis for marine bioeconomy. Bioresour Technol 243:867–873

Ghaderi H, Pishvaee MS, Moini A (2016) Biomass supply chain network design: an optimization-oriented review and analysis. Ind Crop Prod 94:972–1000

Glasson CRK, Sims IM, Carnachan SM, de Nys R, Magnusson M (2017) A cascading biorefinery process targeting sulfated polysaccharides (Ulvan) from Ulva Ohnoi. Algal Res 27:383–391

Golberg A, Liberzon A (2015) Modeling of smart mixing regimes to improve marine biorefinery productivity and energy efficiency. Algal Res 11:28–32

Golden JS, Handfield RB, Daystar J, McConnell TE (2015) An economic impact analysis of the us biobased products industry: a report to the congress of the United States of America. Ind Biotechnol 11(4):201–209

Haberl H, Erb K-H, Krausmann F, Bondeau A, Lauk C, Müller C, Plutzar C, Steinberger JK (2011) Global bioenergy potentials from agricultural land in 2050: sensitivity to climate change, diets and yields. Biomass Bioenergy 35(12):4753–4769

Hanisak M (1987) Cultivation of Gracilaria and other macroalgae in Florida for energy production. Dev Aquac Fish Sci:191–218

Hannon M, Gimpel J, Tran M, Rasala B, Mayfield S (2010) Biofuels from algae: challenges and potential. Biofuels 1:763–784

Holt TJ (1984) The development of techniques for the cultivation of Laminariales in the Irish Sea. Ph.D, University of Liverpool, p 266

Hughes AD, Kelly MS, Black KD, Stanley MS (2012) Biogas from macroalgae: is it time to revisit the Idea? Biotechnol Biofuels 5(1):86

Ingle K, Vitkin E, Robin A, Yakhini Z, Mishori D, Golberg A (2017) Macroalgae biorefinery from Kappaphycus Alvarezii: conversion Modeling and performance prediction for India and Philippines as examples. Bio Energy Res:1–11

International Energy Agency (2011) World energy outlook

Jung KAA, Lim S-RR, Kim Y, Park JMM (2013) Potentials of Macroalgae as Feedstocks for Biorefinery. Bioresour Technol 135:182–190

Keasling JD, Chou H (2008) Metabolic engineering delivers next-generation biofuels. Nat Biotechnol 26:298–299

Keswani C, Singh SP (eds) (2019) Intellectual property issues in microbiology. Springer, Singapore. 425 pages, ISBN:9789811374654

Korzen L, Abelson A, Israel A (2015a) Growth, protein and carbohydrate contents in Ulva rigida and gracilaria bursa-pastoris integrated with an offshore fish farm. J Appl Phycol 23:543–597

Korzen L, Peled Y, Shamir SZ, Shechter M, Gedanken A, Abelson A, Israel A (2015b) An economic analysis of bioethanol production from the marine Macroalga Ulva (Chlorophyta). Technology 03(02n03):114–118

Korzen L, Pulidindi IN, Israel A, Abelson A, Gedanken A (2015c) Marine integrated culture of carbohydrate rich Ulva rigida for enhanced production of bioethanol. RSC Adv 5(73):59251–59256

Kostas ET, White DA, Cook DJ (2017) Development of a bio-refinery process for the production of speciality chemical, biofuel and bioactive compounds from Laminaria digitata. Algal Res 28(May):211–219

Kraan S (2013) Mass-cultivation of carbohydrate rich macroalgae, a possible solution for sustainable biofuel production. Mitig Adapt Strateg Glob Chang 18(1):27–46. http://www.ask-force.org/web/Global-Warming/Kraan-Mass-cultivation-carbyhodrate-Macroalgae-2013.pdf

Kraan S, Guiry MD (2001) Phase II: strain hybridisation field experiments and genetic fingerprinting of the edible brown seaweed Alaria Esculenta 18(18)

Kumar S, Sahoo D (2017) A comprehensive analysis of alginate content and biochemical composition of leftover pulp from brown seaweed Sargassum wightii. Algal Res 23:233–239

Kumar S, Gupta R, Kumar G, Sahoo D, Kuhad RC (2013) Bioethanol production from Gracilaria Verrucosa, a Red Alga, in a biorefinery approach. Bioresour Technol 135:150–156

Langlois J, Sassi J-F, Jard G, Steyer J-P, Delgenes J-P, Hélias A (2012) Life cycle assessment of biomethane from offshore-cultivated seaweed. Biofuels Bioprod Biorefin 6(4):387–404

Laurens LML, Chen-Glasser M, McMillan JD (2017) A perspective on renewable bioenergy from photosynthetic algae as feedstock for biofuels and bioproducts. Algal Res 24(March):261–264

Lee SK, Chou H, Ham TS, Lee TS, Keasling JD (2008) Metabolic engineering of microorganisms for biofuels production: from bugs to synthetic biology to fuels. Curr Opin Biotechnol 19:556–563

Lehahn Y, Ingle KN, Golberg A (2016) Global potential of offshore and shallow waters macroalgal biorefineries to provide for food, chemicals and energy: feasibility and sustainability. Algal Res 17:150–160

Lirasan T, Twide P (1993) Fourteenth international seaweed symposium. In: Chapman ARO, Brown MT, Lahaye M (eds) Fourteenth international seaweed symposium developments in hydrobiology, vol 85. Springer, Dordrecht, pp 353–355

Liu D, Keesing JK, Xing Q, Shi P (2009) World's largest macroalgal bloom caused by expansion of seaweed aquaculture in China. Mar Pollut Bull 58(6):888–895

Liu D, Keesing JK, Dong Z, Zhen Y, Di B, Shi Y, Fearns P, Shi P (2010) Recurrence of the world's largest green-tide in 2009 in Yellow Sea, China: *Porphyra Yezoensis* aquaculture rafts confirmed as nursery for macroalgal blooms. Mar Pollut Bull 60(9):1423–1432

Magnusson M, Carl C, Mata L, de Nys R, Paul NA (2016) Seaweed salt from Ulva: a novel first step in a cascading biorefinery model. Algal Res 16:308–316

Marinho GS, Alvarado-Morales M, Angelidaki I (2016) Valorization of macroalga Saccharina latissima as novel feedstock for fermentation-based succinic acid production in a biorefinery approach and economic aspects. Algal Res 16:102–109

Mhatre A, Gore S, Mhatre A, Trivedi N, Sharma M, Pandit R, Anil A, Lali A (2018) Effect of multiple product extractions on bio-methane potential of marine macrophytic green alga *Ulva lactuca*. Renew Energy 132:742–751

Milledge JJ, Nielsen BV, Bailey D (2016) High-value products from macroalgae: the potential uses of the invasive brown seaweed, Sargassum Muticum. Rev Environ Sci Biotechnol 15(1):67–88

Möller B, Hong L, Lonsing R, Hvelplund F (2012) Evaluation of offshore wind resources by scale of development. Energy 48(1):314–322

Neori A, Chopin T, Troell M, Buschmann AH, Kraemer GP, Halling C, Shpigel M, Yarish C (2004) Integrated aquaculture: rationale, evolution and state of the art emphasizing seaweed biofiltration in modern mariculture. Aquaculture 231:361–391

Notoya M (2010) Production of biofuel by macroalgae with preservation of marine resources and environment. Springer, Dordrecht, pp 217–228

Nunes N, Ferraz S, Valente S, Barreto MC, Pinheiro de Carvalho MAA (2017) Biochemical composition, nutritional value, and antioxidant properties of seven seaweed species from the Madeira archipelago. J Appl Phycol 29(5):2427–2437

Olanrewaju SO, Magee A, Kader ASA, Tee KF (2017) Simulation of offshore aquaculture system for macro algae (seaweed) oceanic farming. Ships and Offshore Structures 12(4):553–562

Palatnik RR, Zilberman D (2017) Economics of natural resource utilization – the case of macroalgae. In: Pinto A, Zilberman D (eds) Modeling, dynamics, optimization and bioeconomics II. Springer, pp 1–21

Park JH, Yoon JJ, Park HD, Lim DJ, Kim SH (2012) Anaerobic digestibility of algal bioethanol residue. Bioresour Technol 113:78–82

Patarra RF, Paiva L, Neto AI, Lima E, Baptista J (2011) Nutritional value of selected macroalgae. J Appl Phycol 23(2):205–208

Peralta-Yahya PP, Ouellet M, Chan R, Mukhopadhyay A, Keasling JD, Lee TS (2011) Identification and microbial production of a terpene-based advanced biofuel. Nat Commun 2:483

Peteiro C, Freire Ó (2012) Outplanting time and methodologies related to mariculture of the edible Kelp Undaria Pinnatifida in the Atlantic Coast of Spain. J Appl Phycol 24:1361–1372

Peteiro C, Sánchez N, Dueñas-Liaño C, Martínez B (2014) Open-sea cultivation by transplanting young Fronds of the Kelp Saccharina Latissima. J Appl Phycol 26:519–528

Pezoa-Conte R, Leyton A, Anugwom I, von Schoultz S, Paranko J, Mäki-Arvela P, Willför S et al (2015) Deconstruction of the green alga Ulva Rigida in ionic liquids: closing the mass balance. Algal Res 12:262–273. Elsevier

Pimentel D (2012) Global economic and environmental aspects of biofuels. CRC Press, Boca Raton

Pimentel M, Pimentel MH (2008) Food, energy, and society. CRC Press, Boca Raton
Postma PR, Cerezo-Chinarro O, Akkerman RJ, Olivieri G, Wijffels RH, Brandenburg WA, Eppink MHM (2017) Biorefinery of the macroalgae *Ulva Lactuca*: extraction of proteins and carbohydrates by mild disintegration. J Appl Phycol:1–13
Potts T, Du J, Paul M, May P, Beitle R, Hestekin J (2012) The production of butanol from Jamaica Bay macro algae. Environ Prog Sustain Energy 31:29–36
Prabhu M, Chemodanov A, Gottlieb R, Kazir M, Nahor O, Gozin M, Israel A, Livney YD, Golberg A (2019) Starch from the sea: the green macroalga Ulva ohnoi as a potential source for sustainable starch production in the marine biorefinery. Algal Res 37:215–227
Reith JH, Deurwaarder EP, Hemmes K, Biomassa E, Curvers APWM, Windenergie E (2005) BIO-OFFSHORE Grootschalige Teelt van Zeewieren in Combinatie Met Offshore Windparken in de Noordzee. https://library.wur.nl/WebQuery/wurpubs/347698
Ricardo R, Neori A, Valderrama D, Reddy CRK, Cronin H, Forster J (2015) Farming of seaweeds. In: Seaweed sustainability. Elsevier, pp 27–57
Roels OA, Laurence S, Vanhemelryck L (1979) The utilization of cold, nutrient-rich deep ocean water for energy and mariculture. Ocean Manag 5:199–210
Roesijadi AG, Copping A, Huesemann M (2008) Techno-economic feasibility analysis of offshore seaweed farming for bioenergy and biobased products. https://arpa-e.energy.gov/sites/default/files/Techno-Economic%20Feasibility%20Analysis%20of%20Offshore%20Seaweed%20Farming%20for%20Bioenergy%20and%20Biobased%20Products-2008.pdf
Roesijadi G, Jones SBB, Snowden-Swan LJ, Zhu Y (2010, September) Macroalgae as a biomass feedstock: a preliminary analysis. Dep. Energy under Contract DE-AC05-76RL01830 by Pacific Northwest Natl. Lab, pp 1–50. http://sailing-sea-farm.com/onewebmedia/PNNL-19944.pdf
Sahoo D, Kumar S, Elangbam G, Devi SS (2012) Biofuel production from algae through integrated biorefinery. Sci Algal Fuels 25:215–230
Sanderson JC, Dring MJ, Davidson K, Kelly M, Culture S (2012) Yield and bioremediation potential of Palmaria Palmata (Linnaeus) Weber & Mohr and Saccharina Latissima (Linnaeus) C.E. Lane, C. Mayes, Druehl & G.W. Saunders adjacent to fish farm cages in Northwest Scotland. Aquaculture 354–355:128–135
Seghetta M, Hou X, Bastianoni S, Bjerre A-B, Thomsen M (2016a) Life cycle assessment of macroalgal biorefinery for the production of ethanol, proteins and fertilizers – a step towards a regenerative bioeconomy. J Clean Prod 137:1158–1169
Seghetta M, Marchi M, Thomsen M, Bjerre AB, Bastianoni S (2016b) Modelling biogenic carbon flow in a macroalgal biorefinery system. Algal Res 18:144–155
Seghetta M, Romeo D, D'Este M, Alvarado-Morales M, Angelidaki I, Bastianoni S, Thomsen M (2017) Seaweed as innovative feedstock for energy and feed – evaluating the impacts through a life cycle assessment. J Clean Prod 150:1–15
Singh HB, Jha A, Keswani C (eds) (2016) Intellectual property issues in biotechnology. CABI, Wallingford. 304 pages, ISBN-13:9781780646534
Star-coliBRi (2011) European biorefinery joint strategic research roadmap for 2020. https://cordis.europa.eu/project/rcn/93170/reporting/en
Steen EJ, Kang Y, Bokinsky G, Hu Z, Schirmer A, McClure A, Del Cardayre SB, Keasling JD (2010) Microbial production of fatty-acid-derived fuels and chemicals from plant biomass. Nature 463:559–562
Stichnothe H, Meier D, de Bari I (2016) Biorefineries: industry status and economics. Dev Glob Bioeconomy:41–67. https://rdm.pure.elsevier.com/it/publications/biorefineries-industry-status-and-economics
Suutari M, Leskinen E, Fagerstedt K, Kuparinen J, Kuuppo P, Blomster J (2015) Macroalgae in biofuel production. Phycol Res 63(1):1–18
Szetela EJ, Krascella NL, Blecher WA, Christopher GL (1976) Evaluation of a marine energy farm concept. Am Chem Soc, Div Fuel Chem, Prepr.; (United States) 19:4
Taheripour F, Hertel TW, Tyner WE, Beckman JF, Birur DK (2010) Biofuels and their by-products: global economic and environmental implications. Biomass Bioenergy 34:278–289

Trivedi N, Baghel RS, Bothwell J, Gupta V, Reddy CRK, Lali AM, Jha B (2016) An integrated process for the extraction of fuel and chemicals from marine macroalgal biomass. Sci Rep 6:30728

Troell M, Joyce A, Chopin T, Neori A, Buschmann AH, Fang JG (2009) Ecological engineering in aquaculture – potential for integrated multi-trophic aquaculture (IMTA) in marine offshore systems. Aquaculture 297(1–4):1–9

Valderrama D, Cai J, Hishamunda N, Ridler N, Neish IC, Hurtado AQ, Msuya FE, Krishnan M, Narayanakumar R, Kronen M et al (2015) The economics of kappaphycus seaweed cultivation in developing countries: a comparative analysis of farming systems. Aquac Econ Manag 19(2):251–277

van den Burg S, Stuiver M, Veenstra F, Bikker P, López Contreras A, Palstra A, Broeze J, Jansen H, Jak R, Gerritsen A, et al (2013) A triple P review of the feasibility of sustainable offshore seaweed production in the North Sea. https://library.wur.nl/WebQuery/wurpubs/442638

van der Wal H, Sperber BLHMHM, Houweling-Tan B, Bakker RRCC, Brandenburg W, López-Contreras AM (2013) Production of acetone, butanol, and ethanol from biomass of the green seaweed Ulva Lactuca. Bioresour Technol 128:431–437

Van Hal JW, Huijgen WJJ, López-Contreras AM (2014) Opportunities and challenges for seaweed in the biobased economy. Trends Biotechnol 32:231–233

van Oirschot R, Thomas J-BE, Gröndahl F, Fortuin KPJ, Brandenburg W, Potting J (2017) Explorative environmental life cycle assessment for system design of seaweed cultivation and drying. Algal Res 27:43–54

Wahlström N, Harrysson H, Undeland I, Edlund U (2018) A strategy for the sequential recovery of biomacromolecules from Red Macroalgae Porphyra Umbilicalis Kützing. Ind Eng Chem Res 57(1):42–53

Wargacki AJ, Leonard E, Win MN, Regitsky DD, Santos CNS, Kim PB, Cooper SR, Raisner RM, Herman A, Sivitz AB et al (2012) An engineered microbial platform for direct biofuel production from brown macroalgae. Science 335:308–313

Wei N, Quarterman J, Jin Y-S (2013) Marine macroalgae: an untapped resource for producing fuels and chemicals. Trends Biotechnol 31(2):70–77

Xie EY, Liu DC, Jia C, Chen XL, Yang B (2013) Artificial seed production and cultivation of the edible brown alga Sargassum Naozhouense Tseng et Lu. J Appl Phycol 25(2):513–522

Yokoyama S, Jonouchi K, Imou K (2007) Energy production from marine biomass : fuel cell power generation driven by methane produced from seaweed. Int J Marine Environ Sci 1(4):320–323

Yuan Y, Macquarrie DJ (2015) Microwave Assisted step-by-step process for the production of fucoidan, alginate sodium, sugars and biochar from Ascophyllum nodosum through a biorefinery concept. Bioresour Technol 198:819–827

Zhang H, Liu Q, Cao Y, Feng X, Zheng Y, Zou H, Liu H, Yang J, Xian M (2014) Microbial production of sabinene–a new terpene-based precursor of advanced biofuel. Microb Cell Factories 13:20

Zilberman D, Lu L, Reardon T (2019) Innovation-induced food supply chain design. Food Policy, Elsevier 83(C):289–297

11 Integrated Bio-cycles System for Sustainable and Productive Tropical Natural Resources Management in Indonesia

Cahyono Agus

Abstract

Indonesia is known as the emerald of the equator which has high values of natural resources, although it also has a huge disaster risk. It's because it is located in strategic areas, namely, (1) the equator, (2) Ring of Fire, and (3) earth plates of Eurasia, Pacific, and Indo-Australia. Tropical ecosystem has high temperature, rainfall, moisture, light intensity, and rapid organic cycling along a year, so they have the highest biodiversity and net primary production in the world. As a part of the Ring of Fire, the land becomes more fertile because it is always supplied by new volcanic materials which contain a lot of important nutrients. The earth plate resulted in the accumulation of valuable mine minerals. Tropical natural resources have a high potential resource but still have less economical values, because it is still under mismanagement that is not based on natural norms. We have to shift paradigm from natural resource-based development to knowledge-based development. New paradigm from extraction to empowerment of natural resource will give new challenge to shift from red-green economic to blue economic concept that should be more smart, global, focused, and futuristic for sustainable development. An education of sustainable development system should be developing with a strong culture and values of humanity that educate the head, heart, and hand, respectively. Development of Integrated Bio-cycles System (IBS) in a closed-to-nature ecosystem will manage land resources (soil, mineral, water, air, microclimate) and biological resources (flora, fauna, human) that could have more high added value in environment, economic, socioculture, and health aspects. This integrated farming system has multifunction and multiproduct that produce food, feed, fiber, fertilizer, energy, water, oxygen, medicine, mystic, and tourism for sustainable and productive tropical natural resource management.

C. Agus (✉)
Faculty of Forestry, Universitas Gadjah Mada, Yogyakarta, Indonesia
e-mail: cahyonoagus@gadjahmada.edu

C. Keswani (ed.), *Bioeconomy for Sustainable Development*,
https://doi.org/10.1007/978-981-13-9431-7_11

Keywords
Integrated system · Knowledge-based education · Natural resources · Sustainable development · Tropical ecosystem

11.1 Introduction

Indonesia is the largest archipelagic state with 16,056 islands, located in Southeastern Asia between the Indian Ocean and the Pacific Ocean. With a total population of 261,890,900 as of 2017, it covers a total area of 1,916,862 km^2 (BPS 2018). Indonesia is the third most populous democracy in the whole world and has a tropical climate. Indonesia is known as the emerald of the equator which has high values of natural resources, although also has a huge disaster risk. It's because it is located in strategic areas, namely, (1) the equator, (2) Ring of Fire, and (3) earth plates of Eurasia, Pacific, and Indo-Australia.

Indonesia's natural resources are a universal inheritance that is obtained for free or gratis without anyone creating, making, planting, and maintaining and not being disturbed for millions of years. Wet tropical forest has the highest productivity in the world because based on its natural nature it has optimal climate and life cycle. Tropical ecosystem has high temperature, rainfall, moisture, light intensity, and rapid organic cycling along a year, so they have highest biodiversity and net primary production in the world. Agus (2018) reported that the high net primary production in moist tropical forest ecosystem was more supported by the rapid organic cycling than their low fertility weathered acid soil.

Indonesia is known as part of the Ring of Fire; nearly 40% of the world's geothermal potential is located in Indonesia. Suharmanto et al. (2015) reported that about 252 geothermal sites in Indonesia spread following the path of volcanic formation which stretches from Sumatra, Java, Nusa Tenggara, Sulawesi, to Maluku, with a total potential of about 27 GWe. Geothermal energy is a renewable energy and environmentally friendly; this large potential needs to upgrade the contribution to fulfill the domestic energy need which is able to reduce Indonesia's dependence on fossil energy sources which are depleting. As a part of the Ring of Fire, the land becomes more fertile because it is always supplied with new volcanic materials which contain a lot of important nutrients.

The earth plate resulted in the accumulation of valuable mine minerals. Rona (1973) reported that the site of economically important mineral deposits was found to have originated at a divergent tectonic plate boundary. The development of divergent plate boundaries may also create a habitat favorable for the accumulation of oil, a finding that would open immense possibilities for petroleum resources in the deep ocean basin (Rona 1973). Hydrothermal processes have concentrated the majority of known metallic sulfide ore bodies along convergent lithospheric plate boundaries originally at continental margins.

Potter et al. (2012) reported that net photosynthetic accumulation of carbon by plants, also known as net primary production (NPP), provides the energy that drives most biotic processes on Earth. NPP represents much of the organic matter that is consumed by microbes and animals. Climate controls on NPP fluxes are an issue of central relevance to society, mainly because of concerns about the extent to which NPP in managed ecosystems can provide adequate food and fiber for a growing human population. Potter et al. (2012) estimated that global terrestrial NPP increased by +0.14 Pg C over the time period of 2000–2009, due almost entirely to a strong upward trend in the Northern Hemisphere. Cramer et al. (1999) reported that predicted terrestrial NPP for the globe in 2009 was 50.05 Pg C, a total carbon flux in the middle of the range of previous vegetation NPP predictions of between 44 and 66 Pg C per year for the period 1982–1998.

Tropical ecosystems, because of their high productivity (871–1098 g C ha^{-1} $year^{-1}$) and extensive footprint on the Earth's surface (2377–3561 g C ha^{-1} $year^{-1}$), comprise nearly half of global NPP and GPP (Gough 2011). Temperate ecosystems and croplands are also a substantial fraction of global terrestrial primary production, accounting for roughly a quarter of global NPP and GPP. Global estimates of terrestrial NPP range from 48.0 to 69.0 Pg (= Petagrams or 1015 g) C $year^{-1}$, with global terrestrial GPP estimated at 121.7 Pg C $year^{-1}$ or approximately double global NPP on land.

Tropical natural resources have a high potential resource but still have less economical values, because it is still under mismanagement that is not based on natural norms (Agus 2018). Intensive deforestation and overexploitation of open-pit mining have become a major cause of land degradation and severe local-global environmental damages (Agus et al. 2018). The overexploitation of the natural resources in tropical ecosystem has caused a big problem on economic, environmental, and sociocultural aspect (Agus 2018).

Indonesia experienced an oil boom in the 1970s, a timber boom in the 1980s, and a coal minerals boom in the 2000s. Indonesia has three times the opportunity to exploit natural resources as the backbone of the country's economy, but it actually causes enormous environmental and sociocultural damage. Now, Indonesia should empower their human and natural resources for more sustainable development management. We have to shift paradigm from natural resource-based development to knowledge-based development. New paradigm from extraction to empowerment of natural resource will give new challenge to shift from red-green economic to blue economic concept that should be more smart, global, focused, and futuristic for sustainable development (Agus 2018).

Tropical rainforests are often called the "lungs of the planet" because they produce oxygen, which helps regulate carbon dioxide levels in the atmosphere (Agus 2018), and have a big effect on global warming. People in the world in the past few decades have put in place attention to Indonesia's tropical forests, especially to deeply prevent damage to Indonesia's tropical forests that have an impact on global warming (Widiaryanto 2018). The loss of tropical forests is thought to be equivalent to emissions excluded from burning fuel oil by developed countries. Maintaining tropical forests is also an important issue in achieving sustainable development goals or often referred to with sustainable development goals (SDGs). Globally, tropical forests are only 2% of the land but where 50% of plant species live in the world. Widiaryanto (2018) reported that Indonesia's forests face serious problems and proposed "redesign of Indonesian sustainable forestry management."

11.2 Natural Resource Management

Indonesia is endowed with rich natural resources, with the mining sector being one of the largest industries in the country. The nation's mineral resource exports are dominated by natural gas and crude petroleum. The key natural resources of Indonesia include silver, coal, fertile soils, natural gas, petroleum, gold, bauxite, tin, copper, timber, and nickel (World atlas 2019; Thomas 2012). The mining industry in Indonesia was 11.9% of its GDP, which is about USD 1.139 trillion as of 2011, with foreign investment in the mining sector exceeding USD 2.2 billion (Thomas 2012). However, most of the minerals are exported in a raw or semi-processed state to industrialized nations.

Indonesia has abundant mineral resources, including tin, gold, natural gas, coal, nickel, and copper. Silver, bauxite, and petroleum are also available in smaller quantities. Globally, Indonesia is a leading producer of copper, nickel, and gold, a leading exporter of liquefied natural gas (LNG), and the second largest producer of tin (Thomas 2012; World atlas 2019). Indonesia's chief industrial mineral is cement, and many cement companies in the country have planned expansions in order to increase the capacity of cement production.

Indonesia is among the leading producers of liquefied natural gas (LNG) in the world. The oil and natural gas accounted for over 90% of the USD 10 billion in exports in 1989, while the nation had 5.14 billion barrels of proven oil reserves (World atlas 2019). The nation boasts of 67.5 trillion cubic feet of established natural gas reserves within its borders and probable reserves estimated to be about 12 trillion standard cubic feet. In 2010, the Natuna natural gas project in Indonesia accounted for 25% of the country's overall gas reserves that are commercially recoverable.

Indonesia's Grasberg mine is the third largest copper mine in the world and the largest gold mine in terms of output and size (Thomas 2012). The nation's copper reserves are found around the Grasberg and Ersberg areas in Papua Island which are home to 15 billion tons of proven and probable reserves of copper. World atlas (2019) reported that the total production of gold was about 13,227 lbs in 1989 and 1,444,000 oz in 2009. Currently, the nation is the eighth largest producer of gold in the world with an annual output of about 100 tons as of 2013 (World atlas 2019).

Indonesia has an estimated 740 million tons of proven tin reserves and is a major exporter of the tin mineral to 31,500 tons for the year 1989 (Thomas 2012). In 2010, Indonesia was a leading global supplier of tin, and the country exported 85% of its tin production. In the same year, the production of tin decreased by 6.1% to about 43,000 t due to several reasons that include severe weather conditions that hit the mining operations in the country, illegal mining activities, and declining reserves (Thomas 2012).

Indonesia has an estimated 4.2 billion tons of proven coal reserves and an additional 12.9 billion tons in inferred reserves (World atlas 2019). In 2010, Indonesia produced about 257 Mt. of coal and was globally ranked as the second largest coal exporter (Thomas 2012). The country's leading coal-producing region is the Kalimantan Island, and the Central Kalimantan province contains about 1400 Mt. of high-quality metallurgical coal reserves. The production of coal experienced a decline in the 1970s due to the promotion of subsidized petroleum. The drop in production leads to efforts to encourage its use in local cement and electric plants, and by 1990, the total production had risen to 11 million tons (World atlas 2019).

Indonesia also has the region's largest exploitable tropical forests, namely, in Papua and Kalimantan, which support a healthy timber industry. The rainforests, which are the world's third largest, are home to 29,000 species of plants and 3000 species of animals (World atlas 2019). The timber industry has witnessed steady growth since the 1960s with legitimate and illegal loggers targeting specific tree species such as teak and meranti which is in high demand due to its reddish easily workable wood that is also considerably lightweight. Vennes and plywood that are produced locally are either consumed locally or exported. The excessive exploitation of forests has led to massive deforestation and substantial environmental degradation. The rate of environmental degradation is further accelerated by government-sanctioned conversion of tracts of forest into agricultural fields.

Despite being blessed with abundant amounts of natural resources, about 11.2% of Indonesia's population lives below the poverty line (World atlas 2019). The nation's level of development also lags behind compared to that of resource-poor Singapore, Taiwan, and South Korea. Academics around the world studying similar phenomenon across the world in resource-rich countries call this the "natural resource phenomenon." The phenomenon describes the contradictory relationship between a nation's economic performances and natural resource wealth.

11.3 Tropical Forest Land Use Changes

According to the UN Food and Agriculture Organization (FAO), the global forest area shrank by an annual average of 3.3 million hectares between 2010 and 2015, with most losses in the tropics (EURedd 2019). Forest loss and degradation affect forest-dependent communities, wild species, and the global climate. Deforestation and forest degradation are mainly driven by a switch to agriculture to satisfy a growing demand for a few commodities – mainly palm oil, soy, cattle (beef and leather) – and to respond to the demand for bioenergy (fuel-wood, charcoal).

Potter et al. (2012) reported that non-deforestation threats to tropical forest biodiversity can be considered in seven distinct categories: (a) selective extraction of plants, (b) selective extraction of animals, (c) biological invasion, (d) fragmentation, (e) climate change, (f) changing atmospheric composition, and (g) changing tree turnover rates. Each phenomenon merits serious consideration as an agent of change in the ecology of tropical forests. Nevertheless, the seven threats are extremely diverse and defy simple classification. Indeed, most of the seven threats are affected by other threats, including deforestation, in a complex and poorly understood web of feedback effects.

Palm oil is Indonesia's top exported commodity and has become a key economic driver for the country. However, it has also been often associated with the loss of Indonesia's valuable forests. Carlson et al. (2012) reported that from 1990 to 2010, almost 90% of lands converted to oil palm were forested (47% intact, 22% logged, 21% agroforests). By 2010, 87% of total oil palm area (31,640 km^2) occurred on mineral soils, and these plantations contributed 61–73% of 1990–2010 net oil palm emissions (0.020–0.024 GtC $year^{-1}$). Oil palm would then occupy 34% of lowlands outside protected areas. Allocated oil palm leases represent a critical yet undocumented source of deforestation and carbon emissions.

The oil palm is the world's most valuable oil crop and supplies >30% of world vegetable oil production. Oil palm produces the highest of oil per hectare (3.68 tonnes/ha/year) compared to rapeseed (0.59), sunflower seed (0.42), and soybean (0.36) (Corley and Tinker 2007). With palm oil production increasing by more than 50% in the last decade of the twentieth century and set to double in the next 20 years, it has never before been so important to understand the history, use, and cultivation of this fascinating crop.

Agus (2017) stated Indonesian tropical palm oil controversies, namely: (1) land use changes from tropical rainforests; (2) complete land clearing (3) land/forest fires; (4) monoculture; (5) intensive cultivation, open organic cycles; (6) aromatic organic matter, lignin materials, difficult to be decomposed; (7) global trade competition; (8) land degradation issues; (9) global warming issues; (10) sustainable management issues; (11) high economic values; and (12) driving the development of the region. We need to develop a closed-to-nature ecosystem that could manage tropical land resources (soil, mineral, water, air, microclimate) and biological resources (flora, fauna, human) that could have more high added value in environment, economic, socioculture, and health aspects.

11.4 Blue Bioeconomy

Grumbine (1994) points out ten important themes of ecosystem management to reach sustainability. These are hierarchy context, ecological boundaries, biological integrity, data collection, monitoring, adaptive management, interagency cooperation, organizational change, humans embedded in nature, and values. These themes cover all aspects of ecosystem management and indicate the need for sustainable resource management in the present as well as the future.

Education for Sustainable Development (ESD) empowers people to change the way they think and work toward a sustainable future. UNESCO aims to improve access to quality education on sustainable development at all levels and in all social contexts, to transform society by reorienting education and help people develop knowledge, skills, values, and behaviors needed for sustainable development. It is about including sustainable development issues, such as climate change and biodiversity into teaching and learning. Individuals are encouraged to be responsible actors who resolve challenges, respect cultural diversity, and contribute to creating a more sustainable world. UNESCO (2019) reported that with a world population of 7 billion people and limited natural resources, we, as individuals and societies, need to learn to live together sustainably. We need to take action responsibly based on the understanding that what we do today can have implications on the lives of people and the planet in the future.

MEXT (2019) stated that ESD is an education that fosters the leadership needed to build a sustainable society. Development of personal character, self-reliance, judgment, and responsibility is the kind of humanity to be fostered, as are individuals who value relationships and connectedness through an awareness of their relationships with other people, with society, and with the natural environment. Hence, ESD not only engages in activities that address various issues related to the environment, peace, human rights, and so on but does so in an interdisciplinary, holistic way that includes environmental, economic, societal, and cultural perspectives. Sinakou et al. (2018) stated that the most often chosen prioritized understanding of sustainable development according to which two or three of the dimensions of the concept (environment, society, economy) are seen as separated to each other and less often in an integrated way.

Pauli (2009) reported that oceans cover 72% of the surface of our blue planet and constitute more than 95% of the biosphere. He proposed "The Blue Economy" as a developing world that is relevant to all coastal states and countries with an interest in waters beyond national jurisdiction. The Blue Economy paradigm constitutes a sustainable development framework for developing countries addressing equity in access to, development of, and sharing of benefits from marine resources, offering scope for reinvestment in human development and the alleviation of crippling national debt burdens. The Blue Economy espouses the same desired outcome as the Rio+20 green economy initiative, namely, "improved human well-being and social equity, while significantly reducing environmental risks and ecological scarcities" (UNEP 2013), and it endorses the same principles of low carbon, resource efficiency, and social inclusion, but it is grounded in a developing world context and fashioned to reflect the circumstances and needs of countries whose future resource base is marine (Pauli 2009).

Pauli (2009) said that sustainable sourcing and usage of local raw materials promotes utilization of "blue" low energy options to realize efficiency and benefits as opposed to the business as usual "brown" scenario of high energy, low employment, and industrialized development models. The natural world is made up of the physical environment, its mineral components and biodiversity at all three levels (genetic, species, ecosystem) are intrinsically interconnected, and the more diverse and productive the natural system, the greater the degree of interconnectivity. The Blue Economy approach will set in place the policies, legislation, infrastructure, and incentives to facilitate the transition to a low-carbon economy utilizing all the tools at its disposal including the ocean's enormous potential for renewable energy (wind, wave, tidal, thermal, and biomass) generation.

Economic development strategies that are socially and environmentally sustainable have enjoyed growing interest (Serageldin 1996) in recent decades.

Fuentes-Saguar et al. (2017) reported that the bio-based economy will be crucial in achieving a sustainable development, covering all ranges of natural resources. The bioeconomy comprises several economic sectors, academic disciplines, and areas of policy. It encompasses the production of renewable biological resources and the conversion of these resources and waste streams into value-added products such as food, feed, bio-based products, and bioenergy. In this way, the bioeconomy is grouped into different sectors of the economy that produce, process, and reuse renewable biological resources (agriculture, forestry, fishing, chemicals, food, bio-based materials, and bioenergy). It is therefore of great interest to analyze the possible impacts of sectoral policies at national or regional level, as well as cross-sectoral policies (environment, climate change, the circular economy, waste, industrial policies, innovation, regional policies, etc.) related to dealing with new social challenges such as increasing food demand and climate change (Singh et al. 2016; Singh et al. 2019).

Etu (2019) stated that bioeconomy is the management and utilization of biological natural resources and equally leveraging upon these resources to create bio-based products and services. It is characterized by renewable bio-based natural resources, environmentally friendly clean technology, and the efficient recycling of materials. Finland's bioeconomy encompasses the following industries: agriculture, food, forestry, wood, paper, construction, chemical, pharmaceutical, renewable energy, water purification/distribution, as well as services including nature tourism, berries, hunting, and fishing.

The bioeconomy covers all sectors and systems that rely on biological resources – animals, plants, microorganisms, and derived biomass, including organic waste – as well as their functions and principles (European Commission 2018). It includes and interlinks land and marine ecosystems and the services they provide, all primary production sectors that use and produce biological resources (agriculture, forestry, fisheries, and aquaculture), and all economic and industrial sectors that use biological resources and processes to produce food, feed, bio-based products, energy, and services. Furthermore, European Commission (2018) reported five objectives from the original Bioeconomy Strategy: (1) ensuring food and nutrition security, (2) managing natural resources sustainably, (3) reducing dependence on nonrenewable resources, (4) mitigating and adapting to climate change, and (5) strengthening competitiveness and creating jobs.

The forest-based sector plays a central role in a bioeconomy: they provide material (wood and non-wood), bioenergy, and a wealth of other regulating and cultural ecosystem services (Wolfslehner et al. 2016). Bioeconomy development increases the interest in forest resources from many sides. First, bioeconomy increases the demand for forest goods and services and therefore also increases economic opportunities for the sector. This list of opportunities is long, including bioenergy, wood construction, packaging products, chemicals, textiles, etc. Second, there are growing requests for forest land for other ecosystem services, e.g., for biodiversity, carbon sequestration, recreation, and effects on human health. Bioeconomy Strategy has five main societal challenges which offer great potential and challenges for the forest-based sector, while not explicitly referring to it: (1) ensuring food security, (2) managing natural resources sustainably, (3) managing natural resources sustainably, (4) mitigating and adapting to climate change, and (5) creating jobs and increasing competitiveness.

11.5 Social Forestry

Forests cover almost one third of the world's land area, and nearly all are inhabited by indigenous and rural communities who have customary rights to their forests and have developed ways of life and traditional knowledge that are attuned to their forest environments (Chao 2012). These communities have been managing the environment through their own systems based on traditional knowledge, practices, rules, and beliefs for generations ("customary use"). Throughout Southeast Asia, some 140 million people are dependent on forests for their livelihoods and have developed their own systems of managing resources based on traditional knowledge, practices, rules, and beliefs for generations ("customary use") (Chao 2012).

Indonesia has an ambitious goal of giving forest-dependent communities access to 12.7 million hectares of forests through social forestry permits (CIFOR 2019). The massive project, launched in 2014, covers five different forest types, including community, community plantation, village, partnership, and customary forests. The program is also the government's way of addressing tenurial conflicts, bringing more justice into forest resource utilization. For local communities, full ownership rights and, in some areas, recognition of their customary institutions are key.

Social forestry program in Indonesia has entered a new era under President Joko Widodo. Community forestry or social forestry (henceforth referred collectively as SF) programs have become new modes of forest management empowering local managers and, hence, allowing integration of diverse local practices and support of local livelihoods. The government has been giving legal access to communities living in and around forests to manage forest resources through five management schemes: community forests, village forests, community plantation forests, partnership, and customary forests (MCA-Indonesia 2019). Through these schemes, community-based forest resources management can be legal and ensure its sustainability in the long run, which is one of the principles of sustainable forest management. Institutions which create and enforce property rights can control resource degradation and improve both economic and ecological efficiency (Poudel et al. 2016). Effective institutional arrangements can provide solutions to economic problems such as the tragedy of the commons, resource depletion, resource encroachment, illegal trading, etc. The government has granted forest management access of 1,053,477 ha for 239,341 families, up to August 2017 (MCA-Indonesia 2019).

11.6 Sustainable Agriculture

Brodt et al. (2011) stated that agriculture has changed dramatically since the end of World War II. Food and fiber productivity has soared due to new technologies, mechanization, increased chemical use, specialization, and government policies that favored maximizing production and reducing food prices. These changes have allowed fewer farmers to produce more food and fiber at lower prices. Sustainable agriculture integrates three main goals – environmental health, economic profitability, and social equity. A variety of philosophies, policies, and practices have contributed to these goals, but a few common themes and principles weave through most definitions of sustainable agriculture. An agroecosystem and food system perspective is essential to understanding sustainability. Agroecosystems are envisioned in the broadest sense, from individual fields to farms to ecozones. Food systems, which include agroecosystems plus distribution and food consumption components, similarly span from farmer to local community to global population.

Conservation of resources critical for agricultural productivity also means taking care of soil so that it maintains its integrity as a complex and highly structured entity composed of mineral particles, organic matter, air, water, and living organisms (Brodt et al. 2011). Farmers interested in long-term sustainability often prioritize caring for the soil, because they recognize that a healthy soil promotes healthy crops and livestock. Maintaining soil functioning often means a focus on maintaining or even increasing soil organic matter. Soil organic matter functions as a crucial source and sink for nutrients, as a substrate for microbial activity, and as a buffer against fluctuations in acidity, water content, contaminants, etc. Furthermore, the buildup of soil organic matter can help mitigate the increase of atmospheric CO_2 and therefore climate change. Another important function of soil organic matter is inducing a better soil structure, which leads to improved water penetration, less runoff, better drainage, and increased stability, thereby reducing wind and water erosion.

Due to a high reliance on chemical fertilizers, agroecosystem functioning has been disconnected from the internal cycling of key plant nutrients (Brodt et al. 2011). The recycling of nutrient (at the farm and regional scale), improving efficiencies of fertilizer applications, and relying on organic nutrient sources (animal and green manures) are important elements of sustainable agriculture. Recycling of nutrients is facilitated by a diversified agriculture in which livestock and crop production are more spatially integrated. Extensive mixed crop-livestock systems, particularly in developing countries, could significantly contribute to future agricultural sustainability and global food security (Brodt et al. 2011).

11.7 Integrated Bio-cycles Farming System

Our blue planet that consists of a blue ocean of 72% and a blue sky over 95% should be supported by blue earth. A red economy that is merely economic-oriented has resulted in environmental and life damage. A green economy is oriented toward environmental and healthy value, although it actually was expensive and dangerous. The concept of the blue economy was developed by Gunter Pauli of ZERI Foundation in 2009 (Agus 2018) through the acceleration of natural cycle processes with the empowerment of land (land, water, and mineral) and biological (plants, animals, and humans) resources with added value of economy, environment, socioculture, technology, and sustainable management (Table 11.1). The blue economy offers

Table 11.1 Key characteristics of various types of sustainable agricultural system

Low input/ integration	Organic farming	Biodynamic agriculture	Agroforestry	Integrated biocycle
Integration of advantageous natural processes	Integration of land, environment, and human health	Management of organisms that optimize quality of land, plants, animals, and human health	Integration of wood and herbal plants	Integration of agriculture and nonagriculture sectors
Adding environmental values	Natural fertilizer, Environmental values	Economic values	Environmental values	Value of environment, esthetics, and economics
Plant rotation	Plant rotation, diversification, and ideal space	Plant rotation, diversification, and ideal space	Spatial diversitas tipe crop	Rotation and diversity of plants
Impact of minimum land management	Adequacy of N through N fixation	Adequacy of N through N fixation, special preparation for improving land quality and living plants	Plant variation and pastoral system	Artificial and functional biotechnology, nanotechnology, and probiotics
The use of chemical fertilizer	Prohibition on treatment of plants and fertilizer	Prohibition on treatment of plants and fertilizer	Fertilization of agricultural plants, the use of cycle in forest plants	Management of closed organic cycle and integration of crop, moisture, nutrient, and pest management
The use of pesticide	Management of traditional animals	Management of traditional animals		Management of integrated bio-protection and ecosystem health management
General principle	Principle of grouping units	Principle of grouping units	General principle	Landscape ecological management, agropolitan concept
Specific management of plants	Specific management of plants	Specific management of plants	Specific management of plants	Specific management of plants
Semi-traditional	Natural	Integrated	Traditional	Holistic and integrated
Stockdale and Cookson (2003) and Chan (2006)	IFOAM (1998)	Koepft et al. (1976)	Stockdale and Cookson (2003)	Agus (2010, 2018)

efficient investment, increased innovation, increased funding, job creation, social capital development, and entrepreneurial stimulation. Performed by the utilization of waste and abandoned goods to be food, energy and employment thus turn poverty to be sustainable development and scarcity to be availability. The blue economy has provided new, creative, and innovative opportunities, clean and dignified.

Tropical bio-geo-resource has high biomass productivity but still has less economical values. Integrated Bio-cycle Farming System (IBFS) is an alternative system which harmoniously combines agricultural sectors and nonagricultural aspects, on landscape ecological management. The cycle of energy, organic matter and carbon, water, nutrient, production, crop, and money, was managed through 9R (reuse, reduce, recycle, refill, replace, repair, replant, rebuild, reward) to get optimal benefits for the farmer, community, agriculture, and global environment (Agus 2018). The system has multifunction and multiproduct (food, feed, fuel, fiber, fertilizer, pharmacy, edutainment, ecotourism). They will meet with the expected basic need for daily, monthly, yearly, and decade's income at short-, medium-, and long-term periods. IBFS was expected to provide additional benefits for farmers with small, medium, and big capital, through the recycling of organic waste into renewable resources to produce high-value production, such as organic fertilizer (liquid and solid), animal feed, and sources of biogas energy (Agus 2018).

Integrated Bio-cycle Farming System (IBFS) is an alternative system of agriculture which harmoniously combines agricultural sectors, such as agriculture, horticulture, plantation, animal husbandry, fisheries, and forestry with nonagricultural aspects, such as settlements, agroindustry, tourism, and industry, which are managed based on landscape ecological management under one integrated area. Agus (2018) developed IBFS through ICM (Integrated Crop Management), INM (Integrated Nutrient Management), IMM (Integrated Soil Moisture Management), and IPM (Integrated Pest Management). The system should collaborate and develop networking system between ABCG (academic, business, community, and government) with economic, environmental, and sociocultural approach as a characteristic of Education for Sustainable Development (Agus 2010). This model facilitates the learning needed to maintain and improve our quality of life and the quality of life for generations. It is about equipping individuals, communities, groups, businesses, and government to live and act sustainably, as well as giving them an understanding of the environmental, social, and economic issues involved. Integrated farming could support for better sustainable life and environment.

The IBFS has comprehensive characteristics compared to other integrated farming system, namely, low input farming, organic farming, biodynamic farming, and agroforestry system (Table 11.1). The key characteristics of IBFS developed in UGM University Farm are (1) an integration of agriculture and nonagriculture sector; (2) value of environment, esthetics, and economics; (3) rotation and diversity of plants; (4) artificial and functional biotechnology, nanotechnology, and probiotic; (5) management of closed organic cycle and integration in an integrated area among ICM, IPM, IMM, INM, and IVM; (6) management of integrated bio-protection and ecosystem health management; (7) landscape ecological management, agropolitan concept; (8) specific management of plant; and (9) holistic and integrated system (Agus 2010, 2018).

IBFS is expected to be one alternative solution for improving land productivity, program development, environmental conservation, and rural development in an integrated management (Agus 2018). They will meet with the expected basic need at short-, medium-, and long-term for food, clothing, and shelter. Thus, IBFS could provide income at daily, monthly, yearly, and decade's term for farmers. The role of micro-, meso-, and macroorganisms on biogeochemical and nutrient cycling in increasing of land productivity is very important. Microorganisms are able to provide essential nutrients to plants through both mutualistic symbiosis and nonsymbiosis. Agus et al. (2004) showed that the ability of N mineralization in tropical soil is three to five times higher than that available in the soil. Meanwhile, legume cover crops could supply nitrogen 9–27 times higher than that available in the soil (Agus et al. 2003). Biotechnology including bio-artificial and functional nanotechnology will greatly enhance the success of integrated biocycle farming in tropical region.

Geo-ecotourism offers the efficiency of investment, increased creative innovation, increased funding, job creation, social capital development, and stimulation of the socioentrepreneurship in community (Cahyanti and Agus 2017). Productive and conservative natural resource management will become the new phenomenal interested object of geo-ecotourism. Indonesia is a country rich in natural and cultural resources. The area of this country consists of more than 17,000 islands and small islands that stretch along 6400 km from west to east, and about 3000 km from north and south, and therefore naturally suggest the high diversities. Butarbutar and Soemarno (2013) reported that Indonesia suggests the high potencies of ecotourism attractions, including the nature ecosystems and its resources, nature biodiversity, and traditional values spread in various regions of archipelagoes. However, management of these ecotourism attractions is actually undeveloped in many tourism destinations. Its activities are really still limited to certain nature areas. Ecotourism is an industry that is based on the natural environment sustainability and the success of promoting ecotourism programs related to flora, fauna, and their ecosystems. The presence of ecotourism in the era of sustainable and tourism development mission should have minimum negative impacts, both on the environment resources and on sociocultural local values. Ecotourism activities were more oriented on the utilization of natural resources, the natural ecosystems, and have not been polluted yet.

Agus (2018) and Cahyanti and Agus (2017) developed the golden agro-production that intends to produce multiple products in an entity of land, and the products represent the real "gold" that has been ignored and given less value, namely, "brown gold" (timbers), "yellow gold" (grains rich of carbohydrate necessary for human life), and "black gold" (organic fertilizer, compost, etc.) in addition to "blue gold" (biomass and biogas energies), "green gold" (green vegetables, fodder, environment, temperature, and humidity), "white gold" (milk, fish, food), "red gold" (animal protein of cattle meat, pork, chicken, ducks, etc.), "transparent gold" (water for life and oxygen), "colorful gold" of herbal medicine, and "magic gold" of spiritual and tourism activities, that play very important role in maintaining human health and dignity human life.

IBFS was expected to provide additional benefits for farmers with small, medium, and big capital, through the recycling of organic waste into renewable resources to produce high-value production, such as organic fertilizer (liquid and solid), animal feed, and sources of biogas energy (Agus 2010). That will be a good prospect that organic farming can provide sustainable economic, environment, and sociocultural aspect.

11.8 Conclusion

Indonesia's natural resources are a universal inheritance that is obtained for free or gratis. Indonesia is located in a strategic area which causes high values of natural resources, although also has a huge disaster risk, namely, (1) the equator, (2) Ring of Fire, and (3) earth plates of Eurasia, Pacific, and Indo-Australia. Tropical ecosystem has the high temperature, rainfall, moisture, light intensity, and rapid organic cycling along a year, so they have highest biodiversity and net primary production in the world. Tropical natural resources have a high potential resource but still have less economical values, because it is still under mismanagement that is not based on natural norms. The overexploitation of the natural resources in tropical ecosystem has caused severe land degradation and environmental damages. We have to shift paradigm from natural resource-based development to knowledge-based development. New paradigm from extraction to empowerment of natural resource will give new challenge to shift from red and green economic to blue economic concept that should be more smart, global, focused, and futuristic for sustainable development. Development of Integrated Bio-cycles System (IBS) in a closed-to-nature ecosystem will manage land resources (soil, mineral, water, air, microclimate) and biological resources (flora, fauna, human) that could have more high added value in environment, economic, socioculture, and health aspects. This integrated farming system has multifunction and multiproduct that produce food, feed, fiber, fertilizer, energy, water, oxygen, medicine, mystic, and tourism for sustainable and productive tropical natural resource management.

References

Agus C (2010) Organic matter management in UGM University Farm for sustainable food, feed, fiber and fertilizer. In: Proceeding seminar on the 16th annual international sustainable development research conference, Hong Kong

Agus C (2017) Integrated bio-cycle system management on palm oil plantation. Keytalk paper on seminar oil palm: land quality and climate change. Faculty of Forestry UGM Yogyakarta, 19 September 2017

Agus C (2018) Development of blue revolution through integrated bio-cycles system on tropical natural resources management. In: Leal Filho W, Pociovălişteanu D, Borges de Brito P, Borges de Lima I (eds) World sustainability series: towards a sustainable bioeconomy: principles, challenges and perspectives. Springer, Cham, pp 155–172. https://link.springer.com/chapter/10.1007/978-3-319-73028-8_9

Agus C, Karyanto O, Hardiwinoto S, Haibara K, Kita S, Toda H (2003) Legume cover crop as a soil amendment in short rotation plantation of tropical forest. J For Environ 45(1):13–19

Agus C, Karyanto O, Kita S, Haibara K, Toda H, Hardiwinoto S, Supriyo H, Na'iem M, Wardana W, Sipayung M, Khomsatun, Wijoyo S (2004) Sustainable site productivity and nutrient management in a short rotation *Gmelina arborea* plantation in East Kalimantan, Indonesia. New For J 28:277–285

Agus C, Primananda E, Faridah E, Wulandari D, Lestari T (2018) Role of arbuscular mycorrhizal fungi and *Pongamia Pinnata* for revegetation of tropical open-pit coal mining soils. Int J Environ Sci Technol (IJEST) 15(11):1–11. https://doi.org/10.1007/s13762-018-1983-5

BPS (2018) Statistical yearbook of Indonesia 2018. BPS Statistik Indonesia, Jakarta. 719 pp

Brodt S, Six J, Feenstra G, Ingels C, Campbell D (2011) Sustainable agriculture. Nat Educ Knowl 3(10):1

Butarbutar R, Soemarno S (2013) Environmental effects of ecotourism in Indonesia. J Ind Tour Dev Std 1(3):97–107

Cahyanti PAB, Agus C (2017) Development of landscape architecture through geo-eco-tourism in tropical karst area to avoid extractive cement industry for dignified and sustainable environment and life. IOP Conf Ser: Earth Environ Sci 83:012028

Carlson KM, Curran LM, Asner GP, Pittman AMD, Triggand SN, Adeney JM (2012) Carbon emissions from forest conversion by Kalimantan oil palm plantations. Nat Clim Chang. https://doi.org/10.1038/nclimate1702. (2012)

Chan GL (2006) Integrated farming system. http://www.scizerinm.org/chanarticle.html. Access 5 Sept 2012

Chao S (2012) Forest peoples: numbers across the world. Forest Peoples Program, Moreton-in-Marsh. http://www.forestpeoples.org/sites/fpp/files/publication/2012/05/forest-peoplesnumbers-across-world-final_0.pdf

CIFOR (2019) Taking stock of Indonesia's social forestry program. https://forestsnews.cifor.org/58344/taking-stock-of-indonesias-social-forestry-program?fnl=en. Access 24 Jan 2019

Corley RHV, Tinker PB (2007) The oil palm, 4th edn. Blackwell Science Ltd., Malden. 578 pp. https://doi.org/10.1002/9780470750971

Cramer W, Kicklighter DW, Bondeau A, Iii B, Moore CG, Nemry B, Ruimy A, Schloss AL (1999) Comparing global models of terrestrial net primary productivity (NPP): overview and key results. Glob Chang Biol 5:1–15

Etu K (2019) Bio economy. http://kainuunetu.fi/bio-economy. Access 23 Jan 2019

EuRedd (2019) Deforestation and forest degradation. http://www.euredd.efi.int/deforestation. Access 2 Feb 2019

European Commission (2018) Bioeconomy: the European way to use our natural resources Action plan 2018. Brussels, 26 pp

Fuentes-Saguar PD, Mainar-Causapé AJ, Ferrari E (2017) The role of bioeconomy sectors and natural resources in EU economies: a social accounting matrix-based analysis approach. Sustainability 9:2383. https://doi.org/10.3390/su9122383

Gough CM (2011) Terrestrial primary production: fuel for life. Nat Educ Knowl 3(10):28

Grumbine RE (1994) What is ecosystem management? Conserv Biol 8(1):27–38. https://doi.org/10.1046/j.1523-1739.1994.08010027.x

IFOAM (International Federation of Organic Agriculture Movements) (1998) Basic standard for organic production and processing. IFOAM, Tholey-Thelei

Koepft HH, Pettersson DD, Schaumann DD (1976) Biodynamic agriculture. Anthroposophic Press, Spring Valey

MCA-Indonesia (2019) New Era of social forestry: for people's welfare. http://www.mca-indonesia.go.id/en/our_news/news/new_era_of_social_forestry_for_peoples_welfare-1212. Access 24 Jan 2019

MEXT (2019) ESD (Education for Sustainable Development). http://www.mext.go.jp/en/unesco/title04/detail04/sdetail04/1375695.htm. Access 24 Jan 2019

Pauli G (2009) Blue economy concept paper. The ZERI Foundation, Singapore, 13 pp

Potter C, Klooster S, Genovese V (2012) Net primary production of terrestrial ecosystems from 2000 to 2009. Clim Chang 115:365. https://doi.org/10.1007/s10584-012-0460-2

Poudel KL, Johnson TG, Tewari R (2016) Property rights and sustainable natural resource management. Environ Manag Sustain Dev 5(2):30–40. https://doi.org/10.5296/emsd.v5i2.9304
Rona PA (1973) Plate tectonics and mineral resources. Sci Am 229(1):86–95
Serageldin I (1996) Sustainable development: from theory to practice. Finan Dev 33(4):3
Sinakou E, Pauwa JB, Goossens M, Van Petegemc P (2018) Academics in the field of education for sustainable development: their conceptions of sustainable development. J Clean Prod 184:321–332. https://doi.org/10.1016/j.jclepro.2018.02.279
Singh HB, Jha A, Keswani C (eds) (2016) Intellectual property issues in biotechnology. CABI, Wallingford. 304 pages, ISBN-13: 9781780646534
Singh HB, Keswani C, Singh SP (2019) Intellectual property issues in microbiology. Springer-Nature, Singapore, 425 pages. ISBN: 9789811374654
Stockdale EA, Cookson WR (2003) Sustainable farming systems and their impact on soil biological fertility-some case studies. In: Abbott LK, Murphy DV (eds) Soil biological fertility. A key to sustainable land use in agriculture. Kluwer Academic Publisher, Dordrecht, pp 225–2392003
Suharmanto P, Fitria AN, Ghaliyah S (2015) Geothermal energy potential as source of alternative energy power plant. Renewable energy and energy conversion conference and exhibition (The 2nd Indo EBTKE-CONEX 2013). KnE 1:119–124. https://doi.org/10.18502/ken.v1i1.325
Thomas GP (2012) Indonesia: mining, minerals and fuel resources. AZOmining. https://www.azomining.com/Article.aspx?ArticleID=101
UNEP (2013) Green economy definition. http://www.unep.org/greeneconomy/AboutGEI/
UNESCO (2019) Education for sustainable development. https://en.unesco.org/themes/education-sustainable-development. Access 24 Jan 2019
Widiaryanto P (2018) Suing foresters creating redesign forestry development towards Indonesia 2045. Indonesian forest glory discussion paper at the anniversary of the faculty of forestry UGM Yogyakarta, October 26, 2018 (in Indonesian)
Wolfslehner B, Linser S, Pülzl H, Bastrup-Birk A, Camia A Marchetti M (2016) Forest bioeconomy – a new scope for sustainability indicators. From science to policy 4. European Forest Institute, Joensuu Finland
World Atlas (2019) What are the major natural resources of Indonesia?. https://www.worldatlas.com/articles/what-are-the-major-natural-resources-of-indonesia.html. Access 19 Feb 2019

Biosynthesized Secondary Metabolites for Plant Growth Promotion

12

April S. Gislason, W. G. Dilantha Fernando, and Teresa R. de Kievit

Abstract

By 2050, the global population is predicted to expand to 9.8 billion people, requiring 70% more food than we are consuming today. At the same time, crop losses are increasing due to resistance to pesticides and restrictions on the use of these products because they are harmful to humans and the environment. The need for alternative and sustainable crop management strategies has prompted interest in plant growth-promoting pseudomonads and the secondary metabolites they produce. These strains, however, have not been widely embraced by the farming community because they have failed to yield consistent results in the field. To take advantage of the plant growth-promoting abilities of pseudomonads, there has been increased interest in the direct application of the biosynthesized secondary metabolites. This review describes the benefits and difficulties associated with pseudomonad-based bioinoculants and recent advances toward commercial application of biosynthetic secondary metabolites from these organisms.

Keywords

Plant growth promotion · Pseudomonads · Secondary metabolites

A. S. Gislason · T. R. de Kievit (✉)
Department of Microbiology, University of Manitoba, Winnipeg, MB, Canada
e-mail: Teresa.Dekievit@umanitoba.ca

W. G. Dilantha Fernando
Department of Plant Science, University of Manitoba, Winnipeg, MB, Canada

C. Keswani (ed.), *Bioeconomy for Sustainable Development*,
https://doi.org/10.1007/978-981-13-9431-7_12

12.1 Introduction

The effects of global warming and increased resistance to currently used antimicrobials, pesticides, and herbicides have been taxing on food crop productivity (Lamberth et al. 2013; Sparks and Nauen 2015; Singh et al. 2014, 2017; van den Bosch et al. 2011). Meanwhile, the global population is increasing, and it is predicted that food requirements will exceed production limits by 2050 (Godfray et al. 2010). As a result, there is an urgent need to find ways to increase food crop capacity worldwide. Conventional methods to maintain crop productivity include the application of synthetic chemicals to provide nutrients and inhibit phytopathogens, as well as the use of cultivars that are resistant to biotic and abiotic stresses (Keswani et al. 2016; Ram et al. 2018; Singh et al. 2016a, b). However, resistant cultivars are not always available, leading to the heavy reliance on the use of pesticides (Tilman 1998). This practice is not sustainable, and concern over pesticide use is growing as they can contaminate the ecosystem, exerting detrimental effects on human, animal, and plant health for decades (Pimentel and Levitan 1986; Arias-Estévez et al. 2008). As a result, regulatory constraints on the use of these chemicals have been put in place by governmental agencies in many countries.

Despite such regulations, pesticides currently approved for use in agriculture have been shown to contain toxic oxidized petroleum distillates and heavy metal pollutants (Defarge et al. 2018; Seralini 2015). In addition to health and environmental concerns, the use of conventional agrochemicals decreases biodiversity in the rhizosphere, resulting in the loss of the microbial-associated functional groups (Giller et al. 1997; Tsiafouli et al. 2015) resulting in reduced soil fertility (Tilman 1998; Glover-Amengor and Tetteh 2009; Aktar et al. 2009). This is concerning because a biologically diverse rhizosphere is more resistant to change in the face of stressors such as drought, soil salinity, extreme temperatures, and invasion by pathogens (Giller et al. 1997; Nelson and Spaner 2010; Brussaard et al. 2007).

The application of agrochemicals also diminishes the genotypic traits responsible for beneficial plant-microbe interactions. A notable example is that long-term fertilization with nitrogen changes soil functionality by altering the composition of the bacterial rhizosphere community (Ramirez et al. 2012; Bai et al. 2010). These changes lead to the evolution of rhizobia that are less mutualistic (Weese et al. 2015). Agrochemical application not only reduces the viability of plant-beneficial rhizobacteria (Singh and Wright 2002) but can also inhibit their plant growth-promoting capabilities (Madhaiyan et al. 2006; Parvaze et al. 2005; Ahemad and Khan 2009). The ubiquitous need for chemical supplements in agriculture is understandable considering that plant domestication coincided with use of biologically homogenous soils containing chemical amendments. The development of high-yield crops has allowed for an increased dependency on these amendments for plants to survive abiotic and biotic stressors (Matson et al. 1997). The result is a vicious cycle, where the loss of functional groups within the soil causes a greater need for fertilization and the use of pesticides and herbicides. Alternate sustainable agricultural practices are needed such that the growing food demand can be met in an eco-friendly manner (Keswani et al. 2014; Keswani 2015; Singh et al. 2019a, b).

12.2 Plant Growth-Promoting Rhizobacteria

It has been known for decades that certain soils are capable of controlling fungal and bacterial diseases of plants. It was later determined that the presence of plant growth-promoting rhizobacteria (PGPR) suppresses plant diseases and promotes plant growth (Kloepper and Schroth 1980). PGPRs are indigenous soil microbes, representing approximately 2–5% of bacteria present in the rhizosphere (Antoun and Prévost 2005). PGPRs promote plant growth directly by facilitating increased nutrient assimilation by the plant, producing exogenous phytohormones that stimulate plant growth, and increasing resistance to biotic and abiotic stressors by inducing systemic resistance (ISR) (Glick 2012; Olanrewaju et al. 2017). Indirect PGP is mediated through a process known as biocontrol. In this case, bacteria inhibit pathogens that cause plant disease by competing for nutrients, occupying niches to the exclusion of pathogens, and producing inhibitory secondary metabolites (SMs) (Glick 2012; Olanrewaju et al. 2017).

12.3 Pseudomonads as Plant Growth-Promoting Rhizobacteria

Of all the bacteria in the rhizosphere, pseudomonads are typically enriched compared to bulk soil and are associated with diverse plant species (Bulgarelli et al. 2012; Coleman-Derr et al. 2016; Haney et al. 2015; Mendes et al. 2011; Niu et al. 2017). *Pseudomonas* species are the most studied PGPRs and are considered to be one of the most effective biological control agents (Kloepper and Schroth 1980; Haas and Défago 2005). These bacteria effect their biocontrol activity through the production of diverse SMs, including pyrrolnitrin (PRN), phenazines (PHZ), 2,4-diacylphloroglucinol (DAPG), pyoluteorin (PLT), hydrogen cyanide (HCN), exoprotease, chitinase and siderophores, exopolysaccharides, and lipopeptides (O'Sullivan and O'Gara 1992; Sarniguet et al. 1995; Haas and Keel 2003; Nandi et al. 2015, 2016a; Selin et al. 2012; Berry et al. 2010). Pseudomonads also play an important role in plant growth promotion by producing phytohormones and SMs that increase nutrient acquisition by the plant (Meyer and Abdallah 1978; Cox et al. 1981; Gügi et al. 1991; Burini et al. 1994; Richardson and Hadobas 1997; Mandal and Kotasthane 2014; Egamberdiyeva 2005). In addition to direct inhibition of bacterial, fungal, insect, and nematode phytopathogens, pseudomonads provide crop protection by inducing systemic resistance (ISR) (Haas and Défago 2005; Bakker et al. 2007).

ISR is mediated through plant reception of bacterial surface structures or bacterial exo-products (Bakker et al. 2007; Jankiewicz and Kołtonowicz 2012). Plant growth and defense against phytopathogens are negatively correlated due to regulatory cross talk between the two processes (Denancé et al. 2013; Huot et al. 2014). Thus, plants have evolved sophisticated methods to balance growth and defense where perception of ISR-eliciting signals induces the plant to become primed (Herms and Mattson 1992; Baldwin 2001; Walling 2009). These priming stimuli trigger direct changes in the plant that are required for enhanced defense while maintaining a low fitness cost (Conrath et al. 2006; Balmer et al. 2015; Pieterse et al. 2014). Priming can involve changes in physiology, transcription, metabolism, and chromatin, which result in an enhanced response by the plant when threatened (Mauch-Mani et al. 2017).

In addition to protecting against pathogens, ISR can shield plants from abiotic factors (Beckers and Conrath 2007; Theocharis et al. 2012), and conversely, abiotic factors can induce resistance to biotic stresses (Frost et al. 2008; Engelberth et al. 2003; Ramarathnam et al. 2011; Bakker et al. 2007). While ISR can provide protection against multiple pathogens, the effect of ISR is often dependent on the bacterial strain, plant, phytopathogen, and the presence of abiotic stressors (Pieterse et al. 2014; Haney et al. 2015, 2018; Haney and Ausubel 2015). Since biotic and abiotic stresses are perceived in parallel, plants that are "primed" could unnecessarily invoke defense responses when triggered by these signals, resulting in fitness cost to the plant (Mauch-Mani et al. 2017). This might contribute to the variability associated with PGP in field applications of pseudomonads. While loss of microbial viability is believed to be the underlying cause, failure to produce the essential PGP factors could also be involved.

12.4 Regulation of Secondary Metabolites in Pseudomonads

Originally, PGPR research focused on identifying the exo-metabolites produced by PGPR. Much is now known about the mechanisms employed by PGPR strains, and attention has shifted to analyzing the regulation and the mechanism of action of the SMs within the context of the ecosystem (Glick 2012). Pseudomonad PGPR are capable of expressing a diverse set of SMs, the expression of which is dependent on specific cues encountered in each environment. The producer senses plant root exudates and soil composition, as well as competitors and cooperators present in the surrounding environment, and SMs are expressed accordingly.

Since devoting energy and precursors to the synthesis of SMs is energetically costly, their expression is not always economical; as such, mechanisms are in place to tightly control their expression. In general, SM expression in pseudomonads is dictated by the interplay between a major regulatory network, including the Gac (global activator of antibiotic and cyanide) system and other regulatory elements (Kay et al. 2005; Laville et al. 1992; Chancey et al. 1999). To achieve repression of SMs, regulator of secondary metabolites (RsmA) proteins bind to an obscure ribosomal binding site of mRNAs involved in SM synthesis, preventing translation (Lapouge et al. 2007). Relief of this repression is achieved by induction of the Gac two-component system (Heeb and Haas 2001), which includes a membrane-bound sensor kinase, GacS, and a cytoplasmic response regulator, GacA. Upon activation by an unknown signal, GacS undergoes autophosphorylation and phosphotransfer to GacA, which can then activate transcription of the noncoding small RNAs (sRNAs), RsmX, RsmY, and RsmZ (Heeb et al. 2002). These sRNAs titrate out the RsmA proteins, allowing expression of mRNAs involved in SM biosynthesis. Two orphan sensor kinases, RetS and LadS, also interact with the Gac system, further modulating the expression of SMs (Goodman et al. 2004, 2009; Ventre et al. 2006; Workentine et al. 2009). In addition, SM expression is mediated through the activity of the alternative sigma factor RpoS, in response to the phase of growth, nutrient limitation, or stress (Sarniguet et al. 1995; Jorgensen et al. 1999; Venturi 2003; Suh et al. 1999). The efficiency in which RpoS binds to RNA polymerase is mediated by the alarmone, ppGpp (Potrykus and Cashel 2008), which is produced after activation to the stringent response in response to growth arrest or nutrient limitation (Kvint et al. 2000).

The expression of SMs is also subject to population-density-based regulation by acyl-homoserine lactone (AHL)-based quorum sensing (QS), such as the PhzI/R system (Selin et al. 2012; Fuqua et al. 1994; Bassler 2002). This is not always the case, however, as some pseudomonads, such as *P. protegens* CHA0, do not possess AHLs (Whistler et al. 1998; Bull et al. 2001) and there is no change in antifungal activity in an AHL-deficient strain of *P. brassicacearum* DF41, which uses QS (Berry et al. 2014). Cross talk between the regulators themselves is common, further complicating our understanding of SM control (Haas and Défago 2005). In the biocontrol strain *P. chlororaphis* PA23, novel regulators, the signals they perceive and how they contribute to PGP continue to be identified, highlighting the complexity of SM regulation existing in a single strain (Klaponski et al. 2014; Shah et al. 2016; Nandi et al. 2016b; Poritsanos et al. 2006; Selin et al. 2012, 2014; Manuel et al. 2012).

The environmental signals perceived as well as the genetic composition of a particular strain dictate the ability to colonize and express SMs (Lynch 1990; Haney and Ausubel 2015). Intriguingly, the responses to environmental signals within these regulatory networks can be divergent, even between closely related strains. For example, in *P. chlororaphis* strains PA23, 30–84, and O6, PHZ production is positively regulated by the Gac system (Kang et al. 2007; Poritsanos et al. 2006; Chancey et al. 1999), while in *Pseudomonas aeruginosa* M18, PHZ is repressed by Gac (Ge et al. 2004). Additionally, RpoS can regulate the expression of SMs positively and negatively or have no effect, depending on the strain and SM in question (Heeb et al. 2005; Sarniguet et al. 1995; Whistler et al. 1998; Ge et al. 2006; Girard et al. 2006; Park et al. 2011). Further complicating matters, the production of different SMs is often coordinated, where the production of one represses the synthesis of another. In *P. protegens* CHA0, 2,4-DAPG and PLT each positively regulates their own biosynthesis but represses the synthesis of the other (Schnider-Keel et al. 2000; Baehler et al. 2005). SMs can also regulate the production of another SM indirectly. In *P. chlororaphis* PA23, inactivating the PHZ biosynthetic enzyme, PhzC, results in a lack of negative feedback of the shikimate pathway, which supplies the precursor for PHZ biosynthesis. Since the shikimate pathway is involved in PRN biosynthesis, production of this antibiotic is increased (Selin et al. 2010). Presumably this enables the cell to dedicate resources to the production of the SM that gives the organism a competitive advantage in that context.

Because the mechanisms of adaptation that have evolved over time are specific to the environment from which the PGPR was isolated, it is understandable that the SM profile will vary in response to the abiotic and biotic signals encountered

(Callaway and Maron 2006). This is a complex situation considering the amounts different allelochemicals produced by competitors, predators, and plant hosts. In general, the success of a PGPR is dependent on how closely the application environment resembles the origin of isolation (Lynch 1990; Glick et al. 2007; Suprapta 2012).

Phase variation is an additional issue for the application of pseudomonas as PGPR. Phase variation is a natural phenomenon that occurs in pseudomonads in the rhizosphere due to spontaneous mutation in the Gac two-component system (van den Broek et al. 2005a; Chancey et al. 2002). In pseudomonads, mutations in either *gacS* or *gacA* repress SMs involved in biocontrol (van den Broek et al. 2003; Hassan et al. 2010; Poritsanos et al. 2006; Berry et al. 2010), inhibit biofilm production, and also increase expression of negatively regulated SMs, such as siderophores (Nandi et al. 2016a; Poritsanos et al. 2006). Gac^- mutants can accumulate in high numbers when *Pseudomonas* species are cultured in nutrient-rich media (Song et al. 2016; Duffy and Défago 2000; Bull et al. 2001; Chancey et al. 2002; van den Broek et al. 2005a). A reduced metabolic load likely provides a growth advantage to Gac^- mutants (van den Broek et al. 2005a). While concerns about the decreased efficacy of bioinoculants containing a high proportion of Gac^- mutants were raised, the rate of mutation can be controlled through mineral supplementation or culture dilution (Duffy and Défago 2000). Intriguingly, two *Pseudomonas* strains have been shown to be capable of switching back to the wild-type phenotype (Han et al.1997; van den Broek et al. 2005b). Whether this mechanism is employed by pseudomonads in the rhizosphere or is strain-specific is not known.

12.5 Biosynthesized SMs

Previous studies provide evidence for the utility of applying SMs directly as single formulations, as complex mixtures, or in combination with PGPR. Initially, supernatants and crude extracts of PGP were found to protect plants against fungal pathogens, prevent predation, and promote plant growth, demonstrating the utility of SM application for PGP (Sayyed and Patel 2011; Jousset et al. 2006; Ongena et al. 2005; Kloepper et al. 1980b). More recent findings indicate that the addition of microbial SMs to PGPR formulations enhances their performance (Morel et al. 2015; Marks et al. 2013). With this in mind, the global microbial product market is predicted to be worth over USD 250 billion by 2023, with a compound annual growth rate of approximately 8.7% (Market Research Future 2016). Although large-scale production of microbial SMs has dominated in the pharmaceutical, biotechnological, clinical, and academic industry sectors, it represents a much smaller market share in the agricultural sector. This is despite the fact that the utility of natural products in agriculture has been well established and many commercial agrochemicals are natural products or synthetic chemicals based on natural products (Sparks et al. 2017). Typically, small companies and academia have been driving bioformulation development. Recently, however, many of these niche bioformulation companies have been acquired by larger companies, leading to rapid growth of the biostimulant and biopesticide market (Agrow Biologicals 2018). Globally, the sale of biostimulants is predicted to reach USD 3 billion by 2021 with an annual growth rate of 10–12% (Agrow Biostimulants Report 2017). Similarly, sales of biopesticides, which represent the fastest-growing sector with an annual growth rate of 16–17%, are projected to increase from USD 3 billion in 2018 to USD 11 billion by 2025 (Agrow Global Biopesticide Regulations 2018).

Although the market for bioformulations is improving, there is still reluctance by growers to implement these products due to the cost of repeated application, shorter shelf life, and fear of inconsistent results in the field. As mentioned earlier, successful PGP in the field can be hampered by reduced viability (particularly for Gram-negative bacteria) and fluctuating SM expression by the producing organism. To circumvent these shortcomings, direct application of biosynthesized SMs, individually or in combination, provides a way to take advantage of bacterial PGP characteristics directly. Recent advances in biosynthetic platforms and instrumentation for detection and purification have made biosynthesis of microbial SMs a realistic objective (Nikel et al. 2014; Pandey et al. 2001; Robinson et al. 2001). Furthermore, technological advances in formulations and chemical modifications can increase the activity of biosynthesized SMs deemed too unstable for field application (Wang et al. 2011; Zhang et al. 2010; Chen et al. 2011). A number of bacterially synthesized SMs have been commercialized as eco-friendly bioformulations, including pesticides such as fenpiclonil and fludioxonil (Ciba-Geigy) (Ligon et al. 2000) and biostimulants such as lipochitooligosaccharides (Ratchet™, Torque™; Monsanto BioAg, http://www.monsantobioag.com). While bacterially synthesized SM formulations currently represent a small fraction of the market share, they are becoming more attractive to users. The existing biosynthesized agrochemical market consists mainly of plant extracts, for which an inconsistent supply chain is problematic (Dunham Trimmer 2017). Moreover there is increasing demand on growers to implement more sustainable, green alternatives to chemicals, which cannot always be easily replaced with conventional bioformulations. A notable example is surfactants, which are widely used in agriculture as a component of fertilizer and micronutrient formulations, biostimulant and crop protection agents, as well as herbicide spray solutions. Synthetic surfactants have been the main contributor to the agricultural surfactants market, which is predicted to be worth USD 1.88 billion by 2022 (Markets and Markets 2016). However, increasing restrictions have limited the growth of the synthetic surfactant market (Markets and Markets 2016), which could provide a niche for the implementation of microbially biosynthesized surfactants.

Since large-scale production of many bacterial SMs is already operational in other industries, utilization of these resources by the agriculture sector seems inevitable. A void in the agricultural market space due to decreased production of synthetic chemicals, along with more amenable registration guidelines for bioformulations (Dunham Trimmer 2017), is likely to result in biosynthesized SMs coming to the forefront of the market. Pseudomonads, in particular, have long been considered a rich source of SMs (Ligon et al. 2000). In the following sections, recent advances toward the commercialization of SMs produced by PGP pseudomonads are described.

12.6 Secondary Metabolites from Pseudomonads

12.6.1 Pyrrolnitrin

The successful commercialization of PRN derivatives is an archetypical example highlighting the potential of biosynthesized SMs involved in PGP. PRN was first discovered as an SM of *P. pyrrocinia* over half a century ago (Arima et al. 1964). Since then, it has been identified as an SM of numerous other pseudomonads (Sarniguet et al. 1995; Haas and Keel 2003). The production of PRN is considered to play a key role in the biocontrol activity of pseudomonads against a broad spectrum of agriculturally important fungi such as *Botrytis cinerea*, *Rhizoctonia solani*, and *Sclerotinia sclerotiorum* (Fernando et al. 2005a). In the biocontrol strain *P. chlororaphis* PA23 which is capable of producing numerous SMs, PRN is the main compound implicated in the antagonism of *S. sclerotiorum* (Selin et al. 2010).

Although biosynthesized PRN had activity against fungal phytopathogens in greenhouse assays (Howell and Stipanovic 1979), it was not considered suitable for use in agriculture due to its degradation by sunlight. However, two stable PRN derivatives, fenpiclonil and fludioxonil, which displayed good efficacy and environmental safety were developed by Ciba-Geigy as agricultural fungicides (Ligon et al. 2000). The use of these products in seed and foliar sprays for over three decades has resulted in the development of negligible resistance (Kilani and Fillinger 2016). Additionally, resistance mechanisms developed under laboratory conditions are associated with fitness defects in the pathogen, translating into a lack of resistance in the field (Kilani and Fillinger 2016).

12.6.2 Phenazines

PHZs are nitrogen-containing heterocyclic pigments that display a broad spectrum of activity against bacteria, fungi, and plants (Mavrodi et al. 2006). The diverse characteristics of PHZs result from substitution of functional groups on two benzene rings flanking the central nitrogen moiety (Buus Laursen and Nielsen 2004). The production and regulation of PHZs are best studied in pseudomonads (Pierson and Pierson 2010). The PHZ, phenazine-1-carboxylic acid (PCA), is the precursor for the production of other PHZ compounds, including phenazine-1-carboxamide (PCN), 1-hydroxyphenazine (1-OH-PHZ), 2-hydroxyphenazine (2-OH-PHZ), pyoverdine or pseudobactin (PVD), pyocyanin (PYO), and aeruginosin A and B (Mavrodi et al. 2001; Hassan and Fridovich 1980; Kerr et al. 1999; Herbert and Holliman 1969). In pseudomonads, the biosynthesis of PCA is dependent on the presence of precursors shunted from a branch point in the shikimate pathway and the highly conserved biosynthetic operon, *phzABCDEFG* (Haas and Keel 2003; Chin-A-Woeng et al. 2003; Mavrodi et al. 1998). Additional genes linked to the biosynthetic cluster are responsible for the production of PHZ derivatives using PCA as a precursor (Mavrodi et al. 2001; Delaney et al. 2001; Chin-A-Woeng et al. 2001).

In the rhizosphere, biocontrol, environmental fitness, and biofilm formation are all associated with PHZ production (Maddula et al. 2008; Selin et al. 2010). PHZs may also promote plant growth through the solubilization of iron in the soil (Hernandez et al. 2004). PHZs are known inhibitors of the agriculturally important pathogenic fungi *Gaeumannomyces graminis* var. *tritici* and *Fusarium oxysporum* but have been shown to have no effect on *S. sclerotiorum* (Thomashow and Weller 1988; Selin et al. 2010; Chin-A-Woeng et al. 1998).

Biosynthesized PHZ derivatives are more effective biocontrol agents than chemically synthesized PHZs (Nansathit et al. 2009). Yet, the large-scale production of PHZs for commercial applications has been restricted due the low yield from wild-type strains and the chemical instability of products (Schisler and Slininger 1997; Cheluvappa 2014). This is no longer the case, as genetic manipulation of the regulatory elements of *Pseudomonas* strains has generated PCA-, 2-OH-PHZ-, and PCN-overproducing mutants (Du et al. 2013; Liu et al. 2016; Bilal et al. 2017; Yao et al. 2018). Shenqinmycin™, a commercial PCA derivative biosynthesized from *Pseudomonas*, was marketed in China in 2011 (Huang et al. 2004; He and Xu 2011). Shenqinmycin™ is a "green," environmentally friendly formulation that is effective for use against *Gaeumannomyces graminis* var. *tritici*, *Fusarium* spp., *Pythium* spp., and *Xanthomonas oryzae* pv. *oryzae* while maintaining low toxicity in humans and animals (Zhao et al. 2018). The commercialization of other PHZ derivatives that suffer from low stability will likely benefit from nanoemulsion formulations, as has been demonstrated for biosynthesized pyoluteorin (Chen et al. 2011).

12.6.3 Siderophores

Despite the fact that iron is one of the most abundant elements in the soil, it is predominantly in a form that cannot be directly assimilated by plants and microorganisms (Lindsay 1979). In this situation, PGP pseudomonads are capable of producing siderophores such as pyoverdine (PVD) and pyochelin (PCL) that have a high affinity for ferric iron (Meyer and Abdallah 1978; Cox et al. 1981). In the rhizosphere, PVD and PCL bind much of the iron present, sequestering it from competing organisms, such as phytopathogenic fungi (Kloepper et al. 1980a; Becker and Cook 1988). Pseudomonads have receptors that specifically bind the siderophore-iron complexes, facilitating iron uptake and assimilation (Hohnadel and Meyer 1988; Bakker et al. 1990; Magazin et al. 1986). Plants also have mechanisms for uptake of siderophore-iron complexes, allowing the plant to use the iron upon release (Duijff et al. 1994a, b; Crowley et al. 1988). In this way, siderophores mediate biocontrol, by reducing the availability of iron to phytopathogens, as well as PGP, by supplying plants with iron. In addition to iron, siderophores form complexes with other essential elements, providing more efficient nutrient uptake by plants and the producing microorganisms (Bellenger et al. 2008; Braud et al. 2009a, b; Kraemer 2004).

Soils can accumulate heavy metals produced from industry, mine tilling, atmospheric deposition, and the application of chemical fertilizers and pesticides (Wuana and Okieimen 2011). Iron deficiency in plants is more pronounced in soils contaminated with heavy metals (Wallace et al. 1992; Romheld and Marschner 1984; Mishra and Kar 1974). The presence of heavy metals induces pseudomonad production of siderophores, which are also capable of sequestering toxic metals (Höfte et al. 1993; Sinha and Mukherjee 2008; Braud et al. 2009a). These complexes, however, do not enter the cell efficiently (Braud et al. 2009b). In the presence of heavy metals, siderophores play an important role in facilitating the uptake of iron for microorganisms and plants, resulting in increased tolerance of the bacteria (Rajkumar et al. 2010; Braud et al. 2009c, 2010).

It has been suggested that siderophores could provide an environmentally friendly alternative to pesticides (Schenk et al. 2012). In addition to chelating metals, siderophores have been shown to induce systemic resistance in plants (Leeman et al. 1996). Siderophores have diverse applications including cancer therapy, rust removal, corrosion resistance, biosensors, bioremediation, and use as antimicrobial compounds (Schalk et al. 2011; Ahmed and Holmström 2014; de Carvalho et al. 2011; Saha et al. 2016). There is increasing interest in the utility of biosynthesized siderophores and their large-scale production from *Pseudomonas* species (Saha et al. 2016; Nikel et al. 2014; Shaikh et al. 2016; Nagendra Prabhu and Bindu 2016). Formulations containing siderophores chelated to essential elements could provide an ecologically friendly alternative to current mineral fertilizers to provide crops with micronutrients that are more efficiently assimilated (Vansuyt et al. 2007; Dimkpa 2016; Valencia 2018; Shenker and Chen 2005).

12.6.4 Volatile Organic Compounds

Volatile organic compounds (VOCs) are low molecular weight lipophilic molecules that act as intra- and interspecies signaling molecules over long and short distances (Kanchiswamy et al. 2015 and Tyc et al. 2015). Bacterial VOCs include alkanes, alkenes, alcohols, esters, ketones, terpenoids, and nitrogen- and sulfur-containing compounds (Lemfack et al. 2014, 2018; Schulz and Dickschat 2007). Bacteria ubiquitously produce VOCs, and although more than 1000 bacterial VOCs have been identified, the functions of less than 10% of these molecules have been elucidated (Piechulla et al. 2017). Interest in agricultural application of VOCs was initiated by observations that they were responsible for biocontrol activity in PGPR (Fernando et al. 2005b). Since then, VOCs have gained attention for their diverse application in PGP, including plant development and growth, inducing plant immunity and direct inhibition of phytopathogens (reviewed by Kanchiswamy et al. 2015 and Schulz-Bohm et al. 2017).

A myriad of VOCs are produced by pseudomonads that have broad-spectrum activity against plant pathogens. For example, hydrogen cyanide (HCN) production is highly conserved in pseudomonads (Castric 1975). HCN inhibits metalloenzymes such as cytochrome c oxidases (Blumer and Haas 2000) and contributes to the broad-spectrum nematocidal and antifungal activity of PGP pseudomonads (Flaishman et al. 1996; Voisard et al. 1989; Gallagher and Manoil 2001; Nandi et al. 2015, 2016a). Dimethyl disulfide (DMDS) produced by *P. protegens* inhibits plant-pathogenic agrobacteria (Dandurishvili et al. 2011). The activity of DMDS exemplifies that a single VOC can have diverse functions. In addition to inhibiting pathogens directly and by inducing systemic resistance in plants (Huang et al. 2012), DMDS improves plant growth by acting as a supply of reduced sulfur (Meldau et al. 2013). The VOC 2,3-butanediol has a high potential for commercial development. 2,3-Butanediol produced by *P. chlororaphis* has been shown to promote plant growth, induce systemic resistance, and enhance drought tolerance (Cho et al. 2008; Cortes-Barco et al. 2010; Han et al. 2006). 2,3-Butanediol is a commodity chemical, usually synthesized from petroleum and used as a precursor to manufacture various chemical products (Białkowska 2016). In fact, the annual global market for 2,3-butanediol derivatives is estimated to be ~ 32 million tons, worth ~ USD 43 billion (Köpke et al. 2011). 2,3-Butanediol can be efficiently produced by wild-type and engineered bacteria, and microbial 2,3-butanediol production has attracted attention worldwide (Ji et al. 2011; Białkowska 2016). Recent studies have demonstrated that microbial-biosynthesized 2,3-butanediol using low-value waste products can be a more cost effective and environmentally friendly alternative to chemical synthesis (Harvianto et al. 2018; Köpke et al. 2011; Kandasamy et al. 2016).

Research directed at the application of other bacterial VOCs in agriculture is just beginning (Schulz-Bohm et al. 2017). The main restriction for implementing VOCs in crop treatments has been effective deployment (Kanchiswamy et al. 2015). Despite this, the DMDS-containing formulation, Paladin™, has been successfully commercialized as a novel soil fumigant that targets nematodes and phytopathogens to replace the ozone-depleting compound, methyl bromide (Fritsch 2005; Piechulla and Degenhardt 2014). Additionally, Song and Ryu have demonstrated the benefits of VOCs when applied in open-field conditions (Song and Ryu 2013). VOCs are considered an unexplored area with high potential for novel agricultural products, and these studies demonstrate the viability of implementing VOCs for efficient and sustainable crop production (Piechulla and Degenhardt 2014).

12.6.5 Pyoluteorin

One of the oldest examples of an antifungal phenylpyrrole SM is the chlorinated polyketide antibiotic PLT. Since it was first isolated from *P. aeruginosa* in 1958 (Takeda 1959), PLT production has been identified in many other pseudomonads (Bailey et al. 1973; Maurhofer et al. 1992; Hu et al. 2005). PLT has been shown to contribute to rhizosphere competence of the producing strain (Carmi et al. 1994; Howell and Stipanovic 1980), and it exhibits bactericidal, herbicidal, and fungicidal activities (Takeda 1959). Much interest surrounding the use of PLT stems from its ability to inhibit bacteria, oomycetes, and fungi, including the agriculturally important phytopathogen *Pythium ultimum* (Howell and Stipanovic 1980; Maurhofer et al. 1992, 1994). In *P. protegens* Pf-5, PLT production positively autoregulates its own expression and, in the rhizosphere, induces the expression of the PLT biosynthetic genes in neighboring bacterial cells (Brodhagen et al. 2004). Thus, it is possible that exogenously applied PLT could not only act to inhibit phytopathogens directly but also stimulate natural PGP pseudomonads to produce PLT.

Large-scale production and purification of PLT biosynthesized by engineered pseudomonads have been accomplished (Wang et al. 2011). Manufactured PLT, however, is sensitive to temperature, pH, and UV-visible light irradiation experienced during storage and in the field, reducing its biological activity to ineffective levels (Zhang et al. 2010). However, Chen et al. (2011) developed a novel slow-release nanostructured silica formulation with PLT synthesized from *P. aeruginosa* M18 (Chen et al. 2011). This formulation greatly prolonged the stability and biological activity of PLT, providing strong evidence for the feasibility of biosynthesized PLT.

12.6.6 2,4-Diacetylphloroglucinol

The phloroglucinol DAPG is considered to be one of the most interesting compounds synthesized by pseudomonads. Originally studied for its biocontrol properties against pathogenic fungi and bacteria, it was found to also have antiviral, herbicidal, and antihelmintic capabilities (Haas and Keel 2003; Weller et al. 2007). In pseudomonads, DAPG production is mediated by the activation of six structural genes (*phlA*, *phlC*, *phlB*, *phlD*, *phlE*, and *phlI*) and the activity of three regulatory genes (*phlF*, *phlG*, and *phlH*) (Mandryk-Litvinkovich et al. 2017). DAPG producing pseudomonads have been consistently isolated from monoculture wheat fields in the United States, where they reduce the incidence of take-all disease caused by *Gaeumannomyces graminis* var. *tritici* (Weller et al., 2002, 2007).

The DAPG biosynthetic cluster is conserved in fluorescent pseudomonads (Fuente et al. 2006); however, many species seem to have lost the ability to synthesize DAPG (Moynihan et al. 2009). It has been suggested that the lack of DAPG production by a strain is dependent on crop type (Landa et al. 2006). Furthermore, many *Fusarium* spp. produce the mycotoxin fusaric acid (Brown et al. 2015), which has been shown to repress DAPG production in pseudomonads (Duffy and Défago 1997). These situations suggest a diminished usefulness of exogenously applied pseudomonads for their DAPG-producing capabilities, which could be complemented by the exogenous application of biosynthesized DAPG. This notion is supported by the fact that DAPG can promote the expression of biocontrol traits in other pseudomonads and nonrelated species of PGP bacteria and fungi (Combes-Meynet et al. 2011; Lutz et al. 2004; Brodhagen et al. 2004; Baehler et al. 2005). In addition, DAPG can function as a signal that stimulates ISR and root branching and exudation in plants (Brazelton et al. 2008; Phillips et al. 2004; Iavicoli et al. 2003; Bakker et al. 2007).

The global market for phloroglucinol derivatives is diverse, including numerous pharmaceuticals, cosmetics, textiles, paints, dyes, insecticides, and pesticides, for which there has been an extraordinary number of patents (Singh et al. 2009). Since the first report of microbial biosynthesized DAPG (Nowak-Thompson et al. 1994), genetic approaches have been employed to increase DAPG production by bacteria as well as the tolerance of canonical biosynthetic strains to phloroglucinol metabolites (Achkar et al. 2005; Rao et al. 2013; Cao et al. 2011; Zha et al. 2009; Zhang et al. 2017). The nature of DAPG makes it amenable to producing a broad spectrum of derivatives through chemical modification (Chauthe et al. 2012). Recently, one such derivative was produced with increased stability, an expanded spectrum of antifungal activity, lower toxicity than albesilate, and a broadly used fungicide in citrus fruits (Gong et al. 2016). Such findings highlight the potential for biosynthesized DAPG as a replacement for current pesticides.

12.6.7 Lipopeptides

There is considerable interest in lipopeptides (LPs) produced by *Pseudomonas* spp. due to their antimicrobial, zoosporicidal, and surfactant properties, and they are capable of inducing systemic resistance to phytopathogens (Raaijmakers et al. 2010). LPs are synthesized in stages through modular enzyme systems encoded by large nonribosomal peptide synthetase (NRPS) genes (Marahiel et al. 1997). LPs comprise a linear or cyclic oligopeptide linked to a lipid (Raaijmakers et al. 2010). The structural diversity and bioactivity of LPs are dictated by the structure of the lipid tail as well as the number and arrangement of amino acids in the peptide chain (Raaijmakers et al. 2010). LPs are SMs commonly produced by pseudomonads that contribute significantly to their bioactivity against agriculturally important phytopathogens (Raaijmakers et al. 2010). Notably, *P. brassicacearum* DF41, which is an effective inhibitor of *S. sclerotiorum* (Lib.) de Bary (Savchuk and Fernando 2004), does not contain the antibiotic biosynthetic genes encoding PCA, PRN, PLT, or 2,4-DAPG (Zhang et al. 2006), and its antifungal effect is due to the production of a novel cyclic LP, called sclerosin (Berry et al. 2010, 2012). LPs produced by *Pseudomonas* species are also important for motility, which may provide an advantage for colonization of plants and inhibition of phytopathogens (Andersen et al. 2003; Roongsawang et al. 2003; de Bruijn et al. 2007).

Microbial-produced bio-surfactants are an environmentally safe alternative to petroleum-based surfactants. The global market for microbial bio-surfactants has been estimated to reach USD 17.1 million by 2020 with an overall compound annual growth rate of 4% (Transparency Market Research 2014). A major impediment of bio-surfactant synthesis using fermentation technologies has been the production of foam and growth inhibition (Coutte et al. 2017). However, advancements such as surface-attached fermentation solve the problem of foaming, and cultures grown in biofilms increase the resistance to surfactants (Coutte et al. 2017), as recently demonstrated for rhamnolipids (Chong and Li 2017).

12.6.8 Phosphate Solubilization

Next to nitrogen, phosphorus is the second most limiting soil nutrient required for plant growth (Schachtman et al. 1998). Paradoxically, many soils contain high amounts of these elements, but most are in a form that is not accessible for plant uptake leading to a loss of crop productivity (Vance et al. 2003; Batjes 1997). *Pseudomonas* species are capable of solubilizing organic phosphates through the activity of various enzymes including phosphatase, phytase, and phosphonatase (Richardson and Hadobas 1997; Burini et al. 1994; Gügi et al. 1991). The production of these enzymes is mediated through the *pho* regulon, which is activated in response to low phosphate concentrations (Monds et al. 2006; Silby et al. 2009).

In addition to enzyme-mediated mineralization of organic phosphates, pseudomonads solubilize inorganic phosphates by producing organic acids, mainly gluconic acid and 2-ketogluconic acids (Arcand and Schneider 2006; Rodríguez and Fraga 1999). Due to its stability, low toxicity, and low corrosivity, gluconic acid is used extensively in food and pharmaceutical industries. Gluconic acid is produced by oxidation of glucose by glucose dehydrogenases, which use pyrroloquinoline quinone (PQQ) as a redox cofactor (Liu et al. 1992). Notably, PQQ is directly involved in PGP, likely owing to its antioxidant activity. A PQQ mutant of *P. fluorescens* B16 completely lost its PGP ability in tomato, cucumber, *Arabidopsis*, and hot pepper (Choi et al. 2008). The PGP effects could be restored by either supplying the PQQ biosynthetic genes in trans or providing synthetic PQQ in the absence of bacteria (Choi et al. 2008).

Large-scale production of gluconic acid is commonly achieved through a combination of microbial fermentation and enzymatic treatment of agricultural and industrial by-products (Cañete-Rodríguez et al. 2016). Awareness of the utility of engineered strains of *Pseudomonas* as noncanonical biosynthetic strains is growing (Shen et al. 2017; Loeschcke and Thies 2015; Nikel et al. 2014). Since they are considered to be one of the most powerful phosphate solubilizers, the use of pseudomonads could improve these fermentation strategies (Satyaprakash et al. 2017).

12.6.9 Plant Growth Regulators

Phytohormones regulate the growth and development of plants. Although *Pseudomonas* species are well known for their biocontrol capabilities, production of the phytohormonesindole-3-acetic acid (IAA), cytokinins, and gibberellins is considered to be one of the main mechanisms underlying their PGP effects (Mandal and Kotasthane 2014; Olanrewaju et al. 2017; Kumar Jha and Saraf 2015). In addition to promoting plant growth, cytokinins produced by *Pseudomonas* have been recently reported to induce hormone-mediated biocontrol (ISR) (Großkinsky et al. 2016).

IAA is a key auxin involved in regulating the growth and development of plants. IAA synthesis by PGP pseudomonads is stimulated in response to components in root exudates, including tryptophan, the precursor of IAA (Duca et al. 2014). Like other SMs, IAA production varies in different strains depending on environmental signals such as pH, osmotic stress, nutrient limitation, as well as genetics (Spaepen et al. 2007; Duca et al. 2014). Commercially available synthetic auxins, naphthylacetic acid and indole butyric acid, have played a major role in horticulture, mainly for their ability to promote root formation. While further investigation is needed to elucidate niche and plant-specific responses to exogenously applied phytohormones (Egamberdieva et al. 2017; Shigenaga et al. 2017), evidence supports the utility of formulations containing IAA, cytokinins, and gibberellins in agriculture (Morel et al. 2016).

Drought stress is known to lead to hormonal imbalances in plants (Farooq et al. 2014). This stress results in a rapid decline in endogenous cytokinin levels, causing decreased plant growth (Pospíšilová et al. 2000; Yasmeen et al. 2013). Application of phytohormones has been suggested in situations where a hormonal imbalance is known to mediate growth stress-adaptation responses (Egamberdieva et al. 2017; Yasmeen et al. 2013). This approach is supported by the fact that crop yield under drought conditions can be improved by the application of natural cytokinin derivatives (Yasmeen et al. 2013; Shaukat et al. 2017). Exogenously applied auxin was found to induce a drought-resistant phenotype in *Arabidopsis* (Shi et al. 2014). Moreover, gibberelin-producing pseudomonads and their culture filtrates increase the yield of wheat and chana bean crops (Pandya and Desai 2014) and enhance the growth of soybean plants under salt and drought stress (Kang et al. 2014).

Pseudomonads are good candidates for large-scale biosynthesis of these SMs, as evidenced by attempts of synthetic biologists to artificially introduce many of the characteristics intrinsic to pseudomonads in exogenous microbes (Nikel et al. 2014). The ability to produce gibberellic acid (GA) is widely conserved among pseudomonads (Sharma et al. 2014; Saber et al. 2015; Pandya and Desai 2014; Kapoor et al. 2016). Interestingly, the non-13-hydroxylation GA biosynthetic pathway found in plants is identical to the pathway in bacteria, resulting in the production of bioactive GA (Nagel et al. 2017; Nett et al. 2017). There is compelling evidence that pseudomonads provide high yields of GA using a less expensive and more efficient commercial production approach compared to GA production from *Gibberella fujikuroi* or other fungi (Kapoor et al. 2016).

12.7 Novel SMs: Toward the Future

12.7.1 Genome Mining for Cryptic Biosynthetic Clusters

The reduced costs associated with next-generation sequencing platforms have resulted in an extraordinary increase in the number of sequenced genomes (Metzker 2010). In turn, the availability of these sequences enables genome mining for novel SMs produced by unannotated or cryptic biosynthetic clusters (Saleh et al. 2012; Chiang et al. 2011; Scherlach and Hertweck 2009; Paterson et al. 2017; Vizcaino et al. 2014). Numerous bioinformatic tools have been developed for genome mining, including antiSMASH (antibiotics and Secondary Metabolite Analysis Shell) (Medema et al. 2011; Weber et al. 2015), NP.searcher (Li et al. 2009), NRPS-PKS (Ansari et al. 2004), PRISM (Skinnider et al. 2015), and PKS/NRPS (Bachmann and Ravel 2009). Of these, the antiSMASH tool provides the most comprehensive database of biosynthetic clusters, including ribosomal peptides, terpenes, polyketide synthase (PKS), and NRPS loci (Weber et al. 2015). This program has the added benefit of being user-friendly.

Many techniques have been employed in an attempt to "awaken" these biosynthetic genes (Rutledge and Challis 2015; Ren et al. 2017). In nature, SMs may only be produced in response to environmental signals or by co-culture with other organisms (Cornforth and Foster 2013; Firn and Jones 2003). In the lab, expression of cryptic gene clusters can be induced by varying the growth conditions as in OSMAC (one strain many conditions) (Bills et al. 2008; Rateb et al. 2011) or upon co-incubation with other bacteria, fungi, or predators (Moree et al. 2012; Nandi et al. 2016a). Alternatively, heterologous expression of cryptic biosynthetic clusters can be achieved by expression in a new host with engineered regulatory elements (Ochi and Hosaka 2013). Historically this approach has been hampered by a lack of vectors for cloning large (>40 kb) biosynthetic clusters, but the availability of artificial chromosomes has overcome this barrier (Jones et al. 2013).

The potential for developing new SMs using the approaches described above shows great promise since these natural products are structurally too complex to be generated by chemical synthesis. Novel SMs can also be created through genetic engineering. Through directed evolution, variants of proteins encoded by biosynthetic clusters can be created to produce SMs with desired properties (Rao et al. 2013; Bender et al. 1999). Numerous tools are available for directed evolution as well as methods for detection and isolation of the evolved proteins (Packer and Liu 2015). In pseudomonads, PKS and NRPS enzyme complexes are involved in the synthesis of a myriad of SMs with diverse structures and biological activities (Bender et al. 1999; Raaijmakers et al. 2006, 2010). These large biosynthetic clusters consist of modules that contain multi-domain enzymes involved in stepwise reactions to synthesize peptide (NRPS) or polyketide (PKS) chains (Walsh et al. 2001; Donadio et al. 1991). The modular structure of NRPS and PKS biosynthetic clusters is conducive to rearrangement. Through combinatorial biosynthesis involving rearrangement of modules, inactivation of domains, or feeding specific or synthetic precursors, these genetic loci can be manipulated to create new pathways (Calcott et al. 2014; Sun et al. 2015; Walsh et al. 2001; Du et al. 2001; Hutchinson 1998). In cases where the SM produced has low stability or activity, post-biosynthetic chemical modification can be used for improving the characteristics of the newly created SMs (Gong et al. 2016; Kirschning and Hahn 2012).

12.8 Summary

The studies mentioned in this review highlight the need and feasibility of using biosynthesized SMs for PGP of crops. However this review is not meant to diminish the importance of developing PGP pseudomonads for commercial use. In our view, the ability of pseudomonads to sense and adapt to a dynamic environment is key to their PGP capabilities. The ability of one strain to produce diverse SMs with multiple biological activities is a powerful trait that should be harnessed. Additionally, some of the pseudomonad PGP functions are mediated by intracellular enzymes, which require the physiological conditions present inside the cell for activity. A notable example of this is the ACC (1-aminocyclopropane-1-carboxylate) deaminase (Glick 2014). Although IAA has been shown to promote plant growth, in some situations, IAA can inhibit plant growth by stimulating the production of ethylene, known as a "stress hormone" in plants (Chadwick and Burg 1967, 1970). The resulting ethylene production can amplify the defense response, causing plant senescence or death (Robison et al. 2008). Plant-derived ACC, however, is also exuded through roots and can be taken up by pseudomonads where it is cleaved by ACC deaminase, decreasing the ethylene level in the plant (Honma and Shimomura 1978; Glick et al. 2007; Glick and Bashan 1997). In this situation, the production of bacterial ACC deaminase by PGPR strains is key for stimulating plant growth and increasing plant tolerance to biotic and abiotic stress (Glick 2014).

There is a positive outlook for the widespread implementation of PGP pseudomonads in crop treatments. Advances toward more effective formulations include the creation of slow-release preparations and those containing a consortium of bacteria, rather than a single strain (Grosskopf and Soyer 2014; Leggett et al. 2011; Bashan et al. 2014; Chen et al. 2011). Expression of SMs can be induced in PGPR formulations containing more than one organism, and formulations containing multiple PGPR organisms are generally more successful in the field (Traxler et al. 2013; Tyc et al. 2014; Moree et al. 2012; Abrudan et al. 2015; Bashan et al. 2014). It may even be possible to predict the success of co-inoculation formulas. Exciting work by Paredes et al. demonstrated that the outcome between the plant host and a synthetic microbial community could be predicted (Paredes et al. 2018). Such findings imply that it is possible to rationally design microbial communities to produce the desired outcomes in the plant host.

In addition to applying biosynthesized SMs to fields, providing exogenous sources of these compounds to PGPR formulations would provide the strain the benefit of the SM without the cost of its production, e.g., siderophores for iron assimilation and inhibiting fungal pathogens. This approach would also circumvent unwanted side effects, such as phenotype switching, from the metabolic burden imposed by SM biosynthesis, as seen for PLT in *P. protegens* (Yan et al. 2018).

New technologies are allowing us to elucidate how PGPR and their SMs interact with competitors, phytopathogens, and the host plant, providing a better understanding of niche-specific outcomes (Schenk et al. 2012; Becker et al. 2017; Chan et al. 2016; Haney et al. 2018; Zhao et al. 2015; De Vos et al. 2007; Paredes et al. 2018; Shah et al. 2016; Duke et al. 2017; Klaponski et al. 2014). This information will enhance our ability to predict the effectiveness of formulations, enabling widespread integration of PGP pseudomonads and their SMs into treatment strategies for enhanced crop yields.

References

Abrudan MI, Smakman F, Grimbergen AJ, Westhoff S, Miller EL, van Wezel GP, Rozen DE (2015) Socially mediated induction and suppression of antibiosis during bacterial coexistence. Proc Natl Acad Sci 112:11054–11059

Achkar J, Xian M, Zhao H, Frost JW (2005) Biosynthesis of phloroglucinol. J Am Chem Soc 127:5332–5333

Agrow Biologicals (2018) https://agrow.agribusinessintelligence.informa.com/-/media/agri/agrow/ag-market-reviews-pdfs/supplements/agrow_biologicals_2018_online.pdf. Accessed 5 Mar 2019

Agrow Biostimulants Report (2017) https://store.agribusinessintelligence.informa.com/product/agrow-biostimulants-2017/. Accessed 5 Mar 2019

Agrow Global Biopesticide Regulations (2018) https://store.agribusinessintelligence.informa.com/product/agrow-global-biopesticide-regulations-2018/. Accessed 5 Mar 2019

Ahemad M, Khan MS (2009) Toxicity assessment of herbicides quizalafop-p-ethyl and clodinafop toward rhizobium pea symbiosis. Bull Environ Contam Toxicol 82:761–766

Ahmed E, Holmström SJM (2014) Siderophores in environmental research: roles and applications. Microb Biotechnol 7:196–208

Aktar MW, Sengupta D, Chowdhury A (2009) Impact of pesticides use in agriculture: their benefits and hazards. Interdiscip Toxicol 2:1–12

Andersen JB, Koch B, Nielsen TH, Sørensen D, Hansen M, Nybroe O, Christophersen C, Sørensen J, Molin S, Givskov M (2003) Surface motility in *Pseudomonas* sp. DSS73 is required for efficient biological containment of the root-pathogenic microfungi *Rhizoctonia solani* and *Pythium ultimum*. Microbiology 149:37–46

Ansari MZ, Yadav G, Gokhale RS, Mohanty D (2004) NRPS-PKS: A knowledge-based resource for analysis of NRPS-PKS megasynthases. Nucleic Acids Res 32:W405–W413

Antoun H, Prévost D (2005) Ecology of plant growth promoting rhizobacteria. In: Siddiqui ZA (ed) PGPR: biocontrol and biofertilization. Springer, Dordrecht, pp 1–38

Arcand MM, Schneider KD (2006) Plant- and microbial-based mechanisms to improve the agronomic effectiveness of phosphate rock: a review. An Acad Bras Cienc 78:791–807

Arias-Estévez M, López-Periago E, Martínez-Carballo E, Simal-Gándara J, Mejuto J-C, García-Río L (2008) The mobility and degradation of pesticides in soils and the pollution of groundwater resources. Agric Ecosyst Environ 123:247–260

Arima K, Imanaka H, Kousaka M, Fukuta A, Tamura G (1964) Agricultural and biological chemistry pyrrolnitrin, a new antibiotic substance, produced by *Pseudomonas*. Agric Biol Chem BioI Chern 288:575–576

Bachmann BO, Ravel J (2009) Methods for in silico prediction of microbial polyketide and nonribosomal peptide biosynthetic pathways from dna sequence data. Methods Enzymol 458:181–217

Baehler E, Bottiglieri M, Péchy-Tarr M, Maurhofer M, Keel C (2005) Use of green fluorescent protein-based reporters to monitor balanced production of antifungal compounds in the biocontrol agent *Pseudomonas fluorescens* CHA0. J Appl Microbiol 99:24–38

Bai Y, Wu J, Clark MC, Naeem S, Pan Q, Huang J, Zhang L, Han X (2010) Tradeoffs and thresholds in the effects of nitrogen addition on biodiversity and ecosystem functioning: evidence from inner Mongolia Grasslands. Glob Chang Biol 16:358–372

Bailey DM, Johnson RE, Salvador UJ (1973) Pyrrole antibacterial agents. Compounds related to pyoluteorin. J Med Chem 16:1298–1300

Bakker PAHM, van Peer R, Schippers B (1990) Specificity of siderophores and siderophore receptors and biocontrol by *Pseudomonas* spp. In: Cook RJ, Henis Y, Ko WH, Rovira AD, Schippers B, Scott PR (eds) Biological control of soil-borne plant pathogens. CAB International, Wallingford, pp 131–142

Bakker PAHM, Pieterse CMJ, van Loon LC (2007) Induced systemic resistance by fluorescent *Pseudomonas* spp. Phytopathology 97:239–243

Baldwin IT (2001) An ecologically motivated analysis of plant-herbivore interactions in native tobacco. Plant Physiol 127:1449–1458

Balmer A, Pastor V, Gamir J, Flors V, Mauch-Mani B (2015) The "prime-ome": toward a holistic approach to priming. Trends Plant Sci 20:443–452

Bashan Y, de Bashan LE, Prabhu SR, Hernandez J-P (2014) Advances in plant growth-promoting bacterial inoculant technology: Formulations and practical perspectives (1998–2013). Plant Soil 378:1–33

Bassler BL (2002) Small talk: cell-to-cell communication in bacteria. Cell 109:421–424

Batjes NH (1997) A world dataset of derived soil properties by FAO-UNESCO soil unit for global modelling. Soil Use Manag 13:9–16

Becker JO, Cook JR (1988) Role of siderophore in suppression of *Pythium* species and production of increased-growth response of wheat by fluorescent pseudomonads. Phytopathology 78:778–782

Becker MG, Walker PL, Pulgar-Vidal NC, Belmonte MF (2017) SeqEnrich: a tool to predict transcription factor networks from co-expressed Arabidopsis and *Brassica napus* gene sets. PLoS One 12:e0178256

Beckers GJ, Conrath U (2007) Priming for stress resistance: from the lab to the field. Curr Opin Plant Biol 10:425–431

Bellenger JP, Wichard T, Kustka AB, Kraepiel AML (2008) Uptake of molybdenum and vanadium by a nitrogen-fixing soil bacterium using siderophores. Nat Geosci 1:243–246

Bender C, Rangaswamy V, Loper J (1999) Polyketide production by plant-associated pseudomonads. Annu Rev Phytopathol 37:175–196

Berry C, Fernando WGD, Loewen PC, de Kievit TR (2010) Lipopeptides are essential for *Pseudomonas* sp. DF41 biocontrol of *Sclerotinia sclerotiorum*. Biol Control 55:211–218

Berry CL, Brassinga AK, Donald LJ, Fernando WGD, Loewen PC, de Kievit TR (2012) Chemical and biological characterization of sclerosin, an antifungal lipopeptide. Can J Microbiol 58:1027–1034

Berry CL, Nandi M, Manuel J, Brassinga AK, Fernando WGD, Loewen PC, de Kievit TR (2014) Characterization of the *Pseudomonas* sp. DF41 quorum sensing locus and its role in fungal antagonism. Biol Control 69:82–89

Białkowska AM (2016) Strategies for efficient and economical 2,3-butanediol production: new trends in this field. World J Microbiol Biotechnol 32:200

Bilal M, Guo S, Iqbal HMN, Hu H, Wang W, Zhang X (2017) Engineering *Pseudomonas* for phenazine biosynthesis, regulation, and biotechnological applications: a review. World J Microbiol Biotechnol 33:191

Bills GF, Platas G, Fillola A, Jiménez MR, Collado J, Vicente F, Martín J, González A, Bur-Zimmermann J, Tormo JR et al (2008) Enhancement of antibiotic and secondary metabolite detection from filamentous fungi by growth on nutritional arrays. J Appl Microbiol 104:1644–1658

Blumer C, Haas D (2000) Mechanism, regulation, and ecological role of bacterial cyanide biosynthesis. Arch Microbiol 173:170–177

Braud A, Hannauer M, Mislin GLA, Schalk IJ (2009a) The *Pseudomonas aeruginosa* pyochelin-iron uptake pathway and its metal specificity. J Bacteriol 191:3517–3525

Braud A, Hoegy F, Jezequel K, Lebeau T, Schalk IJ (2009b) New insights into the metal specificity of the *Pseudomonas aeruginosa* pyoverdine-iron uptake pathway. Environ Microbiol 11:1079–1091

Braud A, Jézéquel K, Bazot S, Lebeau T (2009c) Enhanced phytoextraction of an agricultural Cr- and Pb-contaminated soil by bioaugmentation with siderophore-producing bacteria. Chemosphere 74:280–286

Braud A, Geoffroy V, Hoegy F, Mislin GLA, Schalk IJ (2010) Presence of the siderophores pyoverdine and pyochelin in the extracellular medium reduces toxic metal accumulation in *Pseudomonas aeruginosa* and increases bacterial metal tolerance. Environ Microbiol Rep 2:419–425

Brazelton JN, Pfeufer EE, Sweat TA, Gardener BBM, Coenen C (2008) 2,4-Diacetylphloroglucinol alters plant root development. Mol Plant-Microbe Interact 21:1349–1358

Brodhagen M, Henkels MD, Loper JE (2004) Positive autoregulation and signaling properties of pyoluteorin, an antibiotic produced by the biological control organism *Pseudomonas fluorescens* Pf-5. Appl Environ Microbiol 70:1758–1766

Brown DW, Lee S-H, Kim L-H, Ryu J-G, Lee S, Seo Y, Kim YH, Busman M, Yun S-H, Proctor RH et al (2015) Identification of a 12-gene fusaric acid biosynthetic gene cluster in *Fusarium*species through comparative and functional genomics. Mol Plant-Microbe Interact 28:319–332

Brussaard l, de Ruiter PC, Brown GG (2007) Soil biodiversity for agricultural sustainability. Agric Ecosyst Environ 121:233–244

Bulgarelli D, Rott M, Schlaeppi K, Ver Loren van Themaat E, Ahmadinejad N, Assenza F, Rauf P, Huettel B, Reinhardt R, Schmelzer E et al (2012) Revealing structure and assembly cues for Arabidopsis root-inhabiting bacterial microbiota. Nature 488:91–95

Bull CT, Duffy B, Voisard C, Défago G, Keel C, Haas D (2001) Characterization of spontaneous *gacS* and *gacA* regulatory mutants of *Pseudomonas fluorescens* biocontrol strain CHA0. Antonie Van Leeuwenhoek 79:327–336

Burini J-F, Gugi B, Merieau A, Guespin-Michel JF (1994) Lipase and acidic phosphatase from the psychrotrophic bacterium *Pseudomonas fluorescens*: two enzymes whose synthesis is regulated by the growth temperature. FEMS Microbiol Lett 122:13–18

Buus Laursen J, Nielsen J (2004) Phenazine natural products: biosynthesis, synthetic analogues, and biological activity. Chem Rev 104:1663–1686

Calcott MJ, Owen JG, Lamont IL, Ackerley DF (2014) Biosynthesis of novel pyoverdines by domain substitution in a nonribosomal peptide synthetase of *Pseudomonas aeruginosa*. Appl Environ Microbiol 80:5723–5731

Callaway RM, Maron JL (2006) What have exotic plant invasions taught us over the past 20 years? Trends Ecol Evol 21:369–374

Cañete-Rodríguez AM, Santos-Dueñas IM, Jiménez-Hornero JE, Ehrenreich A, Liebl W, García-García I (2016) Gluconic acid: properties, production methods and applications—an excellent opportunity for agro-industrial by-products and waste bio-valorization. Process Biochem 51:1891–1903

Cao Y, Jiang X, Zhang R, Xian M (2011) Improved phloroglucinol production by metabolically engineered *Escherichia coli*. Appl Microbiol Biotechnol 91:1545–1552

Carmi R, Carmeli S, Levy E, Gough FJ (1994) (+)-(S)-dihydroaeruginoic acid, an inhibitor of *Septoria tritici* and other phytopathogenic fungi and bacteria, produced *by Pseudomonas fluorescens*. J Nat Prod 57:1200–1205

Castric PA (1975) Hydrogen cyanide, a secondary metabolite of *Pseudomonas aeruginosa*. Can J Microbiol 21:613–618

Chadwick AV, Burg SP (1967) An explanation of the inhibition of root growth caused by indole-3-acetic acid. Plant Physiol 42:415–420

Chadwick AV, Burg SP (1970) Regulation of root growth by auxin-ethylene interaction. Plant Physiol 45:192–200

Chan AC, Khan D, Girard IJ, Becker MG, Millar JL, Sytnik D, Belmonte MF (2016) Tissue-specific laser microdissection of the *Brassica napus* funiculus improves gene discovery and spatial identification of biological processes. J Exp Bot 67:3561–3571

Chancey ST, Wood DW, Pierson LS (1999) Two-component transcriptional regulation of N-acyl-homoserine lactone production in *Pseudomonas aureofaciens*. Appl Environ Microbiol 65:2294–2299

Chancey ST, Wood DW, Pierson EA, Pierson LS (2002) Survival of GacS/GacA mutants of the biological control bacterium *Pseudomonas aureofaciens* 30-84 in the wheat rhizosphere. Appl Environ Microbiol 68:3308–3314

Chauthe SK, Bharate SB, Periyasamy G, Khanna A, Bhutani KK, Mishra PD, Singh IP (2012) One pot synthesis and anticancer activity of dimeric phloroglucinols. Bioorg Med Chem Lett 22:2251–2256

Cheluvappa R (2014) Standardized chemical synthesis of *Pseudomonas aeruginosa* pyocyanin. Methods X1:67–73

Chen J, Wang W, Xu Y, Zhang X (2011) Slow-release formulation of a new biological pesticide, pyoluteorin, with mesoporous silica. J Agric Food Chem 59:307–311

Chiang Y-M, Chang S-L, Oakley BR, Wang CC (2011) Recent advances in awakening silent biosynthetic gene clusters and linking orphan clusters to natural products in microorganisms. Curr Opin Chem Biol 15:137–143

Chin-A-Woeng TFC, Bloemberg GV, van der Bij AJ, van der Drift KMGM, Schripsema J, Kroon B, Scheffer RJ, Keel C, Bakker PAHM, Tichy H-V et al (1998) Biocontrol by phenazine-1-carboxamide-producing *Pseudomonas chlororaphis* PCL1391 of tomato root rot caused by *Fusarium oxysporum* f. sp. *radicis-lycopersici*. Mol Plant-Microbe Interact 11:1069–1077

Chin-A-Woeng TFC, Thomas-Oates JE, Lugtenberg BJJ, Bloemberg GV (2001) Introduction of the *phzH* gene of *Pseudomonas chlororaphis* PCL1391 extends the range of biocontrol ability of phenazine-1-carboxylic acid-producing *Pseudomonas* spp. strains. Mol Plant-Microbe Interact 14:1006–1015

Chin-A-Woeng TFC, Bloemberg GV, Lugtenberg BJJ (2003) Phenazines and their role in biocontrol by *Pseudomonas* bacteria. New Phytol 157:503–523

Cho SM, Kang BR, Han SH, Anderson AJ, Park J-Y, Lee Y-H, Cho BH, Yang K-Y, Ryu C-M, Kim YC (2008) 2R,3R-butanediol, a bacterial volatile produced by *Pseudomonas chlororaphis*O6, is involved in induction of systemic tolerance to drought in *Arabidopsis thaliana*. Mol Plant-Microbe Interact 21:1067–1075

Choi O, Kim J, Kim J-G, Jeong Y, Moon JS, Park CS, Hwang I (2008) Pyrroloquinoline quinone is a plant growth promotion factor produced by *Pseudomonas fluorescens* B16. Plant Physiol 146:657–668

Chong H, Li Q (2017) Microbial production of rhamnolipids: opportunities, challenges and strategies. Microb Cell Factories 16:1–12

Coleman-Derr D, Desgarennes D, Fonseca-Garcia C, Gross S, Clingenpeel S, Woyke T, North G, Visel A, Partida-Martinez LP, Tringe SG (2016) Plant compartment and biogeography affect microbiome composition in cultivated and native *Agave* species. New Phytol 209:798–811

Combes-Meynet E, Pothier JF, Moënne-Loccoz Y, Prigent-Combaret C (2011) The *Pseudomonas* secondary metabolite 2,4-diacetylphloroglucinol is a signal inducing rhizoplane expression of *Azospirillum* genes involved in plant-growth promotion. Mol Plant-Microbe Interact 24:271–284

Conrath U, Beckers GJM, Flors V, García-Agustín P, Jakab G, Mauch F, Newman M-A, Pieterse CMJ, Poinssot B, Pozo MJ et al (2006) Priming: getting ready for battle. Mol Plant-Microbe Interact 19:1062–1071

Cornforth DM, Foster KR (2013) Competition sensing: the social side of bacterial stress responses. Nat Rev Microbiol 11:285–293

Cortes-Barco AM, Goodwin PH, Hsiang T (2010) Comparison of induced resistance activated by benzothiadiazole, (2R,3R)-butanediol and an isoparaffin mixture against anthracnose of *Nicotiana benthamiana*. Plant Pathol 59:643–653

Coutte F, Lecouturier D, Dimitrov K, Guez J-S, Delvigne F, Dhulster P, Jacques P (2017) Microbial lipopeptide production and purification bioprocesses, current progress and future challenges. Biotechnol J 12:1600566

Cox CD, Rinehart KL, Moore ML, Cook JC (1981) Pyochelin: novel structure of an iron-chelating growth promoter for *Pseudomonas aeruginosa*. Proc Natl Acad Sci U S A 78:4256–4260

Crowley DE, Reid CP, Szaniszlo PJ (1988) Utilization of microbial siderophores in iron acquisition by oat. Plant Physiol 87:680–685

Dandurishvili N, Toklikishvili N, Ovadis M, Eliashvili P, Giorgobiani N, Keshelava R, Tediashvili M, Vainstein A, Khmel I, Szegedi E et al (2011) Broad-range antagonistic rhizobacteria *Pseudomonas fluorescens* and *Serratia plymuthica* suppress *Agrobacterium* crown gall tumours on tomato plants. J Appl Microbiol 110:341–352 de Bruijn I, de Kock MJD, Yang M, de Waard P, van Beek TA, Raaijmakers JM (2007) Genome-based discovery, structure prediction and functional analysis of cyclic lipopeptide antibiotics in *Pseudomonas* species. Mol Microbiol 63:417–428

de Carvalho CCCR, Marques MPC, Fernandes P (2011) Recent achievements on siderophore production and application. Recent Pat Biotechnol 5:183–198

De Vos RC, Moco S, Lommen A, Keurentjes JJ, Bino RJ, Hall RD (2007) Untargeted large-scale plant metabolomics using liquid chromatography coupled to mass spectrometry. Nat Protoc 2:778–791

Defarge N, Spiroux de Vendômois J, Séralini GE (2018) Toxicity of formulants and heavy metals in glyphosate-based herbicides and other pesticides. Toxicol Reports 5:156–163

Delaney SM, Mavrodi DV, Bonsall RF, Thomashow LS (2001) *phzO*, a gene for biosynthesis of 2-hydroxylated phenazine compounds in *Pseudomonas aureofaciens* 30-84. J Bacteriol 183:318–327

Denancé N, Sánchez-Vallet A, Goffner D, Molina A (2013) Disease resistance or growth: the role of plant hormones in balancing immune responses and fitness costs. Front Plant Sci 4:155

Dimkpa C (2016) Microbial siderophores: production, detection and application in agriculture and environment. Endocytobiosis Cell Res 27:7–16

Donadio S, Staver MJ, McAlpine JB, Swanson SJ, Katz L (1991) Modular organization of genes required for complex polyketide biosynthesis. Science 252:675–679

Du L, Sánchez C, Shen B (2001) Hybrid peptide–polyketide natural products: biosynthesis and prospects toward engineering novel molecules. Metab Eng 3:78–95

Du X, Li Y, Zhou W, Zhou Q, Liu H, Xu Y (2013) Phenazine-1-carboxylic acid production in a chromosomally non-scar triple-deleted mutant *Pseudomonas aeruginosa* using statistical experimental designs to optimize yield. Appl Microbiol Biotechnol 97:7767–7778

Duca D, Lorv J, Patten CL, Rose D, Glick BR (2014) Indole-3-acetic acid in plant–microbe interactions. Antonie Van Leeuwenhoek 106:85–125

Duffy BK, Défago G (1997) Zinc improves biocontrol of fusarium crown and root rot of tomato by *Pseudomonas fluorescens* and represses the production of pathogen metabolites inhibitory to bacterial antibiotic biosynthesis. Phytopathology 87:1250–1257

Duffy BK, Défago G (2000) Controlling instability in *gacS-gacA* regulatory genes during inoculant production of *Pseudomonas fluorescens* biocontrol strains. Appl Environ Microbiol 66:3142–3150

Duijff BJ, Bakker PAHM, Schippers B (1994a) Ferric pseudobactin 358 as an iron source for carnation. J Plant Nutr 17:2069–2078

Duijff BJ, De Kogel WJ, Bakker PAHM, Schippers B (1994b) Influence of pseudobactin 358 on the iron nutrition of barley. Soil Biol Biochem 26:1681–1688

Duke KA, Becker MG, Girard IJ, Millar JL, Fernando WGD, Belmonte MF, de Kievit TR (2017) The biocontrol agent *Pseudomonas chlororaphis* PA23 primes *Brassica napus* defenses through distinct gene networks. BMC Genomics 18:467

Dunham Trimmer (2017) Biological control – global market overview. http://wrir4.ucdavis.edu/events/2017_SLR_Meeting/Presentations/GeneralPresentations/1%20Trimmer%20-%20Global%20Biocontrol%20Market%202017.pdf. Accessed 5 Mar 2019

Egamberdieva D, Wirth SJ, Alqarawi AA, Abd Allah EF, Hashem A (2017) Phytohormones and beneficial microbes: essential components for plants to balance stress and fitness. Front Microbiol 8:2104

Egamberdiyeva D (2005) Characterization of *Pseudomonas* species isolated from the rhizosphere of plants grown in serozem soil, semi arid region of Uzbekistan. ScientificWorldJournal 5:501–509

Engelberth J, Alborn HT, Schmelz EA, Tumlinson JH (2003) Airborne signals prime plants against insect herbivore attack. PNAS 101:1781–1785

Farooq M, Hussain M, Siddique KHM (2014) Drought stress in wheat during flowering and grain-filling periods. CRC Crit Rev Plant Sci 33:331–349

Fernando WGD, Nakkeeran S, Zhang Y (2005a) Biosynthesis of antibiotics by PGPR and its relation in biocontrol of plant diseases. In: PGPR: biocontrol and biofertilization. Springer, Dordrecht, pp 67–109

Fernando WGD, Ramarathnam R, Krishnamoorthy AS, Savchuk SC (2005b) Identification and use of potential bacterial organic antifungal volatiles in biocontrol. Soil Biol Biochem 37:955–964

Firn RD, Jones CG (2003) Natural products – a simple model to explain chemical diversity. Nat Prod Rep 20:382–391

Flaishman MA, Eyal Z, Zilberstein A, Voisard C, Haas D (1996) Suppression of *Septoria tritici* blotch and leaf rust of wheat by recombinant cyanide-producing strains of *Pseudomonas putida*. Mol Plant-Microbe Interact 9:642–645

Fritsch J (2005) Dimethyl disulfide as a new chemical potential alternative to methyl bromide in soil disinfestation in France. Acta Hortic 698:71–76

Frost CJ, Mescher MC, Carlson JE, De Moraes CM (2008) Update on plant defense priming against herbivores plant defense priming against herbivores: getting ready for a different battle. Plant Physiol 146:818–824

Fuente L, Mavrodi DV, Landa BB, Thomashow LS, Weller DM (2006) *phlD*-based genetic diversity and detection of genotypes of 2,4-diacetylphloroglucinol-producing *Pseudomonas fluorescens*. FEMS Microbiol Ecol 56:64–78

Fuqua WC, Winans SC, Greenberg EP (1994) Quorum sensing in bacteria: the LuxR-LuxI family of cell density-responsive transcriptional regulators. J Bacteriol 176:269–275

Gallagher LA, Manoil C (2001) *Pseudomonas aeruginosa* PAO1 kills *Caenorhabditis elegans* by cyanide poisoning. J Bacteriol 183:6207–6214

Ge Y, Huang X, Wang S, Zhang X, Xu Y (2004) Phenazine-1-carboxylic acid is negatively regulated and pyoluteorin positively regulated by *gacA* in *Pseudomonas* sp. M18. FEMS Microbiol Lett 237:41–47

Ge Y-H, Pei D-L, Li W-W, Zhao Y-H, Xu Y-Q (2006) Insertional mutation of the *rpoS* gene contributes to alteration in biosynthesis of antifungal agents in *Pseudomonas* sp. M18. Biol Control 39:186–192

Giller KE, Beare MH, Lavelle P, Izac A-MN, Swift MJ (1997) Agricultural intensification, soil biodiversity and agroecosystem function. Appl Soil Ecol 6:3–16

Girard G, van Rij ET, Lugtenberg BJJ, Bloemberg GV (2006) Regulatory roles of *psrA* and *rpoS* in phenazine-1-carboxamide synthesis by *Pseudomonas chlororaphis* PCL1391. Microbiology 152:43–58

Glick BR (2012) Plant growth-promoting bacteria: mechanisms and applications. Scientifica (Cairo) 2012:1–15

Glick BR (2014) Bacteria with ACC deaminase can promote plant growth and help to feed the world. Microbiol Res 169:30–39

Glick BR, Bashan Y (1997) Genetic manipulation of plant growth-promoting bacteria to enhance biocontrol of phytopathogens. Biotechnol Adv 15:353–378

Glick BR, Cheng Z, Czarny J, Duan J (2007) Promotion of plant growth by ACC deaminase-producing soil bacteria. Eur J Plant Pathol 119:329–339

Glover-Amengor M, Tetteh F (2009) Effect of pesticide application rate on yield of vegetables and soil microbial communities. West African J Appl Ecol 12:1–7

Godfray HCJ, Beddington JR, Crute IR, Haddad L, Lawrence D, Muir JF, Pretty J, Robinson S, Thomas SM, Toulmin C (2010) Food security: the challenge of feeding 9 billion people. Science 327:812–818

Gong L, Tan H, Chen F, Li T, Zhu J, Jian Q, Yuan D, Xu L, Hu W, Jiang Y et al (2016) Novel synthesized 2, 4-DAPG analogues: antifungal activity, mechanism and toxicology. Sci Rep 6:32266

Goodman AL, Kulasekara B, Rietsch A, Boyd D, Smith RS, Lory S (2004) A signaling network reciprocally regulates genes associated with acute infection and chronic persistence in *Pseudomonas aeruginosa*. Dev Cell 7:745–754

Goodman AL, Merighi M, Hyodo M, Ventre I, Filloux A, Lory S (2009) Direct interaction between sensor kinase proteins mediates acute and chronic disease phenotypes in a bacterial pathogen. Genes Dev 23:249–259

Großkinsky DK, Tafner R, Moreno MV, Stenglein SA, García de Salamone IE, Nelson LM, Novák O, Strnad M, van der Graaff E, Roitsch T (2016) Cytokinin production by *Pseudomonas fluo-*

rescens G20–18 determines biocontrol activity against *Pseudomonas syringae* in Arabidopsis. Sci Rep 6:23310

Grosskopf T, Soyer OS (2014) Synthetic microbial communities. Curr Opin Microbiol 18:72–77

Gügi B, Orange N, Hellio F, Burini JF, Guillou C, Leriche F, Guespin-Michel JF (1991) Effect of growth temperature on several exported enzyme activities in the psychrotrophic bacterium *Pseudomonas fluorescens*. J Bacteriol 173:3814–3820

Haas D, Défago G (2005) Biological control of soil-borne pathogens by fluorescent pseudomonads. Nat Rev Microbiol 3:307–319

Haas D, Keel C (2003) Regulation of antibiotic production in root-colonizing *Pseudomonas* spp. and relevance for biological control of plant disease. Annu Rev Phytopathol 41:117–153

Han B, Pain A, Johnstone K (1997) Spontaneous duplication of a 661 bp element within a two-component sensor regulator gene causes phenotypic switching in colonies of *Pseudomonas tolaasii*, cause of brown blotch disease of mushrooms. Mol Microbiol 25:211–218

Han SH, Lee SJ, Moon JH, Park KH, Yang KY, Cho BH, Kim KY, Kim YW, Lee MC, Anderson AJ et al (2006) GacS-dependent production of 2R,3R-butanediol by *Pseudomonas chlororaphis* O6 is a major determinant for eliciting systemic resistance against *Erwinia carotovora* but not against *Pseudomonas syringae* pv. *tabaci* in tobacco. Mol Plant-Microbe Interact 19:924–930

Haney CH, Ausubel FM (2015) Plant microbiome blueprints. Science 349:788–789

Haney CH, Samuel BS, Bush J, Ausubel FM (2015) Associations with rhizosphere bacteria can confer an adaptive advantage to plants. Nat Plants 1:15051

Haney CH, Wiesmann CL, Shapiro LR, Melnyk RA, O'Sullivan LR, Khorasani S, Xiao L, Han J, Bush J, Carrillo J et al (2018) Rhizosphere-associated *Pseudomonas* induce systemic resistance to herbivores at the cost of susceptibility to bacterial pathogens. Mol Ecol 27:1833–1847

Harvianto GR, Haider J, Hong J, Van Duc Long N, Shim J-J, Cho MH, Kim WK, Lee M (2018) Purification of 2,3-butanediol from fermentation broth: process development and techno-economic analysis. Biotechnol Biofuels 11:18

Hassan HM, Fridovich I (1980) Mechanism of the antibiotic action pyocyanine. J Bacteriol 141:156–163

Hassan KA, Johnson A, Shaffer BT, Ren Q, Kidarsa TA, Elbourne LDH, Hartney S, Duboy R, Goebel NC, Zabriskie TM et al (2010) Inactivation of the GacA response regulator in *Pseudomonas fluorescens* Pf-5 has far-reaching transcriptomic consequences. Environ Microbiol 12:899–915

He YW, Xu Y (2011) Development and application of a new antifungal pesticide "Shenquinmycin" by genetically modifying the melon rhizosphere-originating strain *Pseudomonas* sp. M18. In: Proceedings of the 2nd Asian PGPR conference, Beijing, P. R. China, 21–24 Aug 2011

Heeb S, Haas D (2001) Regulatory roles of the GacS/GacA two-component system in plant-associated and other gram-negative bacteria. Mol Plant-Microbe Interact 14:1351–1363

Heeb S, Blumer C, Haas D (2002) Regulatory RNA as mediator in GacA/RsmA-dependent global control of exoproduct formation in *Pseudomonas fluorescens* CHA0. J Bacteriol 184:1046–1056

Heeb S, Valverde C, Gigot-Bonnefoy C, Haas D (2005) Role of the stress sigma factor RpoS in GacA/RsmA-controlled secondary metabolism and resistance to oxidative stress in *Pseudomonas fluorescens* CHA0. FEMS Microbiol Lett 243:251–258

Herbert RB, Holliman FG (1969) Pigments of *Pseudomonas* species. Part II. Structure of aeruginosin B. J Chem Soc C Org 0:2517–2520

Herms DA, Mattson WJ (1992) The dilemma of plants: to grow or defend. Q Rev Biol 67:283–335

Hernandez ME, Kappler A, Newman DK (2004) Phenazines and other redox-active antibiotics promote microbial mineral reduction. Appl Environ Microbiol 70:921–928

Höfte M, Buysens S, Koedam N, Cornelis P (1993) Zinc affects siderophore-mediated high affinity iron uptake systems in the rhizosphere *Pseudomonas aeruginosa* 7NSK2. Biometals 6:85–91

Hohnadel D, Meyer JM (1988) Specificity of pyoverdine-mediated iron uptake among fluorescent *Pseudomonas* strains. J Bacteriol 170:4865–4873

Honma M, Shimomura T (1978) Metabolism of 1-aminocyclopropane-1-carboxylic acid. Agric Biol Chem 42:1825–1831

Howell CR, Stipanovic RD (1979) Control of *Rhizoctonia solani* on cotton seedlings with *Pseudomonas fluorescens* and with an antibiotic produced by the bacterium. Phytopathology 69:480–482

Howell C, Stipanovic R (1980) Suppression of *Pythium ultimum*-induced damping-off of cotton seedlings by *Pseudomonas fluorescens* and its antibiotic, pyoluteorin. Phytopathology 70:712–715

Hu HB, Xu YQ, Chen F, Zhang XH, Hur BKI (2005) Isolation and characterization of a new fluorescent *Pseudomonas* strain that produces both phenazine 1-carboxylic acid and pyoluteorin. J Microbiol Biotechnol 15:86–90

Huang Z, Bonsall RF, Mavrodi DV, Weller DM, Thomashow LS (2004) Transformation of *Pseudomonas fluorescens* with genes for biosynthesis of phenazine-1-carboxylic acid improves biocontrol of rhizoctonia root rot and in situ antibiotic production. FEMS Microbiol Ecol 49:243–251

Huang C-J, Tsay J-F, Chang S-Y, Yang H-P, Wu W-S, Chen C-Y (2012) Dimethyl disulfide is an induced systemic resistance elicitor produced by *Bacillus cereus* C1L. Pest Manag Sci 68:1306–1310

Huot B, Yao J, Montgomery BL, He SY (2014) Growth–defense tradeoffs in plants: a balancing act to optimize fitness. Mol Plant 7:1267–1287

Hutchinson CR (1998) Combinatorial biosynthesis for new drug discovery. Curr Opin Microbiol 1:319–329

Iavicoli A, Boutet E, Buchala A, Métraux J-P (2003) Induced systemic resistance in *Arabidopsis thaliana* in response to root inoculation with *Pseudomonas fluorescens* CHA0. Mol Plant-Microbe Interact 16:851–858

Jankiewicz U, Kołtonowicz M (2012) The involvement of *Pseudomonas* bacteria in induced systemic resistance in plants (review). Appl Biochem Microbiol 48:244–249

Ji X-J, Huang H, Ouyang P-K (2011) Microbial 2,3-butanediol production: a state-of-the-art review. Biotechnol Adv 29:351–364

Jones AC, Gust B, Kulik A, Heide L, Buttner MJ, Bibb MJ (2013) Phage P1-derived artificial chromosomes facilitate heterologous expression of the FK506 gene cluster. PLoS One 8:e69319

Jorgensen F, Bally M, Chapon-Herve V, Michel G, Lazdunski A, Williams P, Stewart GSAB (1999) RpoS-dependent stress tolerance in *Pseudomonas aeruginosa*. Microbiology 145:835–844

Jousset A, Lara E, Wall LG, Valverde C (2006) Secondary metabolites help biocontrol strain *Pseudomonas fluorescens* CHA0 to escape protozoan grazing. Appl Environ Microbiol 72:7083–7090

Kanchiswamy CN, Malnoy M, Maffei ME (2015) Chemical diversity of microbial volatiles and their potential for plant growth and productivity. Front Plant Sci 6:151

Kandasamy V, Liu J, Dantoft SH, Solem C, Jensen PR (2016) Synthesis of (3R)-acetoin and 2,3-butanediol isomers by metabolically engineered *Lactococcus lactis*. Sci Rep 6:36769

Kang BR, Han SH, Zdor RE, Anderson AJ, Spencer M, Yang KY, Kim YH, Lee MC, Cho BH, Kim YC (2007) Inhibition of seed germination and induction of systemic disease resistance by *Pseudomonas chlororaphis* O6 requires phenazine production regulated by the global regulator, *gacS*. J Microbiol Biotechnol 17:586–593

Kang S-M, Radhakrishnan R, Khan AL, Kim M-J, Park J-M, Kim B-R, Shin D-H, Lee I-J (2014) Gibberellin secreting rhizobacterium, *Pseudomonas putida* H-2-3 modulates the hormonal and stress physiology of soybean to improve the plant growth under saline and drought conditions. Plant Physiol Biochem 84:115–124

Kapoor R, Soni R, Kaur M (2016) Gibberellins production by fluorescent *Pseudomonas* isolated from Rhizospheric soil of *Malus* and *Pyrus*. Int J Agric Environ Biotechnol 9:193–199

Kay E, Dubuis C, Haas D (2005) Three small RNAs jointly ensure secondary metabolism and biocontrol in *Pseudomonas fluorescens* CHA0. Proc Natl Acad Sci U S A 102:17136–17141

Kerr JR, Taylor GW, Rutman A, Høiby N, Cole PJ, Wilson R (1999) *Pseudomonas aeruginosa* pyocyanin and 1-hydroxyphenazine inhibit fungal growth. J Clin Pathol 52:385–387

Keswani C (2015) Ecofriendly management of plant diseases by biosynthesised secondary metabolites of *Trichoderma* spp. J Brief Ideas. https://doi.org/10.5281/zenodo.15571

Keswani C, Mishra S, Sarma BK, Singh SP, Singh HB (2014) Unravelling the efficient applications of secondary metabolites of various *Trichoderma* spp. Appl Microbiol Biotechnol 98:533–544

Keswani C, Sarma BK, Singh HB (2016) Synthesis of policy support, quality control, and regulatory management of biopesticides in sustainable agriculture. In: Singh HB, Sarma BK, Keswani C (eds) Agriculturally important microorganisms: commercial and regulatory requirement in Asia. Springer, Singapore, pp 167–181

Kilani J, Fillinger S (2016) Phenylpyrroles: 30 years, two molecules and (nearly) no resistance. Front Microbiol 7:2014

Kirschning A, Hahn F (2012) Merging chemical synthesis and biosynthesis: a new chapter in the total synthesis of natural products and natural product libraries. Angew Chem Int Ed 51:4012–4022

Klaponski N, Selin C, Duke K, Spicer V, Fernando WGD, Belmonte MF, de Kievit TR (2014) The requirement for the LysR-type regulator PtrA for *Pseudomonas chlororaphis* PA23 biocontrol revealed through proteomic and phenotypic analysis. BMC Microbiol 14:94

Kloepper J, Schroth M (1980) Plant growth-promoting rhizobacteria and plant growth under gnotobiotic conditions. Phytopathology 2:642–644

Kloepper JW, Leong J, Teintze M, Schroth MN (1980a) Enhanced plant growth by siderophores produced by plant growth-promoting rhizobacteria. Nature 286:885–886

Kloepper JW, Leong J, Teintze M, Schroth MN (1980b) *Pseudomonas* siderophores: a mechanism explaining disease-suppressive soils. Curr Microbiol 4:317–320

Köpke M, Mihalcea C, Liew F, Tizard JH, Ali MS, Conolly JJ, Al-Sinawi B, Simpson SD (2011) 2,3-Butanediol production by acetogenic bacteria, an alternative route to chemical synthesis, using industrial waste gas. Appl Environ Microbiol 77:5467–5475

Kraemer SM (2004) Iron oxide dissolution and solubility in the presence of siderophores. Aquat Sci – Res Across Boundaries 66:3–18

Kumar Jha C, Saraf M (2015) Plant growth promoting rhizobacteria (PGPR): a review. E3 J Agric Res Dev 5:108–119

Kvint K, Farewell A, Nyström T (2000) RpoS-dependent promoters require guanosine tetraphosphate for induction even in the presence of high levels of sigma(s). J Biol Chem 275:14795–14798

Lamberth C, Jeanmart S, Luksch T, Plant A (2013) Current challenges and trends in the discovery of agrochemicals. Science 341:742–746

Landa BB, Mavrodi OV, Schroeder KL, Allende-Molar R, Weller DM (2006) Enrichment and genotypic diversity of *phlD*-containing fluorescent *Pseudomonas* spp. in two soils after a century of wheat and flax monoculture. FEMS Microbiol Ecol 55:351–368

Lapouge K, Schubert M, Allain FH-T, Haas D (2007) Gac/Rsm signal transduction pathway of γ-proteobacteria: from RNA recognition to regulation of social behaviour. Mol Microbiol 67:241–253

Laville J, Voisard C, Keel C, Maurhofer M, Défago G, Haas D (1992) Global control in *Pseudomonas fluorescens* mediating antibiotic synthesis and suppression of black root rot of tobacco. Proc Natl Acad Sci U S A 89:1562–1566

Leeman M, den Ouden FM, van Pelt JA, Dirkx FPM, Steijl H, Bakker PAHM, Schippers B (1996) Iron availability affects induction of systemic resistance to *Fusarium* wilt of radish by *Pseudomonas fluorescens*. J Phytopathol 86:149–155

Leggett M, Leland J, Kellar K, Epp B (2011) Formulation of microbial biocontrol agents – an industrial perspective. Can J Plant Pathol 33:101–107

Lemfack MC, Nickel J, Dunkel M, Preissner R, Piechulla B (2014) mVOC: a database of microbial volatiles. Nucleic Acids Res 42:D744–D748

Lemfack MC, Gohlke B-O, Toguem SMT, Preissner S, Piechulla B, Preissner R (2018) mVOC 2.0: a database of microbial volatiles. Nucleic Acids Res 46:D1261–D1265

Li MH, Ung PM, Zajkowski J, Garneau-Tsodikova S, Sherman DH (2009) Automated genome mining for natural products. BMC Bioinformatics 10:185

Ligon JM, Hill DS, Hammer PE, Torkewitz NR, Hofmann D, Kempf H-J, van Pée K-H (2000) Natural products with antifungal activity from *Pseudomonas* biocontrol bacteria. Pest Manag Sci 56:688–695

Lindsay WL (1979) Iron. In: Chemical equilibria in soils. Wiley-Interscience, New York, pp 128–149

Liu K, Hu H, Wang W, Zhang X (2016) Genetic engineering of *Pseudomonas chlororaphis* GP72 for the enhanced production of 2-hydroxyphenazine. Microb Cell Fact 15:131

Liu S-T, Lee L-Y, Tai C-Y, Hung C-H, Chang Y-S, Wolfram JH, Rogers R, Goldstein AH (1992) Cloning of an *Erwinia herbicola* gene necessary for the gluconic acid production and enhanced mineral phosphate solubilization in *Escherichia coli* HB101: nucleotide sequence and probable involvement in biosynthesis of the coenzyme pyrroloquinoline quinone. J Bacteriol 174:5814–5819

Loeschcke A, Thies S (2015) *Pseudomonas putida*—a versatile host for the production of natural products. Appl Microbiol Biotechnol 99:6197–6214

Lutz MP, Wenger S, Maurhofer M, Defago G, Duffy B (2004) Signaling between bacterial and fungal biocontrol agents in a strain mixture. FEMS Microbiol Ecol 48:447–455

Lynch JM (1990) The rhizosphere. Wiley, Chichester

Maddula VSRK, Pierson EA, Pierson LS (2008) Altering the ratio of phenazines in *Pseudomonas chlororaphis* (*aureofaciens*) strain 30–84: effects on biofilm formation and pathogen inhibition. J Bacteriol 190:2759–2766

Madhaiyan M, Poonguzhali S, Hari K, Saravanan VS, Sa T (2006) Influence of pesticides on the growth rate and plant-growth promoting traits of *Gluconacetobacter diazotrophicus*. Pestic Biochem Physiol 84:143–154

Magazin MD, Moores JC, Leong J (1986) Cloning of the gene coding for the outer membrane receptor protein for ferric pseudobactin, a siderophore from a plant growth-promoting *Pseudomonas* strain. J Biol Chem 261:795–799

Mandal L, Kotasthane AS (2014) Isolation and assessment of plant growth promoting activity of siderophore producing *Pseudomonas fluorescens* in crops. Int J Agric Environ Biotechnol 7:63–67

Mandryk-Litvinkovich MN, Muratova AA, Nosonova TL, Evdokimova OV, Valentovich LN, Titok MA, Kolomiets EI (2017) Molecular genetic analysis of determinants defining synthesis of 2,4-diacetylphloroglucinol by *Pseudomonas brassicacearum* BIM B-446 bacteria. Appl Biochem Microbiol 53:31–39

Manuel J, Selin C, Fernando WGD, de Kievit TR (2012) Stringent response mutants of *Pseudomonas chlororaphis* PA23 exhibit enhanced antifungal activity against *Sclerotinia sclerotiorum* in vitro. Microbiology 158:207–216

Marahiel MA, Stachelhaus T, Mootz HD (1997) Modular peptide synthetases involved in nonribosomal peptide synthesis. Chem Rev 97:2651–2674

Market Research Future (2016) Microbial products market research report –forecast to 2023. (2017–2023). https://www.marketresearchfuture.com. Accessed 5 Mar 2019

Markets and Markets (2016) Agricultural surfactants market by type (non-ionic, anionic, cationic, amphoteric), application (herbicides, fungicides), substrate type (synthetic, bio-based), crop type (cereals & grains, fruits & vegetables), and region – global forecast to 2022. (2016–2022). https://www.marketsandmarkets.com. Accessed 5 Mar 2019

Marks BB, Megías M, Nogueira MA, Hungria M (2013) Biotechnological potential of rhizobial metabolites to enhance the performance of *Bradyrhizobium* spp. and *Azospirillum brasilense* inoculants with soybean and maize. AMB Express 3:21

Matson PA, Parton WJ, Power AG, Swift MJ (1997) Agricultural intensification and ecosystem properties. Science 277:504–509

Mauch-Mani B, Baccelli I, Luna E, Flors V (2017) Defense priming: an adaptive part of induced resistance. Annu Rev Plant Biol 68:485–512

Maurhofer M, Keel C, Schnider C, Voisard C, Hass D, Defago G (1992) Influence of enhanced antibiotic production in *Pseudomonas fluorescens* strain CHA0 on its disease suppressive capacity. Phytopathology 82:190–195

Maurhofer M, Keel C, Haas D, Défago G (1994) Pyoluteorin production by *Pseudomonas fluorescens* strain CHA0 is involved in the suppression of *Pythium* damping-off of cress but not of cucumber. Eur J Plant Pathol 100:221–232

Mavrodi DV, Ksenzenko VN, Bonsall RF, Cook RJ, Boronin AM, Thomashow LS (1998) A seven-gene locus for synthesis of phenazine-1-carboxylic acid by *Pseudomonas fluorescens* 2–79. J Bacteriol 180:2541–2548

Mavrodi DV, Bonsall RF, Delaney SM, Soule MJ, Phillips G, Thomashow LS (2001) Functional analysis of genes for biosynthesis of pyocyanin and phenazine-1-carboxamide from *Pseudomonas aeruginosa* PAO1. J Bacteriol 183:6454–6465

Mavrodi DV, Blankenfeldt W, Thomashow LS (2006) Phenazine compounds in fluorescent *Pseudomonas*spp. biosynthesis and regulation. Annu Rev Phytopathol 44:417–445

Medema MH, Blin K, Cimermancic P, de Jager V, Zakrzewski P, Fischbach MA, Weber T, Takano E, Breitling R (2011) antiSMASH: rapid identification, annotation and analysis of secondary metabolite biosynthesis gene clusters in bacterial and fungal genome sequences. Nucleic Acids Res 39:W339–W346

Meldau DG, Meldau S, Hoang LH, Underberg S, Wünsche H, Baldwin IT (2013) Dimethyl disulfide produced by the naturally associated bacterium *Bacillus* sp. B55 promotes *Nicotiana attenuata* growth by enhancing sulfur nutrition. Plant Cell 25:2731–2747

Mendes R, Kruijt M, de Bruijn I, Dekkers E, van der Voort M, Schneider JHM, Piceno YM, DeSantis TZ, Anderson GL, Bakker PAHM et al (2011) Deciphering the rhizosphere microbiome for disease-suppressive bacteria. Science 332:1097–1100

Metzker ML (2010) Sequencing technologies—the next generation. Nat Rev Genet 11:31–46

Meyer JM, Abdallah MA (1978) The fluorescent pigment of *Pseudomonas fluorescens*: biosynthesis, purification and physicochemical properties. J Gen Microbiol 107:319–328

Mishra D, Kar M (1974) Nickel in plant growth and metabolism. Bot Rev 40:395–452

Monds RD, Newell PD, Schwartzman JA, O'Toole GA (2006) Conservation of the Pho regulon in *Pseudomonas fluorescens* Pf0-1. Appl Environ Microbiol 72:1910–1924

Moree WJ, Phelan VV, Wu C-H, Bandeira N, Cornett DS, Duggan BM, Dorrestein PC (2012) Interkingdom metabolic transformations captured by microbial imaging mass spectrometry. Proc Natl Acad Sci U S A 109:13811–13816

Morel MA, Cagide C, Minteguiaga MA, Dardanelli MS, Castro-Sowinski S (2015) The pattern of secreted molecules during the co-inoculation of alfalfa plants with *Sinorhizobium meliloti* and *Delftia* sp. strain JD2: an interaction that improves plant yield. Mol Plant-Microbe Interact 28:134–142

Morel MA, Cagide C, Castro-Sowinski S (2016) The contribution of secondary metabolites in the success of bioformulations. In: Arora NK et al (eds) Bioformulations: for sustainable agriculture. Springer, New Delhi, pp 235–250

Moynihan JA, Morrissey JP, Coppoolse ER, Stiekema WJ, O'Gara F, Boyd EF (2009) Evolutionary history of the *phl* gene cluster in the plant-associated bacterium *Pseudomonas fluorescens*. Appl Environ Microbiol 75:2122–2131

Nagel R, Turrini PCG, Nett RS, Leach JE, Verdier V, Van Sluys M-A, Peters RJ (2017) An operon for production of bioactive gibberellin A_4 phytohormone with wide distribution in the bacterial rice leaf streak pathogen *Xanthomonas oryzae* pv. *oryzicola*. New Phytol 214:1260–1266

Nagendra Prabhu G, Bindu P (2016) Optimization of process parameters for siderophore production under solid state fermentation using polystyrene beads as inert support. J Sci Ind Res 75:621–625

Nandi M, Selin C, Brassinga AKC, Belmonte MF, Fernando WGD, Loewen PC, de Kievit TR (2015) Pyrrolnitrin and hydrogen cyanide production by *Pseudomonas chlororaphis* strain PA23 exhibits nematicidal and repellent activity against Caenorhabditis elegans. PLoS One 10:1–19

Nandi M, Berry C, Brassinga AKC, Belmonte MF, Fernando WGD, Loewen PC, de Kievit TR (2016a) *Pseudomonas brassicacearum* strain DF41 kills *Caenorhabditis elegans* through biofilm-dependent and biofilm-independent mechanisms. Appl Environ Microbiol 82:6889–6898

Nandi M, Selin C, Brawerman G, Fernando WGD, de Kievit TR (2016b) The global regulator ANR is essential for *Pseudomonas chlororaphis* strain PA23 biocontrol. Microbiol (United Kingdom) 162:2159–2169

Nansathit A, Phaosiri C, Pongdontri P, Chanthai S, Ruangviriyachai C (2009) Synthesis, isolation of phenazine derivatives and their antimicrobial activities. Walailak J Sci Technol 6:79–91

Nelson AG, Spaner D (2010) Cropping systems management, soil microbial communities, and soil biological fertility. In: Lichtfouse E (ed) Genetic engineering, biofertilisation, soil quality and organic farming, Sustainable Agriculture Reviews. Springer, Dordrecht, pp 217–242

Nett RS, Contreras T, Peters RJ (2017) Characterization of CYP115 as a gibberellin 3-oxidase indicates that certain rhizobia can produce bioactive gibberellin A_4. ACS Chem Biol 12:912–917

Nikel PI, Martínez-García E, de Lorenzo V (2014) Biotechnological domestication of pseudomonads using synthetic biology. Nat Rev Microbiol 12:368–379

Niu B, Paulson JN, Zheng X, Kolter R (2017) Simplified and representative bacterial community of maize roots. Natl Acad Sci 114:E2450–E3459

Nowak-Thompson B, Gould SJ, Kraus J, Loper JE (1994) Production of 2,4-diacetylphloroglucinol by the biocontrol agent *Pseudomonas fluorescens* Pf-5. Can J Microbiol 40:1064–1066

O'Sullivan DJ, O'Gara F (1992) Traits of fluorescent *Pseudomonas* spp. involved in suppression of plant root pathogens. Microbiol Rev 56:662–676

Ochi K, Hosaka T (2013) New strategies for drug discovery: activation of silent or weakly expressed microbial gene clusters. Appl Microbiol Biotechnol 97:87–98

Olanrewaju OS, Glick BR, Babalola OO (2017) Mechanisms of action of plant growth promoting bacteria. World J Microbiol Biotechnol 33:197

Ongena M, Jacques P, Touré Y, Destain J, Jabrane A, Thonart P (2005) Involvement of fengycin-type lipopeptides in the multifaceted biocontrol potential of *Bacillus subtilis*. Appl Microbiol Biotechnol 69:29–38

Packer MS, Liu DR (2015) Methods for the directed evolution of proteins. Nat Rev Genet 16:379–394

Pandey A, Szakacs G, Soccol CR, Rodriguez-Leon JA, Soccol VT (2001) Production, purification and properties of microbial phytases. Bioresour Technol 77:203–214

Pandya ND, Desai PV (2014) Screening and characterization of GA3 producing *Pseudomonas monteilii* and its impact on plant growth promotion. Int J Curr Microbiol App Sci 3:110–115

Paredes S, Gao T, Law TF, Finkel OM, Mucyn T, Teixeira PJPL, Salas González I, Feltcher ME, Powers MJ, Shank EA et al (2018) Design of synthetic bacterial communities for predictable plant phenotypes. PLoS Biol 16:e2003962. https://doi.org/10.1371/journal.pbio.2003962

Park JY, Oh SA, Anderson AJ, Neiswender J, Kim J-C, Kim YC (2011) Production of the antifungal compounds phenazine and pyrrolnitrin from *Pseudomonas chlororaphis* O6 is differentially regulated by glucose. Lett Appl Microbiol 52:532–537

Parvaze A, Zaidi A, Khan AA, Khan MS (2005) Effect of phorate on phosphate solubilization and indole acetic acid releasing potentials of rhizospheric microorganisms. Ann Plant Protect Sci 13:139–144

Paterson J, Jahanshah G, Li Y, Wang Q, Mehnaz S, Gross H (2017) The contribution of genome mining strategies to the understanding of active principles of PGPR strains. FEMS Microbiol Ecol 93:fiw249. https://doi.org/10.1093/femsec/fiw249

Phillips DA, Fox TC, King MD, Bhuvaneswari TV, Teuber LR (2004) Microbial products trigger amino acid exudation from plant roots. Plant Physiol 136:2887–2894

Piechulla B, Degenhardt J (2014) The emerging importance of microbial volatile organic compounds. Plant Cell Environ 37:811–812

Piechulla B, Lemfack MC, Kai M (2017) Effects of discrete bioactive microbial volatiles on plants and fungi. Plant Cell Environ 40:2042–2067

Pierson LS, Pierson EA (2010) Metabolism and function of phenazines in bacteria: impacts on the behavior of bacteria in the environment and biotechnological processes. Appl Microbiol Biotechnol 86:1659–1670

Pieterse CMJ, Zamioudis C, Berendsen RL, Weller DM, Van Wees SCM, Bakker PAHM (2014) Induced systemic resistance by beneficial microbes. Annu Rev Phytopathol 52:347–375

Pimentel D, Levitan L (1986) Pesticides: amounts applied and amounts reaching pests. Bioscience 36:86–91

Poritsanos N, Selin C, Fernando WGD, Nakkeeran S, de Kievit TR (2006) A GacS deficiency does not affect *Pseudomonas chlororaphis* PA23 fitness when growing on canola, in aged batch culture or as a biofilm. Can J Microbiol 52:1177–1188

Pospíšilová J, Synková H, Rulcová J (2000) Cytokinins and water stress. Biol Plant 43:321–328

Potrykus K, Cashel M (2008) (p)ppGpp: still magical? Annu Rev Microbiol 62:35–51

Raaijmakers JM, de Bruijn I, de Kock MJD (2006) Cyclic lipopeptide production by plant-associated *Pseudomonas* spp.: diversity, activity, biosynthesis, and regulation. Mol Plant-Microbe Interact 19:699–710

Raaijmakers JM, De Bruijn I, Nybroe O, Ongena M (2010) Natural functions of lipopeptides from *Bacillus* and *Pseudomonas*: more than surfactants and antibiotics. FEMS Microbiol Rev 34:1037–1062

Rajkumar M, Ae N, Prasad MNV, Freitas H (2010) Potential of siderophore-producing bacteria for improving heavy metal phytoextraction. Trends Biotechnol 28:142–149

Ram RM, Keswani C, Bisen K, Tripathi R, Singh SP, Singh HB (2018) Biocontrol technology: eco-friendly approaches for sustainable agriculture. In: Brah D, Azevedo V (eds) Omics technologies and bio-engineering: toward improving quality of life volume II microbial, plant, environmental and industrial technologies. Academic, London, pp 177–190

Ramarathnam R, Fernando WGD, de Kievit TR (2011) The role of antibiosis and induced systemic resistance, mediated by strains of *Pseudomonas chlororaphis*, *Bacillus cereus* and *B. amyloliquefaciens*, in controlling blackleg disease of canola. BioControl 56:225–235

Ramirez KS, Craine JM, Fierer N (2012) Consistent effects of nitrogen amendments on soil microbial communities and processes across biomes. Glob Chang Biol 18:1918–1927

Rao G, Lee J-K, Zhao H (2013) Directed evolution of phloroglucinol synthase PhlD with increased stability for phloroglucinol production. Appl Microbiol Biotechnol 97:5861–5867

Rateb ME, Houssen WE, Harrison WTA, Deng H, Okoro CK, Asenjo JA, Andrews BA, Bull AT, Goodfellow M, Ebel R et al (2011) Diverse metabolic profiles of a *Streptomyces* strain isolated from a hyper-arid environment. J Nat Prod 74:1965–1971

Ren H, Wang B, Zhao H (2017) Breaking the silence: new strategies for discovering novel natural products. Curr Opin Biotechnol 48:21–27

Richardson AE, Hadobas PA (1997) Soil isolates of *Pseudomonas* spp. that utilize inositol phosphates. Can J Microbiol 43:509–516

Robinson T, Singh D, Nigam P (2001) Solid-state fermentation: a promising microbial technology for secondary metabolite production. Appl Microbiol Biotechnol 55:284–289

Robison MM, Griffith M, Pauls KP, Glick BR (2008) Dual role for ethylene in susceptibility of tomato to verticillium wilt. J Phytopathol 149:385–388

Rodríguez H, Fraga R (1999) Phosphate solubilizing bacteria and their role in plant growth promotion. Biotechnol Adv 17:319–339

Romheld V, Marschner H (1984) Mobilization of iron in the rhizosphere of different plant species. Adv Plant Nutr 2:155–204

Roongsawang N, Hase K, Haruki M, Imanaka T, Morikawa M, Kanaya S (2003) Cloning and characterization of the gene cluster encoding arthrofactin synthetase from *Pseudomonas* sp. MIS38. Chem Biol 10:869–880

Rutledge PJ, Challis GL (2015) Discovery of microbial natural products by activation of silent biosynthetic gene clusters. Nat Rev Microbiol 13:509–523

Saber FMA, Abdelhafez AA, Hassan EA, Ramadan EM (2015) Characterization of fluorescent pseudomonads isolates and their efficiency on the growth promotion of tomato plant. Ann Agric Sci 60:131–140

Saha M, Sarkar S, Sarkar B, Sharma BK, Bhattacharjee S, Tribedi P (2016) Microbial siderophores and their potential applications: a review. Environ Sci Pollut Res 23:3984–3999

Saleh O, Bonitz T, Flinspach K, Kulik A, Burkard N, Mühlenweg A, Vente A, Polnick S, Lämmerhofer M, Gust B et al (2012) Activation of a silent phenazine biosynthetic gene cluster reveals a novel natural product and a new resistance mechanism against phenazines. MedChemComm 3:1009

Sarniguet A, Kraus J, Henkels MD, Muehlchen AM, Loper JE (1995) The sigma factor sigma S affects antibiotic production and biological control activity of *Pseudomonas fluorescens* Pf-5. Proc Natl Acad Sci U S A 92:12255–12259

Satyaprakash M, Nikitha T, Reddi EUB, Sadhana B, Vani SS (2017) Phosphorous and phosphate solubilising bacteria and their role in plant nutrition. Int J Curr Microbiol Appl Sci 6:2133–2144

Savchuk S, Fernando DWG (2004) Effect of timing of application and population dynamics on the degree of biological control of *Sclerotinia sclerotiorum* by bacterial antagonists. FEMS Microbiol Ecol 49:379–388

Sayyed RZ, Patel PR (2011) Biocontrol potential of siderophore producing heavy metal resistant *Alcaligenes* sp. and *Pseudomonas aeruginosa* RZS3 vis-à-vis organophosphorus fungicide. Indian J Microbiol 51:266–272

Schachtman DP, Reid RJ, Ayling SM (1998) Phosphorus uptake by plants: from soil to cell. Plant Physiol 116:447–453

Schalk IJ, Hannauer M, Braud A (2011) New roles for bacterial siderophores in metal transport and tolerance. Environ Microbiol 13:2844–2854

Schenk PM, Carvalhais LC, Kazan K (2012) Unraveling plant–microbe interactions: can multi-species transcriptomics help? Trends Biotechnol 30:177–184

Scherlach K, Hertweck C (2009) Triggering cryptic natural product biosynthesis in microorganisms. Org Biomol Chem 7:1753

Schisler DA, Slininger PJ (1997) Microbial selection strategies that enhance the likelihood of developing commercial biological control products. J Ind Microbiol Biotechnol 19:172–179

Schnider-Keel U, Seematter A, Maurhofer M, Blumer C, Duffy B, Gigot-Bonnefoy C, Reimmann C, Notz R, Défago G, Haas D et al (2000) Autoinduction of 2,4-diacetylphloroglucinol biosynthesis in the biocontrol agent *Pseudomonas fluorescens* CHA0 and repression by the bacterial metabolites salicylate and pyoluteorin. J Bacteriol 182:1215–1225

Schulz S, Dickschat JS (2007) Bacterial volatiles: the smell of small organisms. Nat Prod Rep 24:814

Schulz-Bohm K, Martín-Sánchez L, Garbeva P (2017) Microbial volatiles: small molecules with an important role in intra- and inter-kingdom interactions. Front Microbiol 8:2484

Selin C, Habibian R, Poritsanos N, Athukorala SNP, Fernando WGD, de Kievit TR (2010) Phenazines are not essential for *Pseudomonas chlororaphis* PA23 biocontrol of *Sclerotinia sclerotiorum*, but do play a role in biofilm formation. FEMS Microbiol Ecol 71:73–83

Selin C, Fernando WGD, de Kievit TR (2012) The PhzI/PhzR quorum-sensing system is required for pyrrolnitrin and phenazine production, and exhibits cross-regulation with RpoS in *Pseudomonas chlororaphis* PA23. Microbiology 158:896–907

Selin C, Manuel J, Fernando WGD, de Kievit TR (2014) Expression of the *Pseudomonas chlororaphis* strain PA23 Rsm system is under control of GacA, RpoS, PsrA, quorum sensing and the stringent response. Biol Control 69:24–33

Seralini G-E (2015) Why glyphosate is not the issue with roundup: a short overview of 30 years of our research. J Biol Phys Chem 15:1–9

Shah N, Klaponski N, Selin C, Rudney R, Fernando WGD, Belmonte MF, de Kievit TR (2016) PtrA is functionally intertwined with GacS in regulating the biocontrol activity of *Pseudomonas chlororaphis* PA23. Front Microbiol 7:1–14

Shaikh SS, Wani SJ, Sayyed RZ (2016) Statistical-based optimization and scale-up of siderophore production process on laboratory bioreactor. 3 Biotech 6:69

Sharma S, Kaur M, Srivastva DK, Gambhir G (2014) In vitro evaluation of plant growth regulators on tissue culture bioassay produced by *Pseudomonas* species. Int J Agric Environ Biotechnol 7:747–757

Shaukat M, Ahmad A, Rasul F, Khaliq T, Mudassir MA, Yasin M (2017) Inducing drought tolerance in wheat by applying natural and synthetic plant growth promoters. J Plant Nutr Soil Sci 180:739–747

Shen X, Wang Z, Huang X, Hu H, Wang W, Zhang X (2017) Developing genome-reduced *Pseudomonas chlororaphis* strains for the production of secondary metabolites. BMC Genomics 18:715

Shenker M, Chen Y (2005) Increasing iron availability to crops: fertilizers, organo-fertilizers, and biological approaches. Soil Sci Plant Nutr 51:1–17

Shi H, Chen L, Ye T, Liu X, Ding K, Chan Z (2014) Modulation of auxin content in Arabidopsis confers improved drought stress resistance. Plant Physiol Biochem 82:209–217

Shigenaga AM, Berens ML, Tsuda K, Argueso CT (2017) Toward engineering of hormonal crosstalk in plant immunity. Curr Opin Plant Biol 38:164–172

Silby MW, Nicoll JS, Levy SB (2009) Requirement of polyphosphate by *Pseudomonas fluorescens* Pf0-1 for competitive fitness and heat tolerance in laboratory media and sterile soil. Appl Environ Microbiol 75:3872–3881

Singh G, Wright D (2002) In vitro studies on the effects of herbicides on the growth of rhizobia. Lett Appl Microbiol 35:12–16

Singh IP, Sidana J, Bansal P, Foley WJ (2009) Phloroglucinol compounds of therapeutic interest: global patent and technology status. Expert Opin Ther Pat 19:847–866

Singh HB, Keswani C, Ray S, Yadav SK, Singh SP, Singh S, Sarma BK (2014) *Beauveria bassiana*: biocontrol beyond *Lepidopteran* pests. In: Sree KS, Varma A (eds) Biocontrol of *Lepidoteran* pests: use of soil microbes and their metabolites. Springer, Cham, pp 219–235

Singh HB, Jha A, Keswani C (eds) (2016a) Intellectual property issues in biotechnology. CABI, Wallingford. 304 pages, ISBN-13:9781780646534

Singh HB, Sarma BK, Keswani C (eds) (2016b) Agriculturally important microorganisms: commercialization and regulatory requirements in Asia. Springer, Singapore. 336 pages, ISBN-13:978-9811025754

Singh HB, Sarma BK, Keswani C (2017) Advances in PGPR research. CABI, Wallingford. 408 pages, ISBN-9781786390325

Singh HB, Keswani C, Singh SP (eds) (2019a) Intellectual property issues in microbiology. Springer, Singapore. 425 pages, ISBN- 9789811374654

Singh HB, Keswani C, Reddy MS, Royano ES, García-Estrada C (eds) (2019b) Secondary metabolites of plant growth promoting rhizomicroorganisms: discovery and applications. Springer, Singapore. 392 pages, ISBN-978-981-13-5861-6

Sinha S, Mukherjee SK (2008) Cadmium–induced siderophore production by a high cd-resistant bacterial strain relieved cd toxicity in plants through root colonization. Curr Microbiol 56:55–60

Skinnider MA, Dejong CA, Rees PN, Johnston CW, Li H, Webster ALH, Wyatt MA, Magarvey NA (2015) Genomes to natural products PRediction informatics for secondary metabolomes (PRISM). Nucleic Acids Res 43:9645–9662

Song G, Ryu C-M (2013) Two volatile organic compounds trigger plant self-defense against a bacterial pathogen and a sucking insect in cucumber under open field conditions. Int J Mol Sci 14:9803–9819

Song C, Kidarsa TA, van de Mortel JE, Loper JE, Raaijmakers JM (2016) Living on the edge: emergence of spontaneous *gac* mutations in *Pseudomonas protegens* during swarming motility. Environ Microbiol 18:3453–3465

Spaepen S, Vanderleyden J, Remans R (2007) Indole-3-acetic acid in microbial and microorganism-plant signaling. FEMS Microbiol Rev 31:425–448

Sparks TC, Nauen R (2015) IRAC: mode of action classification and insecticide resistance management. Pestic Biochem Physiol 121:122–128

Sparks TC, Hahn DR, Garizi NV (2017) Natural products, their derivatives, mimics and synthetic equivalents: role in agrochemical discovery. Pest Manag Sci 73:700–715

Suh SJ, Silo-Suh L, Woods DE, Hassett DJ, West SE, Ohman DE (1999) Effect of *rpoS* mutation on the stress response and expression of virulence factors in *Pseudomonas aeruginosa*. J Bacteriol 181:3890–3897

Sun H, Liu Z, Zhao H, Ang EL (2015) Recent advances in combinatorial biosynthesis for drug discovery. Drug Des Devel Ther 9:823–833

Suprapta DN (2012) Potential of microbial antagonists as biocontrol agents agianst plant fungal pathogens. J ISSAAS 18:1–8
Takeda R (1959) *Pseudomonas* pigments. III. J Agric Chem Soc Jpn 23:126–130
Theocharis A, Bordiec S, Fernandez O, Paquis S, Dhondt-Cordelier S, Baillieul F, Clément C, Barka EA (2012) *Burkholderia phytofirmans* PsJN primes *Vitis vinifera* L. and confers a better tolerance to low nonfreezing temperatures. Mol Plant-Microbe Interact 25:241–249
Thomashow LS, Weller DM (1988) Role of a phenazine antibiotic from *Pseudomonas fluorescens* in biological control of *Gaeumannomyces graminis* var. *tritici*. J Bacteriol 170:3499–3508
Tilman D (1998) The greening of the green revolution. Nature 396:211–212
Transparency Market Research (2014) Biosurfactants market – global scenario, raw material and consumption trends, industry analysis, size, share and forecasts. 2014–2020. http://www.transparencymarketresearch.com. Accessed 13 June 2018
Traxler MF, Watrous JD, Alexandrov T, Dorrestein PC, Kolter R (2013) Interspecies interactions stimulate diversification of the *Streptomyces coelicolor* secreted metabolome. MBio 4:e00459–e00413
Tsiafouli MA, Thébault E, Sgardelis SP, de Ruiter PC, van der Putten WH, Birkhofer K, Hemerik L, de Vries FT, Bardgett RD, Brady MV et al (2015) Intensive agriculture reduces soil biodiversity across Europe. Glob Chang Biol 21:973–985
Tyc O, van den Berg M, Gerards S, van Veen JA, Raaijmakers JM, de Boer W, Garbeva P (2014) Impact of interspecific interactions on antimicrobial activity among soil bacteria. Front Microbiol 5:567
Tyc O, Zweers H, de Boer W, Garbeva P (2015) Volatiles in inter-specific bacterial interactions. Front Microbiol 6:1412
Valencia JLM (2018) Solid granulated fertilizer formulated with mineral clays, siderophore chelating agents, secondary nutrients and micronutrients. US Patent 20180029945A1, 1 Feb 2018
van den Bosch F, Paveley N, Shaw M, Hobbelen P, Oliver R (2011) The dose rate debate: does the risk of fungicide resistance increase or decrease with dose? Plant Pathol 60:597–606
van den Broek D, Chin-A-Woeng TFC, Eijkemans K, Mulders IHM, Bloemberg GV, Lugtenberg BJJ (2003) Biocontrol traits of *Pseudomonas* spp. are regulated by phase variation. Mol Plant-Microbe Interact 16:1003–1012
van den Broek D, Bloemberg GV, Lugtenberg B (2005a) The role of phenotypic variation in rhizosphere *Pseudomonas* bacteria. Environ Microbiol 7:1686–1697
van den Broek D, Chin-A-Woeng TFC, Bloemberg GV, Lugtenberg BJJ (2005b) Molecular nature of spontaneous modifications in *gacS* which cause colony phase variation in *Pseudomonas* sp. strain PCL1171. J Bacteriol 187:593–600
Vance CP, Uhde-Stone C, Allan DL (2003) Phosphorus acquisition and use: critical adaptations by plants for securing a nonrenewable resource. New Phytol 157:423–447
Vansuyt G, Robin A, Briat J-F, Curie C, Lemanceau P (2007) Iron acquisition from Fe-pyoverdine by *Arabidopsis thaliana*. Mol Plant-Microbe Interact 20:441–447
Ventre I, Goodman AL, Vallet-Gely I, Vasseur P, Soscia C, Molin S, Bleves S, Lazdunski A, Lory S, Filloux A (2006) Multiple sensors control reciprocal expression of *Pseudomonas aeruginosa* regulatory RNA and virulence genes. Proc Natl Acad Sci U S A 103:171–176
Venturi V (2003) Control of *rpoS* transcription in *Escherichia coli* and *Pseudomonas*: why so different? Mol Microbiol 49:1–9
Vizcaino MI, Guo X, Crawford JM (2014) Merging chemical ecology with bacterial genome mining for secondary metabolite discovery. J Ind Microbiol Biotechnol 41:285–299
Voisard C, Keel C, Haas D, Dèfago G (1989) Cyanide production by *Pseudomonas fluorescens* helps suppress black root rot of tobacco under gnotobiotic conditions. EMBO J 8:351–358
Wallace A, Wallace GA, Cha JW (1992) Some modifications in trace metal toxicities and deficiencies in plants resulting from interactions with other elements and chelating agents-the special case of iron. J Plant Nutr 15:1589–1598
Walling LL (2009) Adaptive defense responses to pathogens and insects. Adv Bot Res 51:551–612

Walsh CT, Chen H, Keating TA, Hubbard BK, Losey HC, Luo L, Marshall CG, Miller DA, Patel HM (2001) Tailoring enzymes that modify nonribosomal peptides during and after chain elongation on NRPS assembly lines. Curr Opin Chem Biol 5:525–534

Wang W, Dong H, Zhang J, Xu Y, Zhang X (2011) The production, separation and stability of pyoluteorin: a biological pesticide. In: Stoytcheva DM (ed) IntechOpen. http://www.intechopen.com/books/pesticides-in-the-modern-world-pests-control-andpesticides-exposure-and-toxicity-assessment/the-production-separation-and-stability-of-pyoluteorin-abiological-pesticide. Accessed 13 Jun 2018

Weber T, Blin K, Duddela S, Krug D, Kim HU, Bruccoleri R, Lee SY, Fischbach MA, Müller R, Wohlleben W et al (2015) antiSMASH 3.0—a comprehensive resource for the genome mining of biosynthetic gene clusters. Nucleic Acids Res 43:W237–W243

Weese DJ, Heath KD, Dentinger BTM, Lau JA (2015) Long-term nitrogen addition causes the evolution of less-cooperative mutualists. Evolution (N Y) 69:631–642

Weller DM, Raaijmakers JM, Gardener BBM, Thomashow LS (2002) Microbial populations responsible for specific soil suppressiveness to plant pathogens. Annu Rev Phytopathol 40:309–348

Weller DM, Landa BB, Mavrodi OV, Schroeder KL, De La Fuente L, Blouin Bankhead S, Allende Molar R, Bonsall RF, Mavrodi DV, Thomashow LS (2007) Role of 2,4-diacetylphloroglucinol-producing fluorescent *Pseudomonas* spp. in the defense of plant roots. Plant Biol 9:4–20

Whistler CA, Corbell NA, Sarniguet A, Ream W, Loper JE (1998) The two-component regulators GacS and GacA influence accumulation of the stationary-phase sigma factor sigmaS and the stress response in *Pseudomonas fluorescens* Pf-5. J Bacteriol 180:6635–6641

Workentine ML, Chang L, Ceri H, Turner RJ (2009) The GacS/GacA two-component regulatory system of *Pseudomonas fluorescens*: A bacterial two-hybrid analysis. FEMS Microbiol Lett 292:50–56

Wuana RA, Okieimen FE (2011) Heavy metals in contaminated soils: a review of sources, chemistry, risks and best available strategies for remediation. ISRN Ecol 2011:1–20

Yan Q, Lopes LD, Shaffer BT, Kidarsa TA, Vining O, Philmus B, Song C, Stockwell VO, Raaijmakers JM, McPhail KL et al (2018) Secondary metabolism and interspecific competition affect accumulation of spontaneous mutants in the GacS-GacA regulatory system in *Pseudomonas protegens*. MBio 9:e01845–e01817

Yao R, Pan K, Peng H, Feng L, Hu H, Zhang X (2018) Engineering and systems-level analysis of *Pseudomonas chlororaphis* for production of phenazine-1-carboxamide using glycerol as the cost-effective carbon source. Biotechnol Biofuels 11:130

Yasmeen A, Basra SMA, Wahid A, Farooq M (2013) Improving drought resistance in wheat (*Triticum aestivum*) by exogenous application of growth enhancers. Int J Agric Biol 15:1307–1312

Zha W, Rubin-Pitel SB, Shao Z, Zhao H (2009) Improving cellular malonyl-CoA level in *Escherichia coli* via metabolic engineering. Metab Eng 11:192–198

Zhang Y, Fernando WGD, de Kievit TR, Berry C, Daayf F, Paulitz TC (2006) Detection of antibiotic-related genes from bacterial biocontrol agents with polymerase chain reaction. Can J Microbiol 52:476–481

Zhang J, Wang W, Lu X, Xu Y, Zhang X (2010) The stability and degradation of a new biological pesticide, pyoluteorin. Pest Manag Sci 66:248–252

Zhang R, Cao Y, Liu W, Xian M, Liu H (2017) Improving phloroglucinol tolerance and production in *Escherichia coli* by GroESL overexpression. Microb Cell Factories 16:227

Zhao Y, Zhao J, Zhao C, Zhou H, Li Y, Zhang J, Li L, Hu C, Li W, Peng X et al (2015) A metabolomics study delineating geographical location-associated primary metabolic changes in the leaves of growing tobacco plants by GC-MS and CE-MS. Sci Rep 5:16346

Zhao X, Chen Z, Yu L, Hu D, Song B (2018) Investigating the antifungal activity and mechanism of a microbial pesticide Shenqinmycin against *Phoma* sp. Pestic Biochem Physiol 147:46–50

Potential of Bioeconomy in Urban Green Infrastructure

13

Petra Schneider, Andreas Meyer, and Kay Plat

Abstract

In the course of the general discussion about sustainable urban development, there is currently a growing interest in urban greenery and the concept of green infrastructure, both at the national and international level. At the international level, the EU Green Infrastructure Strategy explicitly includes urban spaces. Urban green infrastructure offers a promising potential for bioeconomic activities. Bioeconomy stands for the structural change from a petroleum-based to a bio-based economy, which combines economic prosperity with ecological and social compatibility. The concept refers to the provision and use of renewable resources such as plants, animals and microorganisms, as well as the prevention of waste. The present contribution analyses the potential of bioeconomy in urban green infrastructure with a focus on a multifunctional biomass production, particularly focused on the production of food and feed through urban agriculture. The contribution discusses the potentials and challenges of urban gardening as well as urban farming approaches.

Keywords

Urban gardening potential · Urban farming potential · Bioeconomy · Urban green infrastructure · Nature-based solutions

P. Schneider (✉) · A. Meyer
Department of Water, Environment, Civil Engineering, and Safety, Working group "Ecological Engineering", University of Applied Sciences Magdeburg-Stendal, Magdeburg, Germany
e-mail: petra.schneider@hs-magdeburg.de

K. Plat
SIKT, project "Greenhub", Leipzig University, Leipzig, Germany

C. Keswani (ed.), *Bioeconomy for Sustainable Development*,
https://doi.org/10.1007/978-981-13-9431-7_13

13.1 Introduction

In 1950, only 28.8% of the world's population lived in cities, compared with 49.9% in 2009. According to estimates by the United Nations Department of Economic and Social Affairs (UN DESA), by 2050, the proportion will continue to rise to almost 70%. Global urbanization poses enormous challenges to sustainable development. The demand for housing, food, infrastructure and energy is growing steadily. Only sustainable urban development can make its contribution to adaptation to climate change, energy and resource efficiency and sustainable mobility. Bio-based approaches can provide important impetus in terms of housing and urban agriculture, as well as the supply of energy or the use of waste.

Bioeconomy is based on the model of nature as a generally stable circular economy in which nothing is lost and everything is reused (Braungart and McDonough 2002). The concept refers to the provision and use of renewable resources such as plants, animals and microorganisms, as well as the prevention of waste. The principle has been known since time immemorial. Bioeconomy stands for the structural change from a petroleum-based to a bio-based economy, which combines economic prosperity with ecological and social compatibility. A social change of values in favour of sustainable forms of production, trade and consumption is also an indispensable component of the bioeconomy. Since the necessary raw materials for the bioeconomy come directly from nature, it also dictates to our economies basic rules, stipulations and limits. The bioeconomy already plays an important role in the food sector. New bio-based processes and ingredients have made products more versatile, healthier, cheaper and more sustainable. This characteristic forms the potential for bioeconomy in urban green infrastructure.

Green infrastructure is more than a summarising term for the green interior of cities with parks, playgrounds, sports fields or cemeteries. With green infrastructure is associated a strategic planning approach that aims to promote green qualities in the city as a whole. Here, individual areas are intertwined so that they complement each other in their effects, and in total new qualities can arise. Instead of monofunctional use, the focus is on multifunctionality. The concept of green infrastructure offers municipalities the opportunity to act strategically and to develop various ecosystem services.

According to the European Commission (2010), green infrastructure describes a strategically planned network of natural and seminatural areas with different spatial features on different scale levels. These biotope networks aim at preserving biodiversity as well as strengthening and regenerating ecosystem functions and the potential for providing ecosystem services based thereon. In principle, the implementation of green infrastructure aims for a sustainable use of nature. Green infrastructure is a network of natural and artificially created urban and rural vegetation and water areas. This positively affects the ecosystem, biodiversity and resilience of the areas and strengthens the health of flora and fauna as well as humans (Naumann et al. 2011). Green infrastructure is conceptually opposed to concepts of grey and brown infrastructure and offers a cost-effective and stable completion to purely dedicated grey infrastructure. Given the high risk of biodiversity loss in Europe due to intensive land use and fragmentation, the concept of green infrastructure is strongly promoted by the EU. Grey infrastructures such as roads, rainwater management systems, pitches as well as also roofs and facades can become a green infrastructure if they hold back and evaporate water, provide shade, become a place of human well-being, and foster biodiversity. Urban green infrastructure aims to connect multiple interests in cities. The green infrastructure's strategic approach is to understand the entire surface of the city as a potential green infrastructure that can and should deliver environmental and economic benefits.

Green infrastructure describes all elements of a network of connected green spaces and creates the spatial basis for the sustainable use of ecosystems and their services (European Commission 2013). Protected areas are integrated into a common system together with the existing landscape. Some of these elements can be reforestation, green bridges, roofs or walls. Strategic spatial planning deliberately gives nature space to promote the preservation of biodiversity and ecosystem services. Green infrastructure is a part of nature-based solutions, that means "actions to protect, sustainably manage, and restore natural or modified ecosystems, that address societal challenges effectively and adaptively, simultaneously providing human well-being and biodiversity benefits" as defined by the International Union for Conservation of Nature (IUCN) (Cohen-Shacham et al. 2016). Setting up multifunctional areas can also be a green infrastructure measure. Different, compatible usage methods are combined. In these areas, for example, the effects of climate change can be reduced and recreational and compensation areas created at the same time. In addition to the multifunctionality, economic arguments for green infrastructure can also be raised. Despite high costs, for example for a flood prevention program, the resulting costs for dike relocation and an associated restoration are relatively low compared to those caused by flood damage (European Commission/DG Environment 2009).

A special role does green infrastructure play in urban areas. Here, the disintegration of the green areas by sealing for traffic and building infrastructure and thus the loss of biodiversity is particularly pronounced (Neßhöfer et al. 2012). However, a variety of ecosystem services can be provided in cities, especially when the green infrastructure concept is pursued. For example, the air quality can be substantially improved through parks and green spaces. Even overgrown house walls can make a major contribution by absorbing the heat that comes from the sunlight on the houses. These green walls are helping to reduce the effect of the urban "heat islands" (Neßhöfer et al. 2012).

The present contribution analyses the potential of bioeconomy in urban green infrastructure with a focus on a multifunctional biomass production, particularly focused on the production of food and feed through urban agriculture. The total amount of biomass produced on a given area in 1 year is called net primary production (Singh et al. 2017). Having in view the growing global population, especially in large cities, agriculture already faces the challenge of using every hectare of arable land as effectively as possible in order to meet the demand for food and biomass. The gap between the growing population and scarcity of cultivated land could close urban agriculture. The contribution presents options for urban agriculture on different scales and assesses their potential in terms of bioeconomy.

13.2 Materials and Methods

The investigation is based on an in-depth literature review, the collection of case studies from the literature as well as field studies in Congo, Ecuador, Germany and Vietnam. These countries were selected to represent the variety of urban agriculture approaches around the world and on different development levels of the countries. The case studies show that everywhere in the world where are urban food production activities, the levels, approaches and scales are quite different. The term urban agriculture has been coined in recent times by Lohrberg (2001), to describe the agricultural use in the so-called intermediate city and in densification areas. The idea, however, goes back to reflections on urban food production in the 1920s by Migge (1929). Urban farming is often used interchangeably with urban gardening, but there is a significant difference in the scale: while urban horticulture is operated by subgroups of the total population for the purpose of self-sufficiency, urban agriculture has the goal, also on a commercial basis, to supply products for the entire population (Lohrberg and Timpe 2011). In addition, as mentioned above, urban agriculture also includes, at least theoretically, the breeding of (small) cattle in urban areas (Lohrberg and Timpe 2011).

The assessment of case study countries was done qualitatively and semi-quantitatively through a SWOT analysis. The SWOT analysis provides a framework concept, taking into account the internal and external influencing factors, thus allowing different aspects of the current state (David 1993; Helms and Nixon 2010). The SWOT analysis considered a method for systemic situation analysis, where S stands for strengths, W for weaknesses, O for opportunities and T for threats. For better illustration, the SWOT analysis is presented in a matrix comprising two dimensions: firstly, the opportunities and risks arising from the external environment and, secondly, the strengths and weaknesses of the scope in terms of resources. An overall overview is created that takes into account both positive and negative aspects. Urban agriculture usually is significantly different from conventional agriculture in rural areas. The following are overall types of urban food production considered in the present analysis: (a) urban gardening and (b) urban farming. Both differ particularly in their scale, dimension and breeding approaches.

13.3 Bioeconomy Options in Urban Green Infrastructure

13.3.1 Urban Gardening

Urban gardening exists as long as there are cities. Urban gardening is currently experiencing a high level of popularity and boom in the face of demographic change and increasing urban migration. Cities have high potential for unused land. Unused

Fig. 13.1 Urban gardening for food production and cultural purposes in Cuenca (Ecuador, left) and Hoa Binh (Vietnam, right)

areas such as rooftops and urban brownfields offer great potential for urban gardens. Apart from the advantage that up to now unused areas are upcycled, urban gardening offers that in a new way a socialization of culture and nature adjusts itself (Biedermann and Ripperger 2017); see Fig. 13.1.

Urban gardening is mostly done for the personal benefit of a person or a community. According to Lewis et al. (2018), there are three major motivations for urban gardeners: well-being, social aspects of gardening and the outputs (like vegetables) (Lewis et al. 2018). Both the contact with nature and the social and economic aspect play a role here. It is possible for city dwellers to experience nature through gardens in the city. The cultivation of vegetables in city gardens can also cover a certain part of the vegetable and fruit needs of the people involved. Urban gardening can contribute to food security (Rosol 2014). This plays a far greater role in developing countries than in developed countries. Food security is achieved when "all individuals can obtain a culturally accepted, wholesome diet through local, nonemergency sources at all times" (Gottlieb and Fisher 1996).

In addition to food security, especially the sociocultural advantage of urban gardens should be mentioned. Urban gardeners usually aim to improve the quality of life for themselves or for a community (Lohrberg and Timpe 2011). Commercial use is usually not associated with urban gardening. Gusted et al. (2017) shows a connection between garden size and usage. Thus, urban gardens are more likely to be referred to as small, micro or meso gardens claiming a subsistence or sociocultural use. Micro gardens are run by individuals or private households. Meso gardens are run often of associations or start-ups. Macro-garden uses in cities are more likely to be commercial and are counted to urban agriculture. Small garden approaches like allotment gardens, school gardens, roof gardens or community gardens are urban gardening (Fig. 13.2).

13.3.1.1 Environmental Potential

Urban gardens provide a habitat for a variety of animals and insects that would often be out of place in the city. The cultivation of plants can have a positive effect on the urban climate. Thus, urban gardening can partially counteract the increasing surface sealing and mitigate urban heat islands.

Strengths	**Weaknesses**
- Improving the quality of life - Intercultural exchange of experience - Contact to nature - Self-sufficiency in fruit and vegetables - Positive effect on city climate	- Space limited - Possible environmental pollution of the areas and thus also the food
Opportunities	**Threats**
- Integration in infrastructure city planning	- Commercialization through Companies

Fig. 13.2 SWOT analysis of urban gardening

13.3.1.2 Economic Potential

Since urban gardening is usually operated without profit, a resulting economic factor can still be named. Urban gardens have a significant share of ecosystem services in cities. For example, people can benefit from the production of food by plants, the pollination of bees and the urban climate regulation through planting, recreational areas and social cohesion (Langemeyer et al. 2016).

13.3.1.3 Social Potential

As mentioned above, urban gardening has a high social potential. People of different cultures and social environments can meet in urban gardens. Urban gardens can also have a positive effect on environmental education. People who would normally not be in contact with food production in a city can learn how to grow food.

Figures 13.3 and 13.4 provide impressions of special forms of urban gardening: community and private gardens in Magdeburg and Dresden (Germany), Fig. 13.3, and vertical gardens in Berlin (Germany) and Cuenca (Ecuador), Fig. 13.4.

13.3.2 Urban Farming

Urban farming is a generic term for various ways of primary food production in urban areas and their immediate environment for their own needs of the respective region (Lohrberg 2001). In addition to urban forms of horticulture, it also includes animal husbandry in urban areas. Urban farming has been rediscovered in recent years due to the following aspects:

Fig. 13.3 Community gardens in Magdeburg and private urban gardening in Dresden (Germany)

Fig. 13.4 Vertical gardens in Berlin (Germany, left) and Cuenca (Ecuador, right)

- Local food production and consumption is one of the ways to reduce transport routes (and to reduce the carbon footprint) (Halweil and Prugh 2002).
- Large-scale urban plant breeding can help to make transport-intensive and energy-intensive material cycles more local and economical by the direct use of (prepurified) wastewater for irrigation or fertilization purposes.
- Decentralization of food production and connected extensification for increasing food safety, although the risk of pollution from urban products might be potentially higher than that of products grown on healthy topsoil in the countryside.
- Growing interest in local food production goes hand in hand with social movements grouped around the knowledge, enhancement or preservation of local specialties (e.g. slow food).
- There is a growing need for food that is produced in an environmentally sound and socially equitable manner, which is often attempted through in-house production or local acquisition (Nairn and Vitello 2010).

13.3.2.1 Horticulture

Urban horticulture covers all methods of cultivation of plants in urban and peri-urban environments. It includes both the cultivation for food and ornamental plants (see Fig. 13.5). According to Orsini et al., urban horticulture is the most competitive branch of urban farming. The low CO_2 emissions and greater transparency of food

Fig. 13.5 Urban horticulture in Dresden, Germany

production reflect the modern urban lifestyle and make urban farming and horticulture particularly interesting (Eigenbrod and Gruda 2015). Urban horticulture ranges from traditional farming methods to highly innovative farming methods, such as controlled environment agriculture (CEA), organic vegetable cultivation in soil, hydroponics, vertical farming, rooftop use and more. For example, "Z-Farming" (zero-acreage farming) is a method where only space in cities is used which are not resealed but are already sealed and not used (like rooftops).

Due to the limited space in the city, vegetables with high water and fertilizer efficiency are particularly profitable (Orsini et al. 2013). These cultures often have a short growth cycle and are called high-value cash crop (de Bon et al. 2016). According to Thornbuch (2015), worldwide, one third of urban areas (regardless of suitability or availability) are required to cover the whole vegetable consumption of urban dwellers. Abegunde (2012) stated that urban horticulture is important in developing countries to boost food and ornamental plants production, provide job opportunities and promote green space development.

Environmental Potential

The current global greenhouse gas emissions caused by conventional agriculture are 20–30% (Eigenbrod and Gruda 2015). This is largely due to the large proportion of animal feed produced for meat consumption. In addition, large amounts of water and energy from fossil fuels are required for the production, processing and transport of conventional agricultural products. Renewable energy sources and circulation systems such as hydroponics or aquaponics can reduce the ecological footprint in agriculture (Ohyama et al. 2008). Here is a great potential for urban horticulture. Hydroponics can also be used, for example, where the soil is contaminated and unusable for direct planting. In recirculating hydroponics, up to 5–10 times less water is used than in conventional agriculture (Caplow 2009). This is a decisive advantage especially in countries with water scarcity.

Economic Potential

The urban horticulture market has grown steadily in recent years. The yields of the crops vary greatly depending on the type of cultivation. Indoor cultivation methods usually have a much higher yield compared to outdoor cultivation. As an example, in 2015, world crop yields of leafy greens averaged 339,800 lbs per acre in indoor farms. At outdoor lettuce, the average yield per acre was only 30,700 lbs (Agrilyst 2015). This is mainly due to the fact that indoor farms can grow food throughout the year and can be grown by, e.g., soilless cultures like hydroponics that promote faster plant growth. For instance, in 2013, the volume of the vertical farming market rose from USD 0.403 billion to USD 1.5 billion by 2016. According to GlobeNewswire (2015), the vertical farming market will reach USD 6.4 billion by 2023. On average, around 25–30% of urban dwellers work in the agricultural sector worldwide (Orsini et al. 2013)

Social Potential

Urban horticulture is mostly done as commercial. Nevertheless, there are also social advantages for city dwellers. Particularly in developing countries, food security plays an important role. With the help of efficient cultivation methods, food can be produced locally. The cultivation of ornamental plants and the integration of horticulture into architecture concepts can also increase the well-being of city dwellers (Specht et al. 2014) (Fig. 13.6).

13.3.2.2 Aquaculture

According to Hubold, aquaculture is the cultivation of a wide variety of aquatic organisms, such as fish, molluscs and crustaceans, in different artificial and natural forms of ponds and containers (Hubold and Klepper 2013). Aquaculture counts as urban farming when it is practised in the urban or peri-urban environment. For example, fish within a city can be cultivated in tanks, ponds, converted rice fields, borrow pits, lakes and reservoirs, multifunctional wetlands, sea or cages (Bunting and Little 2015). Worldwide fish farming is very popular, and the fish consumption

Strengths	Weaknesses
- regional and decentralized food supply in the city - short transport routes reduce CO_2 emissions - contribution to food security - using closed or semi-closed systems reduce the inputs like fertilizer or pesticides	- Possible health risks from polluted areas and the grown foods (Säumel 2012)
Opportunities	**Threats**
- creation of green spaces in the city - positive effects on microclimate of the city	- Overuse and pollution of the areas by intensive agriculture

Fig. 13.6 SWOT analysis of horticulture

increased with a growth rate of 5.8% during 2001–2016 (FAO 2018a, b, c). In 2017, the average annual per capita consumption of seafood worldwide was 20.5 kg (FAO 2018a, b, c). Especially in developing countries, the availability of aquaculture in the city can be a valuable option as a source of food and income as well as a source of high-quality protein (FAO 2011).

Environmental Potential

Urban aquaculture can help to save long transport distances. However, intensive aquaculture can cause increasing environmental damage such as eutrophication of water bodies. High stocking densities, the use of medicines and enormous water consumption entail environmental risks. Circuit-based systems such as aquaponics offer potential for improvement. In aquaponics, nutrients from fish excreta such as nitrate and phosphorus are used as fertilizers for vegetable cultivation. Recirculating aquaculture systems (RAS) or aquaponics have a significantly better ecological footprint than conventional aquaculture, mainly due to lower water consumption. Modern RAS require about 90% less fresh water than conventional aquaculture systems (Timmons and Ebeling 2013).

Economic Potential

According to the FAO (2018), global aquaculture production in 2016 was 110.2 million tonnes with a market volume of USD 243.5 billion. Of this, the share of fish as food is 80 million tonnes or USD 231.6 billion. The remainder is divided between aquatic plants (30.1 million tonnes) and nonfood products (37,900 tonnes) (FAO 2018a, b, c).

Social Potential

Aquaculture is mainly used to produce food. In developing countries, fish is often the only way for people to obtain animal protein. Worldwide about 20 million people are working in the aquaculture sector, in which 85% are from Asia alone and only about 0.5% in Europe (FAO 2018a, b, c). Depending on the urban environment in developing countries, a fishpond can deliver between 200 and 400 g Nil tilapia per square metre every 4–6 months, i.e., 0.4–1.2 kg/m^2 per year, depending on the scale of inputs and management practices (FAO 2011). Aquaculture can create a positive social impact for urban dwellers by contributing to food security and income and employment opportunities (White and Edwards 2015; Bunting and Little 2015) (Fig. 13.7).

13.3.2.3 Agroforestry

The term agroforestry refers to land use systems in which trees or shrubs are combined with arable crops and/or animal husbandry in such a way that environmental and economic benefits are obtained between the various components (Nair 1993). Usually in agroforestry systems, it is distinguished between the combination of (Nair 1985):

Strengths	**Weaknesses**
- Local product - Increasing market of fish products - Can contribute to food security - Simple source for high quality protein - Income and employment	- Pollution of water bodies at intensive aquaculture - Fish feed usually still consists of a large proportion of fish meal or oils from wild fish - Often high investment costs when environment friendly
Opportunities	**Threats**
- International agreements and guidelines	- inadequate waste water strategy

Fig. 13.7 SWOT analysis of aquaculture

Fig. 13.8 Agroforestry system in the Democratic Republic of the Congo

- Trees with arable crops (silvoarable systems)
- Trees with animal husbandry (silvopastoral systems)
- Trees with arable crops and animal husbandry (agrosilvopastoral systems)

Since the age, distribution and arrangement of woody plants can vary, there are many different forms. Typical of all types of agroforestry are deliberately used interactions between woody and arable crops (Fig. 13.8). For example, earlier forests were also used for pig fattening. The orchard is a traditional agroforestry system that can still be found nowadays in Europe. So the meadow next to the fruit growing still serves as pasture or hay. The diversity of agroforestry systems in terms of their design, species composition and management is large and ranges from shifting cultivation and homeguard systems, particularly in the tropics and subtropics, to aquaculture in mangrove forests (Nair 1985), to water protection strands (Vought et al. 1995) and windbreak hedging systems (Brandle et al. 2004) in North America and Canada and to Knicks and short-drive alley cropping systems (Grüenewald et al. 2007), particularly in Central Europe.

Environmental Potential

Across the world, there has been a growing awareness in recent years that intensive agriculture causes many environmental problems such as soil, air and water pollution, food depletion and soil fertility (McNeely and Scherr 2003). Based on this insight, the concept of multifunctionality of agriculture developed. Agroforestry systems protect the soil from erosion by wind and water and can stabilize and improve the yield of annual plants. In addition, the field strips planted with trees form habitats and retreats for plants and animals.

Economic Potential

In agriculture, there is an increasing field of tension between ecological and economic requirements. For Europe, the potential of agroforestry in the extensive EU project SAFE (Silvoarable Agroforestry for Europe) could be demonstrated by means of models (Dupraz et al. 2005). The ecological advantages of perennial woody crops on agricultural land are the increased structural diversity and thus the positive effect on biodiversity (McNeely and Scherr 2003), the aesthetic enhancement of the landscape, reduced erosion and increased protection against water pollution and flooding. The economic advantages are in particular the higher average productivity of silvoarable tree systems compared to the separate cultivation of trees and crops. Initial economic calculations indicate that farmers with agroforestry systems in Europe in the longer term even achieve greater profit than with farming method's traditional agriculture (Dupraz et al. 2005).

Social Potential

In comparison with pure stands, crops with plants of different stature heights use the solar irradiation on the surface more comprehensively. There is a larger photosynthetic surface, so more biomass can be formed. For example, at the trial sites in Southern France, where single-row walnut strips were combined with wheat, a yield could be achieved on 100 ha of arable land, which would have required a comparatively 140 ha in pure stands. The trees took only 5% of the area. There are many possibilities for adapting this type of land use to the demands of farmers but also to the conditions of the respective location. Maintaining the permanent crops takes place primarily in the winter months and is thus in low competition with the working time requirements of other agricultural activities (Fig. 13.9).

13.3.2.4 Urban Beekeeping

About 15–30% of all food produced depends on pollinators (Greenleaf and Kremen 2006). According to the Food and Agriculture Organization of the United Nations (FAO) 2018, the value of agricultural products from pollinators is estimated at between USD 235 and USD 577 billion per year. Among all pollinators, the honeybee (*Apis mellifera*) is the one that most frequently visits crops worldwide (Hung et al. 2018). The number of hives continues to decline due to various interrelated effects such as climate change, parasites, the use of pesticides and the loss of habitats (Sass 2011). On the other hand, beekeeping has been increasing in cities for several years (Fig. 13.10).

Strengths	Weaknesses
- Provision of highly demanded woody bioenergy sources on agricultural land - Improved area-related energy balance and improved nutrient use efficiency - Positive yield effects and higher yield stability in annual crops due to improved microclimate (e.g., by windbreaking tree strips) - Improved protection of arable crops against weather conditions - Promotion of beneficials in the field - Creation of rest or extensive zones in agricultural areas - Creation of wildlife retreats (especially small game)	- Higher establishment costs of agroforestry systems compared to annual crops - Higher effort and higher management costs - Competition between woody plants and crops for light, nutrients, water and habitats with negative effects on plant growth
Opportunities	**Threats**
- Extension of the agricultural product range (product diversification) - Improvement of income function (especially in low-income locations) - Creation of regional markets for agroforestry products and processing - Strengthening rural regions by promoting regional material cycles and regional added value - Sustainable energy supply	- Long-term capital and area bonding through the comparatively slow-growing trees - In special cases: woody roots can penetrate existing soil drainage systems and destroy / clog them up

Fig. 13.9 SWOT analysis of agroforestry

Fig. 13.10 Urban beekeeping in the zoo of Magdeburg (Germany). The beekeeping station is additionally used for educational purposes on ecosystem services

A bee colony always consists of three different groups of bees. There is generally one queen whose only job is to mate and lay eggs. Furthermore, the beehive consists of drones which are always male and whose task is to mate with the queen. The third group are the working bees. They are the ones who collect nectar and pollen from flowers and make wax and honey from them. The workers are always female but are not able to mate and reproduce. Furthermore, the workers are responsible for the care of the brood, the cleaning, the honeycomb building, the search for food and the guarding of the beehive (Cramp 2008).

Environmental Potential

Over 80% of all flowers and 75% of all fruits and vegetables are pollinated (Rose et al. 2016). Pollination thus contributes significantly to biodiversity conservation and has a key role in ecosystem services. Bees play a role in these tasks because they perform the skin part of pollination (Bradbear 2009). As Marinelli (2017) pointed out, urban green spaces today have to fulfil several functions, such as verdant getaways, playgrounds or gathering spots. Green spaces must also be a place for native plants and animals in order to preserve diversity.

Economic Potential

In 2017, the worldwide production of honey was 1,860,712 tonnes and for beeswax 42,307 tonnes and has been rising for years (FAO 2019a, b, c). The export of honey alone had a trade volume of US$ 2.1 billion this year (Workmann 2018). The honey produced and traded comes mainly from commercial production. A single hive produces between 11 and 27 kg per season (BBKA 2019), depending on the local conditions for the bees. In order to make a living, a beekeeper must have many individual hives, which is why the price of non-commercial honey is usually higher than that of commercial honey. However, people in countries such as Germany prefer regional products (Statista 2019a, b) and are willing to spend more money on them.

Social Potential

Urban beekeeping is a hobby rather than a commercial activity due to its low-financial profits. Therefore, the social benefits are the enjoyment and fun of the hobby. This concerns the individual, the family or the community in which the beekeeping is carried out in order to produce healthy food or to sell part of it (GNSW 2000). Further, it can be combined with educational activities, providing cultural ecosystem services (see Fig. 13.11).

13.3.2.5 Insects Farming

Insects have been consumed for thousands of years. Worldwide about 1400 species are known to be used by humans as food (Durst and Shono 2010). Consumption takes place in over 100 countries around the globe, with most known species coming from the African, Asian and American continents (Johnson 2010). Depending on the circumstances, the insects are regarded as a staple food or delicacy. In front of the background of an increasing world population, the FAO assumes that livestock production will increase by 60% compared to today's production (Hanboonsong et al. 2013). In this context, 80% of agricultural land is already being used for livestock farming (FAO 2019a, b, c).

In general, there exist two types to get edible insects. In the past, the insects were harvested in wild, but with the development of farming techniques, the insects will

Strengths	**Weaknesses**
- Pollination services for biodiversity - Sheltered from pesticides - Bee survival is high in cities - Sale of local products - Low space requirement for hives	- Limited number of hives for honey production - Honey production depends on plants in the neighbourhood - Honey production isn't possible during the whole year
Opportunities	**Threats**
- Improve local material cycles - Interacting with other urban gardening activities	- Disturbing for other residents - Allergic reactions due to stitching

Fig. 13.11 SWOT analysis of urban beekeeping

be more and more breeding on farms (Hanboonsong 2013). While insect-eating in Asian countries is very popular and common, in western countries, it is the main focus on feed production for animals (Jansson and Berggren 2015). However, the current state of European countries is changing, and insect breeding could change from animal food to human food. At the beginning of 2018, the EU Regulation 2015/2283 on novel food changed. Since then, it has been permitted on the European market to sell certain insects as food for humans. This has been accompanied by the establishment of start-up companies and increased research. Furthermore, insect breeding isn't only for food production. Bioconversion into energy like biodiesel is feasible as well (Surendra et al. 2016).

Environmental Potential

The feed conversion ratio, this means how much feed is needed to produce a 1 kg increase in weight, is very high in conventional meat production in contrast to insect farming. Smil (2002) wrote that a chicken needs 2.5 kg, a pork 5.0 kg and a cattle 10 kg of feed to increase their own weight about 1 kg. Furthermore, at the end of conventional production for a chicken and a pork, 55% of weight are edible and for beef just 40% of weight. On the other hand, insects need 1.7 kg or less of feed to gain their weight about 1 kg (Collavo et al. 2005). With the focus on land consumption and water use, 1 kg of beef requires 8–14 times more land and 5 times more water compared to mealworms (van Huis and Oonincx 2017). In 2012, Oonincx and de Boer examined the global warming potential (GWP) and found out that in contradiction to mealworms the GWP value for a chicken is 1.3–2.6 times, for pork 1.5–3.8 times and for beef 5.5–12.5 times higher.

Economic Potential

In 2018, the global market value of edible insects is about 406 million US$. According to estimations on Business Wire (2018), the market will reach up to 1.18 billion US$ by 2023. This includes the whole insect itself as well as further processing from insect breeding to meal, protein bars or bake products. Alone the market in the USA is projected to reach 50 million $US for flour, protein bars and snacks (Statista 2019a, b). Tao and Li (2018) found out in this context that 72% of American people definitely or probably would eat insects.

Consumer Behaviour

Food acceptance is controlled by affective, personal, cultural and situational factors, but motives are based mostly on sensory/pleasure considerations and health. Humans are inclined to avoid unfamiliar foods (neophobia), particularly when they are of animal origin.

With these novel foods, humans exhibit both an interest in (obtaining a wide variety of nutrients) and a reluctance to (the possibility that these foods may be harmful or toxic) eating them (the omnivore's dilemma). Neophobic reactions towards novel foods of animal origin may be decreased by lowering individuals' perceptions of their disgusting properties (Van Huis 2012) (Fig. 13.12).

13.3.2.6 Molluscs Farming

As part of aquaculture, molluscs farming is one of the major components in this area together with fish, crustaceans and plants (FAO 2014). Today, if we exclude a small number of land snails, only marine bivalves (mainly oysters, clams, scallops and mussels) are farmed with great success, utilizing methods sometimes old of centuries (Cattaneo-Vietti 2016). Although they have been cultivated for centuries, recent technological advances in the field of mollusc farming have allowed increasing their production significantly (Olivares-Banuelos 2018).

Strengths	**Weaknesses**
- Protein, fat, mineral and vitamin source - Feed conversion ratio - Low GWP and land consumption	- Distaste again edible insects - High energy consumption (heating in cold countries)
Opportunities	**Threats**
- Innovated business markets in western countries - Source for bioenergy	- Resistance to try new food (especially in western countries)

Fig. 13.12 SWOT analysis of insects farming

Environmental Potential
Filter feeding by populations of bivalve molluscs is reviewed with respect to their ability to act as an estuarine filter, increase clarity of coastal waters and facilitate the removal of nitrogen and other nutrients from eutrophic coastal waters. Most species of cultured bivalve molluscs clear particles from waters at rates of 1–4 l/h, and populations of shellfish in healthy assemblages can filter a substantial fraction of the water in coastal estuaries on a daily basis. Actively growing shellfish incorporates nitrogen and other nutrients into their tissues as they grow. On average, 16.8 g of nitrogen is removed from estuaries for every kilogram of shellfish meats harvested (Rice 2001).

Economic Potential
The most heavily traded bivalve mollusc species are mussels, clams, scallops and oysters, and the vast majority are farmed. China is by far the largest exporter of bivalves, exporting almost three times as much as Chile, the second largest exporter, in 2016. China also has significant domestic consumption, although the European Union is the largest single market for bivalves. Bivalves are widely promoted as healthy and sustainable food items, and demand has been rising in recent years. The world farmed food fish production amounted to 54.1 million tonnes of finfish, 17.1 million tonnes of molluscs and 7.9 million tonnes of crustaceans in 2016 (FAO 2018a, b, c). Only for the European Union, the value of molluscs farming reached 902.7 million euros, in 2014 (Eurostat 2018).

Social Potential
If aquaculture is planned as grow-out operations using a feedlot concept, then the benefits to communities are small. However, if aquaculture is planned as community-based development of a highly integrated, local operation, then employment opportunities and the potential for positive community impacts increase dramatically. Aquaculture can play an important economic role by creating new economic niches by generating employment in areas where there are few alternate job choices, by providing local sources of high-quality food and opportunities for attractive investments for local entrepreneurs to invest in the local economy, thereby increasing local control over economic development (White and Edwards 2015) (Fig. 13.13).

13.4 Options for Brownfield Rehabilitation

Most of the gardens are located on former urban brownfields. According to Tobisch (2013), who made a respective investigation in Germany, 69% of the gardens are located on urban brownfields. At first glance, urban gardening proves to be an extremely adaptable instrument that can and will be practiced on inner-city brownfields with various uses and varying periods of non-use. One example is the currently running research project “productive green infrastructure for post-industrial urban renewal” (proGIreg), where “productive green infrastructure for the regeneration of old” refers to industrial cities. Nature-based urban development measures in

Strengths	Weaknesses
- Increasing consumption of seafood - Consumer confidence in local products	- Harvest can´t reach a marketable size - Slow growth
Opportunities	**Threats**
- Efficient biological treatment of water - Export potential	- Needs a constant salinity level - Disease outbreak can destroy the harvest

Fig. 13.13 SWOT analysis of molluscs farming

disadvantaged districts of Dortmund (Germany) is one of the three cities in which the green infrastructure is to be realized, and the other two cities are Turin (Italy) and Zagreb (Croatia).The mammoth project is funded by the European Union with more than 10 million euros. At a conference in Dortmund-Huckarde in September 2018, the official launch of the large-scale project began, which is expected to run until the summer of 2023. Horizon 2020 is the EU's Research Framework Programme, which will provide € 75 billion in research projects between 2014 and 2020.

Many after use ideas are often not implemented on brownfields, because investors have to reckon with hidden costs in an unknown amount because of their constructional-structural defects (Tobisch 2013). Classically, these deficiencies include the sanitation measures like soil and groundwater remediation, foundations or other building residues, contaminated sites that pollute the soil and compensatory measures that investors must carry out in return for their construction. These obstacles are met by urban community gardens with great adaptability (Tobisch 2013). To build a communal garden, the initiatives need space to garden on, either open or undeveloped, as they can be planted in mobile containers without soil. On many areas of the surveyed urban gardens, there are various remnants of pipes, foundations or building components left over from previous uses. As long as there is enough acreage and the gardeners are safe enough, the garden initiatives can react extremely flexibly to these remnants (Tobisch 2013).

13.5 Economic Considerations

As a community-based venture, urban agriculture provides the products, in most cases, directly to the neighbourhood. Small businesses have the potential to stimulate the local economy through job creation and income generation. Beyond this, additional support business has the opportunity to occur on the cultivation, progression and distribution stage (Vitalyst 2017). Due to limited space, urban farmers faced other conditions as conventional farmers that leads to innovative ideas like vertical farming, microgreen operations, aquaponics, etc. in the past with the possibility to create new ideas in the future (Lanarc and Golder 2013).

As its nature, urban agriculture highly depended on local conditions. Business challenges and opportunities can be very different; for instance, sites like Berlin, New York or Vancouver would not compare to Nairobi or Havana (Hallett et al. 2017).

13.5.1 Havana (Cuba)

In Havana, the idea of urban agriculture is based on "production in the community, by the community, for the community". With a total area of 721 km^2, the city used 41% of its land for agriculture production. Thousands of people work either directly in urban farming or in supporting sectors like popular councils, service networks or research institutions (Novo and Murphy 2000). Nowadays, one of the key reasons for entering gardening is a better chance for more income (French et al. 2010) that can reach many times over the average government salary (The Economic Times 2008).

13.5.2 Brooklyn, New York (United States)

In 2009, Gotham Greens was found through inspirations of innovation and technology in agriculture systems. It was the first commercial rooftop greenhouse in the USA and the state of the art in greenhouse facility in 2011. With the opening of the second greenhouse, they integrated a supermarket and expanded greenhouse space. Up to now, they are located in New York and Chicago with a totalling space of 170,000 ft^2 (approx. 15,800 m^2) (GothamGreens 2019).

The two examples of Havana and New York show that under economic considerations, urban farming can be implemented everywhere. However, it must be distinguished in terms of the economic value, meaning on the one hand, the production for self-consumption and on the other hand the production for sale. Both increase the income for the people and lead to a better economic welfare.

In a depressed economy with high unemployment, urban agriculture can create jobs, generate income and promote financial stability (M-NCPPC 2012). According to the Food and Agriculture Organization (FAO 2019a, b, c), 800 million people worldwide practise urban agriculture that helps low-income urban residents to save money. The OECD (2010) estimate that Detroit could generate 200 million US dollars in sales and approx. 5000 new jobs. Vitiello et al. (2010) investigated the urban framing area in Camden, New Jersey, and estimated the value of USD 64,756 by 48 gardens. Another study came to the estimation of USD 4.9 million for the summer vegetable production in Philadelphia (Vitiello and Nairn 2009).

The sample illustrates the economic potential of urban agriculture as well as the potential for multifunctionality. In the frame of green infrastructure, a multifunctional urban agriculture will deliver a large potential of ecosystem services like the following:

- Production of raw materials and food
- Design and maintenance of cultural landscapes
- Design and maintenance of diverse natural habitats and climate mitigation zones including heat buffers
- Provision of equalization areas to agglomerations (for instance, in the form of green belts)
- Provision of space for social life in urban areas

Further, there is a large potential for industrial symbiosis in order to further develop the economic potential. Industrial symbiosis refers to a business collaboration wherein residuals from one enterprise serve as inputs to another. Referring to the circular economy approach, a further scope should be to close all material cycles. A first material cycle analysis for the following options of urban agriculture in the city of Magdeburg was given in Plat et al. (2018):

- Use of fallow land and brownfields with aquaponics
- Use of fallow land and brownfields with organic farming
- Rooftop with aquaponics
- Rooftop with organic farming
- Vertical farming with aquaponics
- Vertical organic farming

According to Plat et al. (2018), the results showed that the variant "vertical farming with aquaponics" has the highest implementation potential for Magdeburg.

A recent investigation by Schneider et al. (2019) underlined the environmental potential of urban agriculture. The study investigated the resource-saving potentials in the frame of industrial symbiosis through insect farming (Berlina et al. 2015). The results showed a resource-saving potential of up to 2 powers of 10 that can particularly be proven regarding the impact category "fossil resource depletion". The economic potential of industrial symbiosis in the frame of urban agriculture is not estimated yet, but the results of Schneider et al. (2019) indicate the resource-saving dimension to be expected.

13.6 Conclusion

A sustainable bioeconomy can potentially help to replace the era of fossil resources and supply a growing world population. The transformation to such a bio-based economy is characterized by economic, ecological and social opportunities but also by risks. The possible potentials, based on the key objectives of the 2030 Agenda and the Sustainable Development Goals (SDGs), and challenges have been illustrated in the present contribution. The World Community has set itself 17 ambitious SDGs with the scope to preserve the earth for future generations and to improve the lives of those who still live in hunger and poverty. It is in the interest of climate protection as well as for humanitarian reasons, across sectors and across the entire production chain, to transform to sustainable food production approaches.

Bioeconomy in the form of urban green infrastructure is one pillar to establish sustainable and multifunctional food production solutions on different scales of a community in developing as well as developed countries. Developing countries have the chance and challenge to include bioeconomy solutions in the frame of urban green infrastructure in the spatial planning and consider multifunctionality already during the urbanization process. But also developed countries have a chance and challenge for transformation in the direction of bio-based solutions during the process of urban renewal. Urban communal gardens can respond very flexibly to obstacles to the reuse of brownfield land and be implemented in places that have no potential for other uses. First of all, urban gardens serve as places for community cultivation of food and the experience of nature. In the beginning, attractiveness plays a minor role from the food growing perspective. Nevertheless, the gardens make the area used more attractive and usually even lead to a higher property value. Also the ecological improvement and upgrading of an area is a central motivation of the garden initiatives. With their focus on the community, the new urban gardens differ in design and outward form of allotment gardens and private gardens. The social component plays a central role in the urban gardening movement. Social motives and the desire for community activities in the horticultural sector are central motivations for the majority of gardeners for participating in the gardens. Although urban community gardens do not focus on the economic development of an area, they not only reactivate fallow land on an ecological and social level but usually also lead to an economic revival.

Urban agriculture and closed resource cycles are by no means short-term phenomena. Corresponding initiatives should therefore be adapted locally and sustainable system solutions developed. Historical crops of cereals or legumes need to be returned to the field to ensure long-term agricultural biodiversity. Here, the bioeconomy can provide opportunities to develop new products from original crops, making it worthwhile to grow them again. In general, defining community gardens as a planning tool for urban development would give basic planning security to projects, giving them the opportunity to develop and exploit their great social and economic potential for the city, as well as its ecology and its inhabitants (Tobisch 2013).

Acknowledgments This paper was prepared in the context of the project "Water Management and Climate Change in the Focus of International Master Programs (WATERMAS)," funded by the Erasmus+ Programme of the European Union. In this regard, this manuscript reflects only the views of the authors; as such, the European Union cannot be held responsible for these views or any future use of them.

References

Abegunde AA(2012) Urban horticulture and community economic development of lagging regions. In: Prof. Alejandro Isabel Luna Maldonado (ed) Horticulture. ISBN: 978-953-51-0252-6. InTech. Available from: http://www.intechopen.com/books/horticulture/urban-horticulture-and-community-economicdevlopment-of-lagging-regions

Agrilyst (2015) Average yield for indoor and outdoor farming worldwide in 2015, by crop type (in 1000 pounds per acre). Statista. Available from https://www.statista.com/statistics/752128/average-yield-for-indoor-and-outdoor-farming-worldwide-by-crop-type/

BBKA (The British Beekeepers Association) (2019) Honey. Retrieved from https://www.bbka.org.uk/honey on 01/23/2019

Berlina A, Lindberg G, Mikkola N, Smed Olsen L, Teräs J 2015 The potential of industrial symbiosis as a key driver of green growth in Nordic regions Johnsen IHG (ed) Nordregio: Stockholm. ISBN: 978-91-87295-34-8

Biedermann A, Ripperger AL (2017) Das Phänomen Urban Gardening. In: Urban Gardening und Stadtentwicklung. BestMasters. Springer Spektrum, Wiesbaden

Bradbear N (2009) Bees and their role in forest livelihoods – a guide to the services provided by bees and the sustainable harvesting, processing and marketing of their products. Food and agriculture Organization of the United nations, Rome

Brandle J, Hodges L, Zhou X (2004) Windbreaks in North American agricultural systems. Agrofor Syst 61:65. https://doi.org/10.1023/B:AGFO.0000028990.31801.62

Braungart M, McDonough W (2002) "Cradle to Cradle", remaking the way we make things. NorthPoint Press, New York

Bunting SW, Little DC (2015) Urban aquaculture for resilient food systems. In: de Zeeuw H, Drechsel P (eds) Cities and agriculture developing resilient urban food systems, Earthscan food and agriculture. Taylor & Francis. ISBN 978-1-13-886058-2, pp 312–335

Business Wire (2018) Edible insects market by type and by product – global opportunity analysis and industry forecast 2018–2023. Retrieved from https://www.businesswire.com/news/home/20180405005878/en/Edible-Insects-Market-Type-Product%2D%2D-Global on 24/01/2019

Caplow T (2009) Building integrated agriculture: philosophy and practice. In: Heinrich Böll Foundation (ed) Urban futures 2030: urban development and urban lifestyles of the future. Heinrich-Böll-Stiftung, Berlin, pp 54–58

Cattaneo-Vietti R (2016) Man and shells – molluscs in the history. Bentham Science Publishers, Sharjah

Cohen-Shacham E, Janzen C, Maginnis S, Walters G (2016) Nature-based solutions to address global societal challenges. ISBN: 978-2-8317-1812-5. https://doi.org/10.2305/IUCN.CH.2016.13.en

Collavo A, Glew RH, Huang Y-S, Chuang L-U, Bosse R, Paoletti MG (2005) House cricket small-scale farming. In: Paoletti MG (ed) Ecological implications of minilivestock: potential of insects, rodents, frogs and sails. Taylor & Francis, Boca Raton

Cramp D (2008) A practical manual of BEEKEEPING – how to keep bees and develop your full potential as an apiarist. Spring Hill House, Oxford

David F (1993) Strategic management, 4th edn. Macmillan Publishing Company, New York

de Bon H, Holmer RJ, Aubry C (2016) Urban horticulture. In: de Zeeuw H, Drechsel P (eds) Cities and agriculture – developing resilient urban food systems. RUAF Foundation and International

Dupraz C, Burgess P, Gavaland A, Graves A, Herzog F, Incoll LD, Jackson N, Keesman K, Lawson G, Lecomte I, Liagre F, Mantzanas K, Mayus M, Moreno G, Palma J, Papanastasis V, Paris P, Pilbeam DJ, Reisner Y, Van Noordwijk M, Vincent G, Werf Van der W (2005) Synthesis of the silvoarable agroforestry for Europe project. INRAUMR system editions. Montpellier, 254 p

Durst PB, Shono K (2010) Edible forest insects: exploring new horizons and traditional practices. In: Durst PB, Johnson DV, Leslie RN, Shono K (eds) Forest insects as food: humans bite back – Proceedings of a workshop on Asia-Pacific resources and their potential for development. Food

and Agriculture Organization of the United Nations, pp 1–4. http://www.fao.org/docrep/012/i1380e/i1380e00.pdf
Eigenbrod C, Gruda (2015) Urban vegetable for food security in cities. A review. Agron Sustain Dev 35:483. https://doi.org/10.1007/s13593-014-0273-y
European Commission (2010) Green infrastructure. Available online: http://ec.europa.eu/environment/nature/ecosystems/index_en.htm
European Commission(2013) Communication from the Commission to the European Parliament, the council, the European Economic and social committee and the committee of the regions: green infrastructure (GI) — enhancing Europe's natural capital. Available online: https://eur-lex.europa.eu/legal-content/EN/TXT/?uri=celex%3A52013DC0249
European Commission/DG Environment (2009) Zielrichtung: eine grüne Infrastruktur in Europa. In: Natura 2000: Newsletter "Natur und Biodiversität" der Europäischen Kommission. Heft 27, S. 3–7
Eurostat (2018) Aquaculture statistics. Retrieved from https://ec.europa.eu/eurostat/statistics-explained/index.php/Aquaculture_statistics#cite_note-7 on 01/28/2019
FAO (2011) The place of urban and peri-urban agriculture (UPA) in national food security programmes. http://www.fao.org/3/i2177e/i2177e00.pdf
FAO (2014) Small-scale aquaponic food production – integrated fish and plant farming. FAO Fisheries and Aquaculture Technical Paper 589. http://www.aquacultuurvlaanderen.be/sites/aquacultuurvlaanderen.be/files/public/aquaponics_fao.pdf
FAO (2018a). The state of world fisheries and aquaculture 2018 – meeting the sustainable development goals. Rome
FAO (2018b) Why bees matter – the importance of bees and other pollinators for food and agriculture. http://www.fao.org/3/I9527EN/i9527en.PDF
FAO (2018c) Average annual per capita consumption of seafood worldwide from 2014 to 2017 (in kilograms). https://www.statista.com/statistics/820953/per-capita-consumption-of-seafood-worldwide/ on 02/15/19
FAO (2019a) FAO's role in urban agriculture. Retrieved 15 March 2019 from http://www.fao.org/urban-agriculture/en/
FAO (2019b) Statistics – honey and beeswax. Retrieved from http://www.fao.org/faostat/en/#data/QL on 01/23/2019
FAO (2019c). Animal production. Retrieved from http://www.fao.org/animal-production/en/ on 24/01/2019
FAO, IFAD, UNICEF, WFP, WHO (2018) The state of food security and nutrition in the world. Building climate resilience for food security and nutrition. Rome
French C, Becker M, Lindsay B (2010) Havana's changing urban agriculture landscape: a shift to the right? J Agric Food Syst Commun Dev 1(2):155–165. https://doi.org/10.5304/jafscd.2010.012.013
GlobeNewswire (2015) Projected vertical farming market worldwide from 2013 to 2023 (in million U.S. dollars). https://www.statista.com/statistics/487666/projection-vertical-farming-market-worldwide/. Accessed 14 Feb 2019, 08:01 PM
GNSW (THE GOVERNMENT OF NEW SOUTH WALES) (2000) Inquiry into beekeeping in urban areas – report. https://www.aph.gov.au/DocumentStore.ashx?id=9f37a963-78f1-499f-8853-31887f4f0c86&subId=206793
GothamGreens (2019) History. Retired 15 March 2019 from http://www.gothamgreens.com/our-history
Gottlieb R, Fisher A (1996) First feed the face: environmental justice and the community food security. Antipode 28:2. Cambridge & Oxford
Greenleaf SS, Kremen C (2006) Wild bees enhance honey bees' pollination of hybrid sunflower. PNAS 103(37):13890–13895. https://doi.org/10.1073/pnas.0600929103
Gruenewald H, Brandt BKV, Schneider BU, Bens O, Kendzia G, Hüttl RF (2007) Agroforestry systems for the production of woody biomass for energy transformation purposes. Ecol Eng 29(4):319–328., ISSN 0925-8574. https://doi.org/10.1016/j.ecoleng.2006.09.012

Gusted E et al (2017) Reflexions of urban gardening in Germany. Challenges Sustain 4(1):63–70. Hannover

Hallett S, Hoagland L, Toner E (2017) Urban agriculture: environmental, economic, and social perspectives. In Janick J (ed) Horticultural reviews, vol 44, XLIV, First Edition. Wiley-Blackwell

Halweil B, Prugh T (2002) Home grown: the case for local food in a global market. Available online: https://books.google.de/books?hl=de&lr=&id=9cef41L_nVEC&oi=fnd&pg=PT3&dq=local+food+co2&ots=eOPKdj3F5J&sig=N18TM1elHaPEq1AxUv8J4k99It8&redir_esc=y#v=onepage&q=local%20food%20co2&f=false

Hanboonsong Y, Jamjanya T, Durst PB (2013) Six-legged livestock: edible insect farming, collection and marketing in Thailand. Food and Agriculture Organization of the United Nations. http://www.fao.org/3/a-i3246e.pdf

Helms MM, Nixon J (2010) Exploring SWOT analysis – where are we now? A review of academic research from the last decade. J Strateg Manag 3(3):215–251

Hubold G, Klepper R (2013) The importance of fishing and aquaculture for the global food security, Thuenen Institute for market analysis (in German)

Hung K-LJ, Kingston JM, Albrecht M, Holway DA, Kohn JR (2018) The worldwide importance of honey bees as pollinators in natural habitats. Proc R Soc B285:20172140. https://doi.org/10.1098/rspb.2017.2140

Jansson A, Berggren A (2015) Insects as food – something for the future? A report from future agriculture. Swedish University of Agricultural Sciences, Uppsala. https://www.slu.se/globalassets/ew/org/centrb/fr-lantbr/publikationer/insects_as_food_2015.pdf

Johnson DV (2010) The contribution of edible forest insects to human nutrition and to forest management: current status and future potential. In: Durst PB, Johnson DV, Leslie RN, Shono K (eds) Forest insects as food: humans bite back – Proceedings of a workshop on Asia-Pacific resources and their potential for development. Food and Agriculture Organization of the United Nations, pp 5–22. http://www.fao.org/docrep/012/i1380e/i1380e00.pdf

Lanarc, Golder (2013) The urban farming guidebook. Planning for the business of growing food in BC's towns & cities. Retrieved from www.refbc.com/sites/.../Urban-Farming-Guidebook-2013.pdf

Langemeyer J, Latkowska MJ, Nicolas Gomez-Baggethun E, Voigt A, Calvet-Mir L, Pourias J, et al. (2016) Ecosystem services from urban gardens. In: Bell S, Fox-Kämper R, Keshavarz N, Benson M, Caputo S, Noori S, Voigt A (eds) Urban allotment gardens in Europe. Routledge, pp 115–141

Lewis O, Home R, Kizos T (2018) Digging for the roots of urban gardening behaviours. Urban For Urban Green. https://doi.org/10.1016/j.ufug.2018.06.012

Lohrberg F (2001) Stadtnahe Landwirtschaft in der Stadt- und Freiraumplanung: Ideengeschichte, Kategorisierung von Konzepten und Hinweise für die zukünftige Planung, Dissertation at RTWH Aachen. Germany (in German)

Lohrberg F, Timpe A (2011) Urban agriculture – new forms of primary production in the city, specialist magazine "Planerin" 5/2. Germany

Marinelli J (2017) Urban refuge: how cities can help rebuild declining bee populations. Yale Eniviron 360. https://e360.yale.edu/features/urban-refuge-how-cities-can-help-rebuild-declining-bee-populations

McNeely JA, Scherr SJ (2003) Ecoagriculture. Strategies to feed the world and save wild biodiversity. Island Press, Washington, DC

Migge L (1929) Grünpolitik der Stadt Frankfurt am Main. In: Der Städtebau. Heft 2, 1929, S. 37–46

M-NCPPC (The Maryland-National Capital Park and Planning Commission) (2012) Urban agriculture: a tool for creating economic development and healthy communities in Prince George's County, MD. Maryland

Nair PKR (1993) An introduction to agroforestry. Kluwer Academic Publishers (in cooperation with the International Centre for Research in Agroforestry), Dordrecht

Nair PKR (1985) Classification of agroforestry systems. Agrofor Syst 3:97. https://doi.org/10.1007/BF00122638

Nairn M, Vitello D (2010) Lush lots. Everyday urban agriculture, cited in Lohrberg & Timpe (2011)
Naumann S, McKenna D, Kaphengst T, Pieterse M, Rayment M (2011) Design, implementation and cost elements of Green Infrastructure projects. Final report to the European Commission, DG Environment, Contract no. 070307/2010/577182/ETU/F.1, Ecologic institute and GHK Consulting
Neßhöfer C, Kugel C, Schniewind I (2012) Ökosystemleistungen im Europäischen Kontext: EU Biodiversitätsstrategie 2020 und "Grüne Infrastruktur" (Helmholtz Zentrum für Umweltforschung – UFZ). In: Hansjürgens B, Neßhöver C, Schniewind I (eds) Der Nutzen von Ökonomie und Ökosystemleistungen für die Naturschutzpraxis. Workshop I: Einführung und Grundlagen. BfN-Skripte 318. S. 22–27
Novo MG, Murphy C (2000) Urban agriculture in the city of Havana: a popular response to a crisis. Retrieved from https://www.ruaf.org/sites/default/files/Havana_1.PDF
OECD (2010) Urban agriculture: good food, good money, good idea!. Retrieved 15 March 2019 from http://oecdinsights.org/2010/09/13/urban-agriculture-good-food-good-money-good-idea/
Ohyama K, Takagaki M, Kurasaka H (2008) Sustain Sci 3:241. https://doi.org/10.1007/s11625-008-0054-0
Olivares-Banuelo TN (2018) How important it is to produce seeds for the aquaculture of bivalve molluscs? Oceanogr Fish 8(3):1–2. https://juniperpublishers.com/ofoaj/pdf/OFOAJ.MS.ID.555740.pdf
Oonincx DGAB, de Boer IJM (2012) Environmental impact of the production of mealworms as a protein source for humans – a life cycle assessment. Plos One 7(12):e51145. https://doi.org/10.1371/journal.pone.0051145
Orsini F, Kahane R, Nono-Womdim R et al (2013) Urban agriculture in the developing world: a review. Agron Sustain Dev 33:695–720
Plat K, Meyer A, Schneider P, Perret K (2018) Potential for sustainable urban food production in a medium scale city in Germany. In: Leal Filho W, Pociovălişteanu D, Borges de Brito P, Borges de Lima I (eds) Towards a sustainable bioeconomy: principles, challenges and perspectives. World Sustainability Series. Springer, Cham, pp 233–260. https://doi.org/10.1007/978-3-319-73028-8_13
Rice MA (2001) Environmental impacts of shellfish aquaculture: filter feeding to control eutrophication. In: Tlusty M, Bengtson D, Halvorson HO, Oktay S, Pearce J, Rheualt R (eds) Marine aquaculture and the marine environment: a meeting for the stakeholders in the northeast. Held Jan. 11–13, 2001 at the University of Massachusetts Boston. Cape Cod Press, Falmouth, pp 77–86
Rose T, Kremen C, Thrupp A, Gemmill-Herren B, Graub B, Azzu N (2016) Policy analysis paper: mainstreaming of biodiversity and ecosystem services with a focus on pollination. Food and agriculture Organization of the United nations, Rome
Rosol M (2014) Ernährungssicherung durch Urban Gardening? Standort 38:220–224. https://doi.org/10.1007/s00548-014-0352-y
Sass J (2011) Why we need bees: nature's tiny workers put food on our tables. NRDC. https://www.nrdc.org/sites/default/files/bees.pdf
Schneider P, Folkens L, Meyer A, Fauk T (2019) Sustainability and dimensions of a Nexus approach in a sharing economy. Sustainability. 2019 11(3):909. https://doi.org/10.3390/su11030909
Singh HB, Sarma BK, Keswani C (eds) (2017) Advances in PGPR research. CABI, Wallingford. 408 pages, ISBN-9781786390325
Smil V (2002) Worldwide transformation of diets, burdens of meat production and opportunities for novel food proteins. Enzym Microb Technol 30(2002):305–311. https://doi.org/10.1016/S0141-0229(01)00504-X
Specht K, Siebert R, Hartmann I et al (2014) Urban agriculture of the future: an overview of sustainability aspects of food production in and on buildings. Agric Hum Values 31:33. https://doi.org/10.1007/s10460-013-9448-4
Statista (2019a) Edible insects – statistics & facts. Retrieved from https://www.statista.com/topics/4806/edible-insects/ on 24/01/2019

Statista (2019b) Level of agreement towards the statement "If available, I prefer regional products" in Germany from 2013 to 2017. Retrieved from https://www.statista.com/statistics/504178/consumer-preference-for-regional-products-germany/on 01/23/2019

Surendra KC, Oliver R, Tomberlin JK, Jha R, Khanal SK (2016) Bioconversion of organic wastes into biodiesel and animal feed via insect farming. Renew Energy 98(2016):197–202. https://doi.org/10.1016/j.renene.2016.03.022

Tao J, Li YO (2018) Edible insects as a means to address global malnutrition and food insecurity issues. Food Qual Saf 2(1):17–26. https://doi.org/10.1093/fqsafe/fyy001

The Economic Times (2008) Cuba's urban farming program a stunning success. Retrieved 15 March 2019 from https://economictimes.indiatimes.com/news/international/cubas-urban-farming-program-a-stunning-success/articleshow/3110698.cms

Thornbuch M (2015) Urban agriculture in the transition to low carbon cities through urban greening. Environmental Science, Ontario

Timmons MB, Ebeling JM (2013) Recirculating aquaculture, 3rd edn. Northeastern Regional Aquaculture Center (NRAC), Michigan State University, Lansing

Tobisch C (2013) Oasen im Beton. Urban Gardening als Instrumentzur Attraktivierung und Belebungvon Brachflächen, diploma thesis at Technische Universität Dortmund

Van Huis A (2012) Potential of insects as food and feed in assuring food security. Annu Rev Entomol. (2013 58:563–583. https://doi.org/10.1146/annurev-ento-120811-153704

Van Huis A, Oonincx DGAB (2017) The environmental sustainability of insects as food and feed. A review. Agron Sustain Dev. (2017 37:43. https://doi.org/10.1007/s13593-017-0452-8

Vitalyst (2017) URBAN FARMING an introduction to urban farming, from types and benefits to strategies and regulations. Phoenix, Arizona

Vitiello D, Nairn M (2009) Community gardening in Philadelphia 2008 harvest report. Penn planning and urban studies. University of Pennsylvania

Vitiello D, Nairn M, Grisso JA, Swistak N (2010) Community gardening in Camden, NJ harvest report: summer 2009. Penn's Center for Public Health Initiatives, Pennsylvania

Vought LBM, Pinay G, Fuglsang A, Ruffinoni C (1995) Structure and function of buffer strips from a water quality perspective in agricultural landscapes. Landsc Urban Plan 31(1–3):323–331., ISSN 0169-2046. https://doi.org/10.1016/0169-2046(94)01057-F

White P, Edwards P (2015) Social impacts of aquaculture. Aquac Manag. Retrieved from http://aquaculture.management/2015/social-impacts-of-aquaculture/ on 01/28/2019

Workmann D (2018) Natural honey exports by country. Worlds top exports. Retrieved from http://www.worldstopexports.com/natural-honey-exporters/ on 01/23/2019

Part III

Pharmaceutical Biotechnology

Vaccines: Biotechnology Market, Coverage, and Regulatory Challenges for Achieving Sustainable Development Goals

14

Cristina Possas, Adelaide Maria de Souza Antunes, Jorge Lima de Magalhães, Flavia Maria Lins Mendes, Mateus Pinheiro Ramos, Juliana De Simone Morais, and Akira Homma

Abstract

This chapter provides an overview, from bioeconomic and global sustainability perspectives, of the main constraints to the current global vaccine innovation system for achieving Sustainable Development Goals – SDGs. Biotechnology market trends, gaps in vaccine coverage against emerging and neglected diseases, and patent protection and regulation are discussed. A structured long-term "public-return-driven" innovation model to overcome vaccine market failure is proposed.

Keywords

Biotechnology market · Emerging and neglected diseases · Sustainable Development Goals · Regulation and patents · Vaccine innovation system

C. Possas (✉) · A. Homma
Bio-Manguinhos, Oswaldo Cruz Foundation, Rio de Janeiro, Brazil
e-mail: cristina.possas@bio.fiocruz.br

A. M. d. S. Antunes
School of Chemistry, Federal University of Rio de Janeiro, Rio de Janeiro, Brazil

National Institute of Industrial Property (INPI), Rio de Janeiro, Brazil

J. L. de Magalhães
Farmanguinhos, Oswaldo Cruz Foundation, Rio de Janeiro, Brazil

F. M. L. Mendes · M. P. Ramos · J. De Simone Morais
School of Chemistry, Federal University of Rio de Janeiro, Rio de Janeiro, Brazil

C. Keswani (ed.), *Bioeconomy for Sustainable Development*,
https://doi.org/10.1007/978-981-13-9431-7_14

14.1 Introduction

Innovative preventive vaccines against emerging and neglected infectious diseases, such as Zika, dengue, chikungunya, influenza, and HIV/AIDS, are examined here from bioeconomics and global sustainability perspectives, aiming to integrate public health and biotechnology market approaches.

Novel vaccines with reduced adverse effects can have an enormous impact on life expectancy and on the quality of life of the global population, significantly reducing government, individual, and business costs (Bloom et al. 2018). Nevertheless, there are significant production, technological development, market, coverage, regulatory, and governance constraints to achieving Sustainable Development Goals (SDGs). In this chapter we examine vaccine biotechnology market and the factors contributing to market failure, discussing policy strategies to optimize science, technology, and innovation (STI) and drastically reduce current constraints to vaccine development (Singh et al. 2016a, b, 2019).

For achieving SDG, it should be noted that only one of these goals, SDG3, refers specifically to vaccines (3.b.1). However, in addition, we have also identified 7 other SDG goals strongly related to vaccines and 6 SDG goals related to vaccine, in a total of 14 vaccine-related goals in 17 SDGs. Two of these goals are related to innovation and technological development of vaccines (SDG9 and SD17). We discuss the main vaccine development challenges for achieving SDG and current technological and regulatory obstacles particularly affecting developing countries. From this perspective, we propose STI governance strategies to overcome these gaps and increase global access to vaccines, focusing on institutional and regulatory perspectives, including intellectual property and ethics. Policy recommendations for vaccine funding and incentives for innovation, development, and production are made. Finally, we emphasize the enormous potential role that access to innovative vaccines can play on global sustainability (Milstien et al. 2007; Possas et al. 2015), benefiting particularly the poorest countries in a global context permeated by sharp social inequalities.

14.2 Vaccines: Global Market Trends

The global market for human vaccines is projected to reach USD 50.42 billion by 2023 from 36.45 billion in 2018 at a CGAR of 6.7% (Markets and Markets 2019) driven by the growing importance of vaccines in public health, reducing healthcare costs and contributing through prevention of diseases toward a more sustainable healthcare system.

Drastic changes in the dynamics of the global vaccine market occurred between 2000 and 2018, with a sharp growth from USD 6 billion in 2000 to USD 33 billion in 2014 (Access to Vaccines Index 2017) and to USD 36.45 billion in 2018 (Markets and Markets 2019), with sales to high-income countries representing about 65% of the total value of this market (Access to Vaccines Index 2017). In Fig. 14.1 we

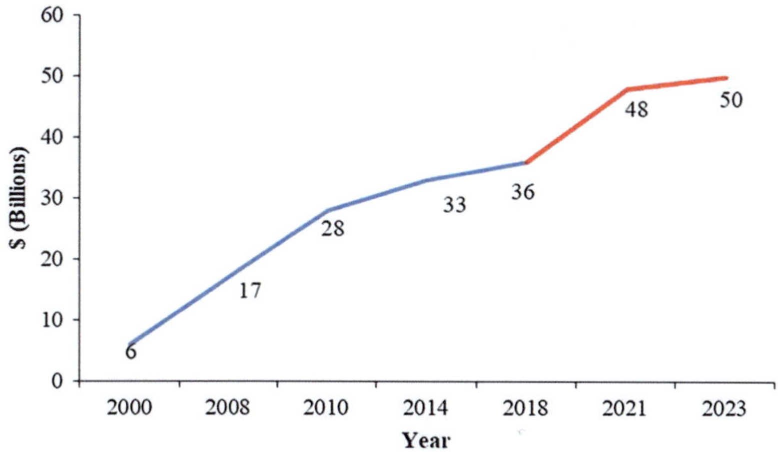

Fig. 14.1 Global human vaccine market growth in USD billion. (Sources: Access to Vaccines Index, 2017 for 2000–2014; Markets and Markets Reports, 2019 for 2018 and forecasts for 2021 and 2023)

Table 14.1 Top 5 pharmaceutical companies: global vaccine revenue market share in 2017

Company	
GlaxoSmithKline Plc.	24.0%
Merck and Co	23.6%
Pfizer	21.7%
Sanofi	20.8%
Novavax	5.9%

Source: Statista (2019)

indicate the evolution of the global human vaccine market from 2000 to 2018 and the forecast for 2023.

Recently, other reports have been released anticipating an even more favorable scenario for the global human vaccines market. A recent study estimated that this market would grow from 32.5 billion in 2015 to 77.5 billion by 2024 (Grand View Research 2018).

These market forecasts also anticipate rising R&D investments in vaccine development projects by the main global players in the vaccine market. Table 14.1 indicates the top 5 pharmaceutical players according to global revenue share in 2017. Pfizer is expected to increase its participation in the market in the next decade due to the success of its pneumococcal vaccine Prevnar 13 and increasing investments in vaccine development.

Other important vaccine players include Emergent Biosolutions, CSL, Inovio Pharmaceuticals, Bavarian Nordic, Mitsubishi Tanabe, Serum Institute of India Pvt. Ltd., ALK-Abelló A/S, Altimmune, Inc., Bharat Biotech International, and MedImmune.

The world vaccine market consists of four segments: Gavi (the Global Alliance for Vaccine and Immunizations), UNICEF, PAHO Revolving Fund (RF), and rest of the world (ROW). About 70 of the lowest-income countries in the world rely on Gavi for funding of some key vaccines (WHO 2017). UNICEF Supply Division (SD) is the procurement agent for most of these countries and for an additional approximately 30 middle-income countries (MICs) (totaling about 100 countries). The PAHO RF provides financial and procurement support to about 40 countries and territories in the Americas. The ROW consists of self-funding and self-procuring countries spanning all income levels and receiving only marginal, mostly indirect, financial, procurement, market shaping, or other related support.

These increasing investments in vaccine innovation, development, and production are guided by the growing global need for preventive vaccines and immunotherapy strategies against cancer, Zika, HPV, HSV, HIV, and a broad range of infectious diseases that currently burden the healthcare system and societies worldwide (WHO 2015, 2018).

This scenario of increasing global demand for vaccines in the next decade is supported by epidemiological indicators: annual burden of new HPV-related cancers worldwide to the tune of 670,000; rise of Zika into a public health emergency with over 86 countries reporting 230,000 cumulative confirmed cases of infection between 2015 and 2018; very high prevalence of HSV which infects approximately 67% of the world population under 50 years of age; continued prevalence of tuberculosis which infects 10 million and takes 1.5 million lives each year despite the progress made toward eliminating the disease; and rise in HIV infections worldwide over 36.9 million (WHO 2018; Global Industry Analysts 2018).

Developments in reverse vaccinology and synthetic vaccinology are expected to help increase the rate of successful vaccine design and development (Sette and Rappuoli 2010). Emerging countries with mandatory immunization programs represent large markets with enormous potential for future growth and expansion. With large population base and relatively high proportion of young children and teen population, emerging markets including China and India represent the fastest growing markets in Asia-Pacific, with this region expected to grow at the fastest rate of 8.2% in the next decade.

14.3 Innovation: From Genomics and Proteomics to Immunome

The pharma industry is rapidly becoming a competitive player in the bioeconomy market, a new global paradigm that will introduce novel technologies such as genomics and proteomics across multiple economic sectors and industries. Immunome, resulting from advances in sequencing technology and a bioinformatics resource, is also contributing to vaccine innovation and development.

These advances in genomics, proteomics, immunome, bioinformatics and new information technologies and their increasing convergence are driving these new market trends. This accelerated innovation scenario is revolutionizing healthcare with new preventive and therapeutic technologies, expected to provide longer, healthier lives to the global population.

Physicians will eventually be able to predict a person's predisposition to a broad range of diseases and intervene appropriately, insert new genes to replace faulty ones, and tailor therapies to an individual's needs and profile.

The immune system is a highly complex system, based on a coordinated expression of a wide array of genes and proteins. One of the major gaps in vaccine innovation, particularly affecting the development of new vaccines against emerging and neglected diseases, is related to the inability of scientists to explain the diversity of individual immune responses and clinical outcomes to the same vaccine and how this diversity relates to innate and acquired immunity.

The Human Immunome, a specific set of genes and molecular structures underlying the response of the immune system to fight disease, is vast and estimated at 100 billion times larger than the Human Genome Project in terms of data output. Because of this scale, scientists have never been able to characterize the core parts by which the immune system responds to pathogens and develops a disease. Only recently, with the dramatic advances in sequencing technologies and bioinformatics, exponentially extending their informational scale, it became possible for the first time for scientists to uncover the complexity of the human immunome (Soto et al. 2019; Briney et al. 2019).

Immunome, a bioinformatics resource, has been conceived for the characterization of the human immune system. It contains information about immunity-related proteins, their domain structure, and the related ontology terms and contains also information about the localization and mechanisms involved in the coding genes.

Determining the core parts of the immune system in the human immunome could drastically transform how we diagnose, prevent, and treat disease through the identification of new biomarkers while enabling highly targeted, computationally designed vaccines and therapies that reduce time and risk of product development.

The Immunome Program of the Human Vaccines Project is sequencing, in a global collaborative 10-year effort, receptors from a group of genetically diverse individuals in several continents and determines the structure and function of a key subset of receptors. Through an open-source procedure, data will be made available to researchers across the world.

In this program, laboratory analyses of biospecimens will be combined with an array of other genetic, lifestyle, and health information provided by volunteers to help researchers to identify individual genetic differences that contribute to diverse immune responses. The initial study will assess immune responses of ten healthy adults (ages 40–80) to a licensed hepatitis B vaccine (considered an ideal model to study human immunological protection), and it is expected that this study will expand to include several hundred people from neonates to the elderly in middle- and low-income countries.

The Immunome Program can thus bring crucial information to the development of more effective vaccines against emerging and neglected infectious diseases. Vaccine manufacturers in developing countries, particularly affected by these diseases, should be actively involved in its international scientific and technological collaborations.

In the near future, sophisticated technology will allow patients to search and manage their medical records, comparing them to current public health information based on individual genomic profiles. Intelligent marketing agents will aggregate patient information from a variety of sources to provide timely and relevant responses. Networks of distributed processing systems will offer new insights into data by mining all enterprise and public data sources. And supercomputing platforms and information management will enable the rigorous manipulation of genomic data.

Advances in remote sensing and artificial intelligence technologies have already created intelligent operating rooms with sensory control mechanisms and devices that transmit health information via telephones and personal digital assistants. The industry can now develop programmable microchips for the subcutaneous delivery of precisely timed doses of drugs and vaccines. Interactive chips with built-in sensors may mimic the body's own regulatory ability.

As the bioeconomy evolves, the industry faces an unprecedented era of opportunity and challenge. Companies recognize that alliances are critical to their future and are now making them a major component of their strategy. Virtual research organizations are now conceived searching to provide discovery technologies to scientists in pharmaceutical enterprises; providing them links to gene database, proteomics database, or high-throughput screening capabilities; and enabling fast and efficient access to vaccine and immunotherapy information.

14.4 Emerging and Neglected Diseases: Challenges for Global Sustainability

The world is facing multiple public health challenges, such as outbreaks of vaccine-preventable diseases, increasing reports of drug-resistant pathogens, climate change, and multiple humanitarian crises. The World Health Organization (WHO) included several emerging and neglected diseases among the ten global threats for 2019 (WHO 2019): influenza, dengue, HIV, and also high-threat pathogens, such as Ebola, several other hemorrhagic fevers, Zika, Nipah, Middle East respiratory syndrome coronavirus (MERS-CoV), and severe acute respiratory syndrome (SARS) and disease X, which represents the need to prepare for an unknown pathogen that could cause a serious epidemic.

To address these and other threats, 2019 started its new 5-year strategic plan: the 13th General Programme of Work. This plan focuses on a triple billion target: ensuring 1 billion more people benefit from access to universal health coverage, 1 billion more people protected from health emergencies, and 1 billion more people in better health and well-being. Reaching this goal will require addressing these threats to health from a variety of angles.

14.4.1 Influenza Pandemic

The world will face another influenza pandemic; the only thing we don't know is when it will hit and how severe it will be. Global defenses are only as effective as the weakest link in any country's health emergency preparedness and response system. WHO is constantly monitoring the circulation of influenza viruses to detect potential pandemic strains: 153 institutions in 114 countries are involved in global surveillance and response.

Every year, WHO recommends which strains should be included in the flu vaccine to protect people from seasonal flu. In the event that a new flu strain develops pandemic potential, WHO has set up a unique partnership with all the major players to ensure effective and equitable access to diagnostics, vaccines, and antivirals (treatments), especially in developing countries, in order to make possible the supply of required vaccines as soon as possible.

14.4.2 Dengue

Dengue, a mosquito-borne disease that causes flu-like symptoms and can be lethal and kill up to 20% of those with severe dengue, has been a growing threat for decades. A high number of cases occur in the rainy seasons of countries such as Bangladesh and India. Now, its season in these countries is lengthening significantly (in 2018, Bangladesh saw the highest number of deaths in almost two decades), and the disease is spreading to less tropical and more temperate countries such as Nepal that have not traditionally seen the disease. An estimated 40% of the world is at risk of dengue fever, and there are around 390 million infections a year. WHO's dengue control strategy aims to reduce deaths from the disease by 50% by 2020.

14.4.3 HIV/AIDS

Since the beginning of the epidemic, more than 70 million people have acquired the HIV infection, and about 35 million people have died. The progress made against HIV has been enormous in terms of getting people tested, providing them with antiretrovirals (22 million are on treatment) and providing access to preventive measures such as a preexposure prophylaxis (PrEP, which is when people at risk of HIV take antiretrovirals to prevent infection).

However, the epidemic continues to rage with nearly a million people every year dying of HIV/AIDS. Today, around 37 million worldwide live with HIV. Reaching more vulnerable people like sex workers, people in prison, men who have sex with men, or transgender people is hugely challenging. Often these groups are excluded from health services. A group increasingly affected by HIV are young girls and women (aged 15–24), who are particularly at high risk and account for one in four

HIV infections in sub-Saharan Africa despite being only 10% of the population. This year, WHO will work with countries to support the introduction of self-testing so that more people living with HIV know their status and can receive treatment (or preventive measures in the case of a negative test result).

14.4.4 Zika and Other "High-Threat Pathogens" Defined as Global Public Health

WHO has included in these ten global threats for 2019 diseases and pathogens that have potential to cause a public health emergency but lack effective treatments and vaccines. This list for priority research and development includes Ebola, several other hemorrhagic fevers, Zika, Nipah, Middle East respiratory syndrome coronavirus (MERS-CoV), and severe acute respiratory syndrome (SARS) and disease X, which represents the need to prepare for an unknown pathogen that could cause a serious epidemic.

14.5 Global Sustainability Initiatives: Decade of Vaccines, MDG and SDG

International recognition of vaccines' impact and increased global demand for vaccines have stressed the need for global strategies to assure timely provision of low-price vaccines (Meissner 2016) through policies supporting free and universal access.

In this scenario, the Decade of Vaccines (DoV) initiative was launched at the World Economic Forum in Davos in 2010, signed by international agencies, such as the World Health Organization (WHO), UNICEF, the US National Institute of Allergy and Infectious Diseases (NIAID), and the Bill & Melinda Gates Foundation, with the mission: "to extend, by 2020 and beyond, the full benefits of immunization to all people, regardless of where they are born, who they are, or where they live." This declaration was supported by a commitment by the Bill & Melinda Gates Foundation to donate USD 10 billion to research and development and to delivering vaccines for the poorest countries.

The DoV initiative gained significant international support and visibility. Two years later, after consultations with DoV stakeholders, including industry groups, a Global Vaccine Action Plan (GVAP) was launched by the 194 member states of the 65th World Health Assembly in May 2012, aiming to deliver universal access to immunization by 2020.

Following the collaborative DoV strategies, the GVAP brought together multiple stakeholders to achieve the ambitious goals of the plan: the leadership of the Bill & Melinda Gates Foundation, Gavi Alliance, UNICEF, US National Institute of Allergies and Infectious Diseases (NIAID), and WHO, mobilizing many partners (governments, health professionals, academia, manufacturers, funding agencies, development partners, civil society, media, and the private sector). If the GVAP is translated into action and resources are mobilized, it is expected that between 24.6 and 25.8 million deaths could be averted by the end of the decade, with gains in billions of dollars in productivity.

Nevertheless, it is important to note that actions and resources will not be sufficient for the success of GVAP if the plan does not conceive a global strategy to support manufacturers in the developing world to overcome the main IPR and regulatory barriers that delay and hinder vaccine development and production.

The Millennium Development Goals (MDGs) for 2000–2015 were incorporated by governments worldwide and had a strong global mobilization power on promoting development and social initiatives, engaging national leaders in elaborating and monitoring these goals (UN 2015). This mobilization was facilitated since the targets were quantifiable and could potentially be attained. Although the two health-related goals, MDG4 (reduce under-5 mortality from 1990 to 2015 by two-thirds) and MDG5 (reduce maternal mortality from 1990 to 2015 by three-quarters), had not been met by 2015 and it is estimated by WHO that 19.4 million infants worldwide are still missing out on basic vaccines, significant progress has been made, with child and maternal mortality approximately halved, with significant global progress.

In sequence to MDGs, the United Nations promoted an in-depth revision of this strategy (UN 2014, 2015) and formulated a new global strategy, Sustainable Development Goals (SDGs) for 2016–2030 with 17 goals, with one of them (SDG3) directly related to health (UN 2016).

The target of SDG3 is to "ensure healthy lives and promote well-being for all at all ages." Its 13 sub-targets include 2 ones that could be met: two-thirds less maternal mortality and a third less noncommunicable disease (NCD) mortality. They also include ending preventable newborn and under-5 deaths and ending HIV/AIDS, tuberculosis, malaria, and neglected tropical diseases, besides other non-vaccine related sub-targets.

In this chapter, we argue that a major component of SDGs is crucial for attaining SDG3 goals and should not be minimized: innovation and technological development of vaccines. We discuss how this component should be incorporated into monitoring the sub-targets of this goal, and we emphasize the need for a new vaccine innovation model based on an expanded role of the state and incentive mechanisms to pharmaceutical companies and public manufactures to correct the current scenario of "market failure" constraining access to vaccines.

It is certainly unacceptable, from ethical and sustainable development perspectives, to simply recognize this "market failure" as a detrimental and inevitable consequence of the rationale of a global market economy. On the contrary, it should be seen as a massive public health failure and a global failure to direct economic development for the benefit of societies (Trouiller et al. 2001).

14.6 Pipeline for Vaccines Against Neglected Diseases: The "Valley of Death"

Although 240 vaccine candidates are in the development pipeline for neglected and emerging infectious diseases mainly affecting the poorest countries such as malaria, dengue, HIV, tuberculosis, and pneumonia, only 2 of them have made it through the pipeline recently and are widely used in these countries: a conjugate vaccine for meningitis serogroup A diseases and a vaccine against Japanese encephalitis virus (Kaslo et al. 2018; WHO 2018).

It has been estimated by these authors that unfortunately much of this promising pipeline could go to waste and fall into the so-called valley of death, failing to move from proof-of-concept to second-phase trial due to lack of market interest in vaccines against these emerging and neglected diseases affecting only the poorest populations in developing countries. No single organization or group is interested in supporting the costly and more complex late-stage clinical trials for neglected diseases that mainly affect the poor nations.

This scenario raises great concern for two reasons. First, around 60% of these vaccine candidates in the development pipeline target the mentioned neglected and emerging infectious diseases, a much higher problem in lower- and middle-income countries (Kaslow et al. 2018). Second, this means a significant waste of global resources in a crucial area for sustainable development, considering that these vaccine candidates received billions of dollars for the first phase of vaccine development from prestigious donors, such as the US National Institutes of Health (NIH), the European Union, the Welcome Trust, and the Bill and Melinda Gates Foundation.

Taking a vaccine candidate from a discovery at the laboratory bench to widespread deployment is a complex, lengthy, and expensive endeavor, with many financial, licensing, and regulatory barriers. No organization or group plans to support the emerging and neglected diseases vaccines from the beginning to end. Therefore, it could take many decades to incorporate these vaccines into the national immunization programs in these poorest countries (Kaslow et al. 2018).

In Table 14.2 we provided a selection of promising projects for vaccines for emerging and neglected infectious diseases affecting the poorest developing countries that could significantly impact on achieving SDG targets.

14.7 Market Failure: From Free Market to Public Health Sustainability

Science and technology have made enormous progress and are now prepared to provide the innovative-intensive vaccines that the poorest populations in the world urgently need. But innovation and discovery are not the major bottleneck, which reside in technological development, production, and timely provision of vaccines to people (Homma et al. 2013).

R&D-based pharmaceutical major industry players are reluctant, due to free-market rationale, to invest in the development of vaccines to treat the major neglected

Table 14.2 Novel innovative vaccines for emerging and neglected diseases that could impact on SDGs – selected promising projects[a]

Dengue	Sanofi CYD-TDV vaccine registration and pricing, entering now the market after 20 years of development. However, due to the evidence showing the serostatus-dependence, WHO has recommended to make pre-screening of all individuals before vaccination, which may complicate the operation for its use. There are at least two other vaccines in development at phase III clinical trials expected to be completed in next 3–4 years
Pneumococcal vaccine	Merck: V114 is being evaluated in two phase 3 clinical trials
	Pfizer: phase 3 clinical trial testing its own next-generation pneumococcal vaccine
RSV vaccine	Novavax: ResVax RSV, a vaccine for protecting infants from RSV via maternal immunization. Phase 3 study
Human papillomavirus (HPV) vaccine	VGX-3100 vaccine for treating cervical dysplasia caused by HPV. Phase 3 clinical study
Malaria	GSK/path: RTS, S malaria vaccine. Registration after 28 years of development
Diarrhea	Takeda pharmaceuticals – phase 2 trial, bivalent norovirus vaccine candidate
	Vaccine was well-tolerated and induced immune responses that persisted for 1 year after vaccination. Following these promising results, one of the vaccine formulations has been selected to move forward to phase 3 study
Influenza	Sanofi Fluzone – marketed
	University of Washington School of Medicine: breakthrough research for development of novel universal DNA influenza vaccine
HIV	National Institute of Allergy and Infectious Diseases (NIAID): VCR01 phase IIb and III
	Target: overcome barriers and to develop a clinically effective vaccine with more than 50% efficacy, improved safety, and good tolerability profile, with reduced adverse effects. This result would be a breakthrough when compared with the previous efficacy of 31% of the HIV vaccine in the former Thailand trial

Sources: Evaluate (2017) and World Health Organization (2018)
[a]Promising projects selected by the authors. Target for HIV vaccines elaborated by the authors

and emerging diseases affecting mainly the poorest nations, since return on their investments cannot be guaranteed.

National and international policies currently support a free-market-based global order, with economic opportunities, rather than global public health needs guiding the direction and rationale of vaccines development.

It is certainly unacceptable, from ethical and sustainable development perspectives, to simply recognize this "market failure" as a detrimental and inevitable consequence of the rationale of a global market economy. On the contrary, it should be seen as a massive public health failure and a global failure to direct economic development for the benefit of societies (Trouiller et al. 2001).

14.8 Vaccine Pipeline: Global Governance and National Strategies

An urgent redefinition of priorities in vaccine development is needed. This strategy cannot rely only on fragmented contributions of researchers, funding agencies, and the pharmaceutical industry. Effective national and international policies need to be urgently conceived to redirect the global economy to address the true public health needs of society (Homma et al. 2013; Røttingen et al. 2017).

"Political will," identified as the need for a strong commitment to prioritize health considerations over economic interests, has been frequently emphasized by policy-makers as a major issue to ensure access to vaccines but is not sufficient. It is necessary to go beyond "political will," with a clear goal in mind and a realistic plan to achieve it. From this perspective, it will be necessary to promote effective global implementation of strategies to accelerate innovation, technological development, and production of new vaccines and to ensure timely global access to them.

Moreover, a global vaccine policy strategy should be conceived to promote the necessary enforcement of regulations and other mechanisms to stimulate vaccine development, production, and global access to these products.

Novel, creative, and effective strategies involving both the public and the private sector are needed to ensure low-price vaccines, accelerating innovation and technological of vaccines against emerging and neglected diseases.

Priority action areas should include:

1. Advocating a preventive vaccines R&D agenda
2. Conceiving capacity-building programs adequate to the conditions of developing countries' manufacturers
3. Promoting technology transfer to public and private manufacturers in emerging countries
4. Elaborating an adapted legal and regulatory framework to increase flexibility and "fast-track" procedures
5. Prioritizing funding for vaccine development
6. Securing availability, accessibility, and distribution of these vaccines

14.9 Global Strategies: Alternative Models for Governance of Vaccine R&D

Consensus is building among the main stakeholders in the global vaccine community that the spiraling costs of risks associated with vaccine R&D are detrimental to global access to these products, particularly in the poorest developing countries. Most of them agree that these vaccine R&D costs should be instead rewarded by means other than financial returns in the market from charging high product prices. Novel mechanisms such as incentives, prizes, and "patent pools" for drugs and vaccine innovation and development have been proposed in the last two decades. There is now vast literature on the subject, claiming for alternative models that should be urgently implemented to meet the increasing global demand for vaccines, particularly in the poorest developing countries.

The main question is: how to conceive a feasible long-term mechanism to minimize these risks faced by pharma companies? Which global organizations should be responsible for this alternative model?

We recommend this new vaccine incentive model should be coordinated by three international organizations: WHO, Gavi, and UNICEF. These organizations would, in collaboration with the main stakeholders, identify from the list of 240 candidates the priority vaccine candidates, identify the funding mechanisms necessary to these candidates to enter the second-phase clinical trials, and specify which organization, or alliance, would be responsible for these selected vaccine candidates from beginning to end.

In this innovative global collaboration strategy, these three leading international organizations should bring together the main players and stakeholders in the vaccine market, with funding agencies such as the Bill & Melinda Gates Foundation, NIH, Welcome Trust, and other organizations as PATH, IAVI, and the International Vaccine Institute in Seoul and vaccine manufacturers, in collaboration with other nongovernmental organizations in order to conceive and implement this alternative long-term model for sustainable development and provision of vaccines which are uncertain business products or require a great amount of public funding to go beyond the initial proof-of-concept phase.

14.10 Priority Setting, Funding, and "Advanced Market Commitment"

A novel global priority-setting strategy, driving adequate implementation, will be necessary to assess the 240 vaccine candidates in the pipeline, trying to identify the most favorable candidates which are uncertain business cases that will require significant public funding to move into second-phase clinical trials.

Funding mechanisms supported by subsidies from governments, such as those of the G20 countries, and philanthropic organizations, such as the Bill & Melinda Gates Foundation, could remedy the market failure threatening vaccine development for LMICs. Gavi already provides one form of subsidy (Gavi 2018a, b). Support to develop vaccines or to make them available during epidemics is also provided by public organizations, such as the Coalition for Epidemic Preparedness Innovations in Oslo and the Biomedical Advanced Research and Development Authority, part of the US Department of Health and Human Services.

Such schemes need to be expanded and rethought to give vaccine developers more certainty and upfront financial backing (Kaddar et al. 2013). For instance, Gavi could commit to purchasing a vaccine before it has been developed, on the condition that the developers meet certain regulatory milestones. At present, the alliance buys vaccines to distribute to LMICs after they have been licensed or recommended by the WHO for general use (Gavi 2012, 2018b).

Only with this kind of leadership will the global community secure vaccines for some of the world's most debilitating diseases.

14.11 Global Governance: Incentives and "Mission-Oriented" Approaches

There is an urgent need for a paradigm shift in global governance of health innovation systems to achieve Sustainable Development Goals (Buse and Hawkes 2015; Possas et al. 2015; Seib et al. 2017; Mazzucato 2018). "Mission-oriented" approaches have been proposed to overcome current constraints in innovation systems (Mazzucato and Penna 2015). Recently, in a new report, "The People's Prescription: Re-imagining Health Innovation to Deliver Public Value" (Mazzucato 2018), the authors call for restructuring research and development innovation systems in order to create, rather than extract, value. It also calls for long-term "mission-oriented" public investment and a public return on this investment. In this report the authors argue that health innovation is about making new treatments and cures available to the people that need them. Profits might be earned but not at the cost of doing what the health system is meant to do: heal. This report is the outcome of result of collaboration between the UCL Institute for Innovation and Public Purpose, STOPAIDS, and Global Justice Now and Just Treatment. The report identifies gaps of the current health innovation system and sets principles for a new model. It proposes concrete policy actions that can be taken in the long term to actively shape and co-create a health system that delivers real public value. The report is structured into two sections. The first is "diagnosis" with chapters on "Problems with the current health innovation system" and "Principles for a health innovation model that delivers public value." The second section, "remedies," includes chapters on "Immediate policy actions: Getting better prices today" and "Transformative proposals: Re-imagining our health innovation system to deliver public value." The report focuses on the unethical and unacceptable current global scenario for health innovation, highly inefficient, with a pharmaceutical industry that makes billions in profits without providing the affordable products that people need.

The report examines all those problems, and then it sets out some key principles of how a "healthy" innovation model for health would work, based on an analysis of case studies from different countries and different contexts, looking at where innovation has been done well. In vaccine development, as in drugs development, there is a tremendous waste of resources because public health is not driving the R&D agenda. We have all the money going into proof-of-concept studies instead of developing public accountability.

14.12 Regulatory Barriers: Intellectual Property and "Fast Track"

The need to provide more flexible and expedite new vaccine products and processes resulting from biotechnology is challenging both developed and developing countries to accelerate the implementation of adequate regulations and intellectual property rights (Crager 2014; Possas et al. 2015).

IPR are granted by the state to individuals, enterprises, or organizations under temporary monopolistic conditions (patents) in order to compensate them for the investments made in their creations/innovations. In industry, a patent is clearly an instrument to guarantee the returns of the investments on R&D through the commercialization of the patented products and through the payment of property rights.

Patents are viewed as a crucial incentive to innovation. Nevertheless, Arrow (1962) recognized in his pioneer theory that in spite of its advantages, the patent system creates a suboptimal situation in economic terms: patents create a monopoly that restricts the diffusion and dissemination of innovation. The argument is that this restriction is temporary (after 20 years the patent protection "falls" to public domain) and is compensated by the fact that the knowledge related to the patent is necessarily published in the moment that the patent is granted.

Nevertheless, several authors have noted the detrimental impacts of the monopoly created by the patent system on health products' innovation, particularly on the development and accessibility to new drugs for neglected and emerging diseases and proposed incentive mechanisms, such as prizes, "patent pools," and awards to compensate this "market failure."

Although in the vaccine sector many intellectual property and market issues affecting price remain unclear, in the current regulatory scenario, the access to new technologies in multipatented vaccines, such as adjuvants for vaccine compositions, remains a main challenge (Possas et al. 2015).

For vaccine manufacturers in emerging countries, access to patent information on vaccine adjuvants is a crucial issue, detrimental to vaccine development. The incorporation of new adjuvants for vaccines which boost the immune response has become crucial to the development of innovative vaccines, as new antigens, with purer and smaller molecules, may have less then optimal immune responses, necessary to vaccine protection for a lengthy period of time.

The malaria vaccine candidate RTS provides a good example of the crucial role new adjuvants can play: this vaccine, based on the *Plasmodium falciparum* sporozoite antigen circumsporozoite protein (CSP), was successful in providing protection against clinical malaria only when combined with a powerful adjuvant (AS02 or AS01). Another example are the tests using hybrid flagelins also in malaria vaccines. Adjuvants have emerged thus as an alternative route for vaccine development with enormous potential in the global market (Mbow et al. 2010). The development of new, powerful, and safe adjuvants is therefore a key component of vaccine research. We present in Table 14.3 some of licensed vaccine adjuvants, with company and class.

Table 14.3 Licensed vaccine adjuvants

Adjuvant	Company	Class	Indications
Alum	Various	Mineral salts	Various
MF59	Novartis	O/W emulsion	Influenza (Fluad)/pandemic flu
ASO3	GSK	O/W emulsion+αtocopherol	Pandemic flu (Pandemrix)
AS04	GSK	MPL+alum	HBV (Fendrix), HPV (Cervarix)
Liposomes	Crucell	O/w emulsion	HAV, Flu (EU)

Source: Mbow et al. (2010)

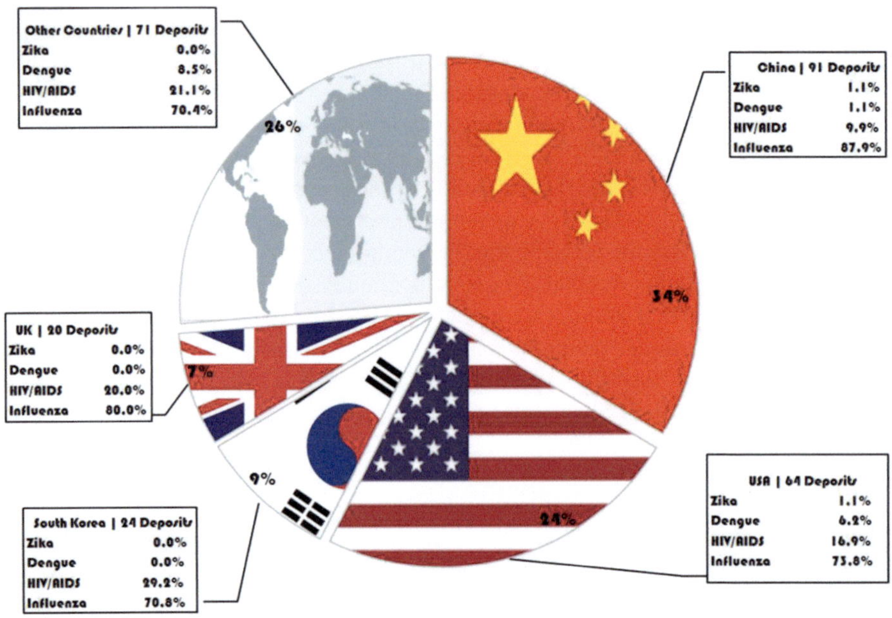

Fig. 14.2 Top countries in patent deposits for adjuvant for vaccine compositions: influenza, HIV/AIDS, dengue, and Zika. (Source: Federal University of Rio de Janeiro School of Chemistry Information System on the Chemical Industry (SIQUIM); Espacenet, Accessed On: January 10, 2019)

Figure 14.2 indicates the countries concentrating patent deposits for adjuvants to vaccine compositions and the diseases related to them (Zika, dengue, HIV/AIDS, influenza), China (34%), the USA (24%), South Korea (9%), and the UK (8%), with these three countries accounting for nearly 75% of all patent deposits. It is also observed that chikungunya had no deposit in the period from 2000 to 2018, as indicated the Espacenet base. It should be stressed that very few deposits are related to Zika and dengue vaccine, only in China and the USA.

Table 14.4 Top countries in patent deposits for adjuvant for vaccine compositions: influenza, HIV/AIDS, dengue, and Zika

Country	Zika	Dengue	HIV/AIDS	Influenza	Total
China	1	1	9	80	91
USA	2	4	11	47	64
South Korea	0	0	7	17	24
UK	0	0	4	16	20
Total	3	5	31	160	199

Source: Federal University of Rio de Janeiro School of Chemistry Information System on the Chemical Industry (SIQUIM); Espacenet (Accessed on: January 10, 2019)

Table 14.4 shows by country and by disease the patent deposits of adjuvants and formulations of vaccines with adjuvants in the world between 2000 and 2018, searched from the base Espacenet.

It also indicates that most of these adjuvant deposits are concentrated in vaccine compositions related to just two diseases, influenza (80%) and HIV/AIDS (15.6%), while Zika, followed by dengue, is more neglected. The increasing risk of a pandemic of influenza, rapidly affecting all continents, might explain the concentration of R&D efforts on vaccine compositions and adjuvants for this disease.

Table 14.5 provides an overview of the patent holders of deposits for vaccine adjuvants against Zika, dengue, and HIV in the 2000–2018 period, listed by country, number of deposits, and deposits with partnerships and partners. This table indicates that the number of partnerships is very low and should be stimulated.

Figure 14.3 shows the temporal evolution of patent deposits in the period 2000–2008 of the three largest patent depositors of adjuvants and formulation of vaccines with adjuvants for the diseases studied, showing that only in the last decade there has been a greater R&D effort.

This figure also indicates the leadership of China and the increasing role played by this country in the development of adjuvants for vaccine compositions.

Finally, it should be noted that in addition to these intellectual property barriers to access to vaccine formulations and vaccine adjuvants, such as confidentiality and constraints to patent information sharing (Possas 2013, Possas et al. 2016), other regulatory obstacles remain also a challenge to vaccine development and access to timely immunization: virtual inexistence in many countries, particularly the developing ones, of expedite and "fast-track" review processes (FDA 2018; U.S. Dept. of HHS et al. 2013) for evaluating priority and emergency projects; lack of flexible regulatory procedures for sharing biospecimens and samples; legal constraints in access to biorepositories and to biobank information; and the difficulties in defining the standard of care to be provided during clinical trials.

Table 14.5 Patent holders for adjuvants for vaccine compositions: influenza, HIV/AIDS, dengue, and Zika 2000–2018

Company, organization, or individual	Country	Total number of deposits	Deposits with partnership	Partners (number of deposits together)
GlaxoSmithKline	UK	29	1	GSK Deutschland (1)
Novartis	Switzerland	17	0	
Chinese Academy of Medical Sciences	China	12	3	Kunming Institute of Botany (2) and Nat. Tsing Hua Univ. (1)
Sanofi Pasteur (incluindo Aventis Pasteur)	France	11	0	Connaught Lab (1)
Genexine	Korea	8	5	Postech Screw Piles (5) and Progen Co. Ltd. (2)
Yebio Bioengineering Co., Ltd	China	8	0	
National Tsing Hua University	Taiwan	5	0	
MORIYAMA MASAMI	Japan	3	0	
Luoyang Pulike Bio-Engineenng Co.Ltd.	China	3	0	
SUN JUAN	China	3	0	
Konkuk University	Korea	3	0	
Tianjin Ringpu	China	3	0	
Pennsylvania University	USA	3	2	Inovio Pharmaceuticals. Inc. (2)
Abbott Biologicals BV	USA	2	0	
Celltrion	Korea	2	0	
Cha Vaccine RES INST CO LTD [KR]	China	2	0	
Istituto Superiore di Sanità	Italy	2	0	
LG Life Sciences	Korea	2	0	
Medeva Holdings BV	Netherlands	2	0	
Nitto Denko Corporation	Japan	2	0	
Novavax, Inc.	USA	2	0	
Fudan University	China	2	0	
South China Agricultural University	China	2	0	
Sun Yat-sem University'	China	2	0	

(continued)

Table 14.5 (continued)

Company, organization, or individual	Country	Total number of deposits	Deposits with partnership	Partners (number of deposits together)
Qinhuangdao Gangyuan Real Estate Group Co.	China	2	0	
West Pharmaceutical Services	USA	2	0	

Source: University of Rio de Janeiro School of Chemistry Information System on the Chemical Industry (SIQUIM) and the European Patent Office (EPO) (2018)

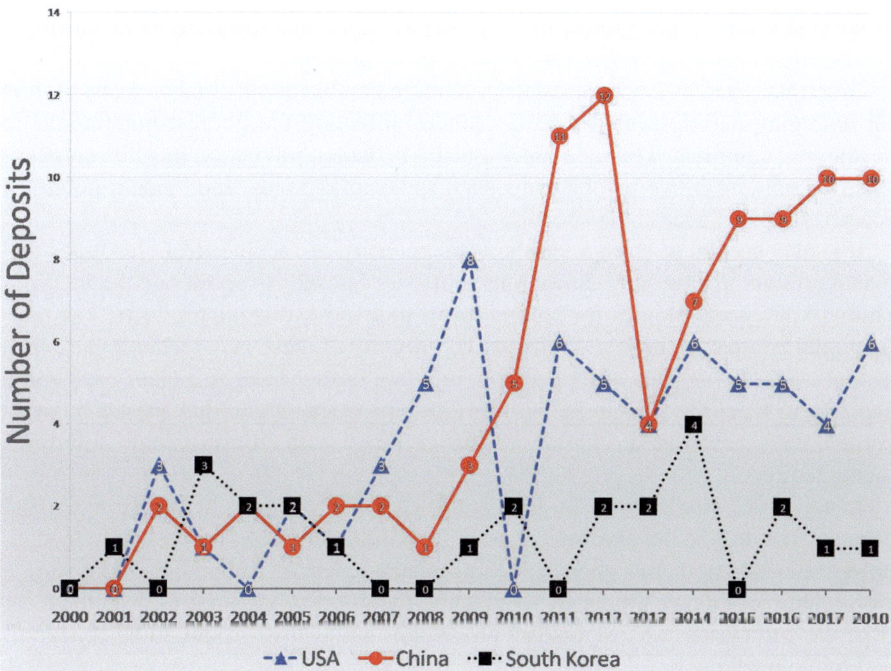

Fig. 14.3 Evolution of patents deposits of top countries 2000–2018 (Zika, dengue, HIV, influenza). (Source: Federal University of Rio de Janeiro School of Chemistry Information System on the Chemical Industry (SIQUIM); Espacenet. Accessed on January 10, 2019)

14.13 Conclusion

We examined here, from bioeconomics and global sustainability perspectives, the current innovation system for vaccine development, providing considerations on how this system should be better designed and implemented to address SDGs' vaccine coverage targets.

Public investment plays a crucial role in biomedical R&D worldwide, but research and development activities supported by governments with public resources are in most cases not directed into targets that have the most public health value, such as vaccines. For this reason, there is increasing awareness and concern in the global vaccine community on the fact that public-supported vaccine products should be shared by the public and not just privately appropriated by pharmaceutical companies.

This scenario evidences a need for conceiving a new innovation governance paradigm for vaccine development compatible with the market rationale, aiming profitable products with social return: an expanded role of the state, with clear procedures for planning, development, regulation, and forecast; a redefinition of the patterns of relationship between private and public players; and finally, a political agreement between the main players and stakeholders on the more adequate mechanisms to balance the risks and the rewards between those players.

Moreover, there is a need to conceive a more flexible intellectual property regime for emerging and neglected diseases mainly affecting the poorest populations in developing countries. The monopoly created by patent protection must be compensated by new incentive mechanisms such as awards, prizes, and "patent pools" to accelerate global access to vaccines.

It is also crucial to strengthen the local capacity of vaccine R&D institutes and manufacturers in emerging developing countries in order to accelerate the incorporation of new technologies for production of innovative vaccine products. The multinational companies have the intellectual property of these new technologies, such as adjuvants for vaccine compositions, but they do not have sufficient production capacity to meet the global demand for these products, a gap that should be overcome with expanded global collaboration with developing countries' manufacturers.

In addition, these players should identify gaps and priorities in infrastructure and capacity building in developing countries' manufacturers in order to facilitate technology transfer agreements with leading pharmaceutical companies and ensure a sustainable long-term supply of vaccines to the poorest populations. Finally, clear expedite and "fast-track" regulatory procedures and pathways should be conceived and implemented.

In other words, it will be necessary for international organizations and pharmaceutical companies to move from a short-term "shareholder-driven innovation model" to a long-term "public-return-driven innovation model", aiming global sustainability and social welfare for meeting the needs of low-income populations while searching for innovative and profitable vaccine products.

References

Access to Vaccines Index (2017) How vaccines companies are responding to calls for greater immunization coverage. Report. Access to Medicines Foundation. https://accesstomedicine-foundation.org/publications/2017-access-to-vaccines-index

Arrow KJ (1962) Economic welfare and the allocation of resources for invention. In: Nelson RR (ed) The rate and direction of inventive activity. Princeton University Press, Princeton, pp 609–625

Bloom DE, Fan VY, Sevilla JP (2018) The broad socioeconomic benefits of vaccination. Sci Transl Med 10:eaaj2345

Buse K, Hawkes S (2015) Health in the sustainable development goals: ready for a paradigm shift? Glob Health 11(13):1–8

Crager SE (2014) Improving global access to new vaccines: intellectual property, technology transfer, and regulatory pathways. Am J Public Health 104(11):e85–e91

Briney B, Inderbitzin A, Joyce C, Burton DR (2019) Commonality despite exceptional diversity in the baseline human antibody repertoire. Nature 566(7744):393–397

European Patent Office (2018) Espacenet patent search. EPO, Brussels, Belgium. Accessed 10 January 2019. Available in https://www.epo.org/searching-for-patents/technical/espacenet.html#tab-1

Evaluate (2017) Evaluate pharma. World preview 2017, outlook to 2022. Evaluate, London/Boston/Akasaka

FDA (2018) Fast track, breakthrough therapy, accelerated approval and priority review for patients and patient advocates, US Food and Drug Administration https://www.fda.gov/forpatients/approvals/fast/ucm405399.htm

Global Alliance for Vaccines and Immunizations (2012) Global Vaccine Action Plan (GVAP). http://www.who.int/immunization/global_vaccine_action_plan/GVAP_doc_2011_2020/en/

Global Alliance for Vaccines and Immunizations (2018a) GAVI sustainable development goals 2018. https://www.gavi.org/about/ghd/sdg/

Global Alliance for Vaccines and Immunizations (2018b) Sustainable development goals monitoring framework. https://www.gavi.org/gavi-brief-on-the-vaccine-indicators

Global Industry Analysts (2018) Human vaccines: market analysis, trends and forecasts. Available in:https://www.strategyr.com/market-report-human-vaccines-forecasts-global-industry-analysts-inc.asp

Grand View Research (2018) Market report on pharmaceuticals. Available in: https://www.grandviewresearch.com/services/market-research-reports

Homma A, Tanuri A, Duarte AJ, Marques E, de Almeida A, Martins R, Possas C (2013) Vaccine research, development, and innovation in Brazil: a translational science perspective. Vaccine 31S:B54–B60

Kaddar M, Schmitt S, Makinen M, Milstien J (2013) Global support for new vaccine implementation in middle-income countries. Vaccine 31(Suppl. 2):B81–B96

Kaslow DC, Black S, Bloom DE, Dalta M, Salisbury D, Rappuoli R (2018) Vaccine candidates for poor nations are going to waste. Nature 564(7736):337–339

Markets and Markets (2019) Vaccine markets by diseases and technologies, report 2019. Available in: https://www.marketsandmarkets.com/Market-Reports/vaccine-technologies-market-1155.html. Accessed 22 Mar 2019

Mazzucato M (October 2018) The people's prescription: re-imagining health innovation to deliver public value. UCL Institute for Innovation and Public Purpose, London

Mazzucato M, Penna C (2015) Mission-oriented finance for innovation: new ideas for investment-led growth. Rowman & Littlefield International, London

Mbow ML, De Gregorio E, Valiante NM, Rappuoli R (2010) New adjuvants for human vaccines. Curr Opin Immunol 22(3):411–416

Meissner HC (2016) Immunization policy and the importance of sustainable vaccine pricing. J Am Med Assoc 315(10):981–982

Milstien J, Gaule G, Kaddar M (2007) Access to vaccine technologies in developing countries: Brazil and India. Vaccine 25(44):7610–7619

Possas CA (2013) Propriété intellectuelle et le SIDA dans les pays en dévéloppement: innovation et accès aux produits pharmaceutiques. In: Possas C, Larouzé B (eds) Propriété intellectuelle et politiques publiques pour l'accès aux antirétroviraux dans les pays du Sud. ANRS, Paris, pp 37–50

Possas C, Antunes AMS, Mendes FML, Schumacher SOR, Martins RM, Homma A (2015) Access to new technologies in multi-patented vaccines: challenges for Brazil. Nat Biotechnol 33:599–603

Possas C, Antunes A, Mendes FML, Menezes Martins R, Homma A (2016) Innovation and intellectual property issues in the "Decade of Vaccines". In: Singh HB, Alok J, Keswani C (eds) Intellectual property issues in biotechnology. CABI International, Wallingford/Boston, pp 181–192

Røttingen JA, Gouglas D, Feinberg M, Plotkin S, Raghavan KV, Witty A, Draghia-Akli R, Stoffels P, Piot P (2017) New vaccines against epidemic infectious diseases. N Engl J Med 376:610–613

Seib K, Pollard AJ, Wals P, Andrews RM, Zho F, Hatchett RJ, Picketing LK, Orestein WA (2017) Policy making for vaccine use as a driver of vaccine innovation and development in the developed world. Vaccine 35(10):380–1389

Sette A, Rappuoli R (2010) Reverse vaccinology: developing vaccines in the era of genomics. Immunity 33:530–541

Singh HB, Jha A, Keswani C (eds) (2016a) Intellectual property issues in biotechnology. CABI, Wallingford. 304 pages, ISBN-13:9781780646534

Singh HB, Jha A, Keswani C (2016b) Biotechnology in agriculture, medicine and industry: an overview. In: Singh HB, Jha A, Keswani C (eds) Intellectual property issues in biotechnology. CABI, Wallingford, pp 1–4

Singh HB, Keswani C, Singh SP (2019) Intellectual property issues in microbiology. Springer, Singapore. 425 pages, ISBN:9789811374654

Soto C, Bombardi RG, Branchizio A, Kose N, Matta P, Sevy AM, Sinkovits RS, Gilchuk P, Finn JA, Crowe JE (2019) High frequency of shared clonotypes in human B cell receptor repertoires. Nature 566(7744):398–402. https://doi.org/10.1038/s41586-019-0934-8

Statista, Global Pharmaceutical Industry-Statistics and Facts (2019) Available in: https://www.statista.com/topics/1764/global-pharmaceutical-industry/. Accessed 22 April 2019

Trouiller P, Torreele E, Olliaro P, White N, Foster S, Wirth D, Pécoul B (2001) Drugs for neglected diseases: a failure of the market and a public health failure? Tropical Med Int Health 6(11):945–951

U.S. Dept. of HHS, FDA, CDER, and CBER (2013) Guidance for industry: expedited programs for serious conditions – drugs and biologics, CENTER FOR DRUG EVALUATION AND RESEARCH, at 9 (June). Available at http://www.fda.gov/downloads/Drugs/GuidanceComplianceRegulatoryInformation/Guidances/UCM358301.pdf

United Nations (2014) Secretary General. The road to dignity by 2030: ending poverty, transforming all lives and protecting the planet. Synthesis report of secretary-general on the post-2015 Sustainable Development Agenda. https://digitallibrary.un.org/record/785641

United Nations (2015) The Millennium Development Goals Report 2015. http://www.un.org/millenniumgoals/2015_MDG_Report/pdf

United Nations Development Programme (2016) 2030 sustainable development goals. United Nations, Geneva

World Health Organization (2015) WHO's vision and mission in immunization and vaccines 2015–2030. WHO, Geneva

World Health Organization (2017) Fact sheet V3P pricing report. WHO, Geneva. www.who.int/immunization/v3p

World Health Organization (2018) Vaccines in pipeline. WHO, Geneva. http://www.who.int/immunization/diseases/en/

World Health Organization (2019) Ten threats to Global Health in 2019. WHO, Geneva

Achieving Sustainable Drug Development Through CSR: Possibility or Utopia

15

Dhanay M. Cadillo Chandler

Abstract

This book chapter looks into corporate social responsibility (CSR) and meta-regulation as legislative technique as a way to implement regulatory incentives of sorts aiming to foster sustainable drug development. The chapter pays particular attention to supplementary protection certificates, the recent manufacturing waiver in the EU, and PRIME. By reviewing these regulatory inccentives, the chapter reflects on the need to include CSR principles in future legislative reforms, which should not focus exclusively in recouping investments but instead in achieving a balance amongst stakeholders' and patients' interests, thus proposing meta-regulation as a means to achieve it.

Keywords

Corporate social responsibility · IP laws · WTO · Pharmaceutical biotechnology · Bioeconomy

15.1 Background

Numerous studies have focused on the advantages and the need to perpetuate intellectual property rights (IPRs) protection within the framework of international trade. Is a well known fact that, in 1994 as part of the mandatory package of treaties ratified with the creation of the World Trade Organization (WTO), the Agreement

The views expressed in the article are those of the author and do not represent any official position by the institution represented by the author.

D. M. Cadillo Chandler (✉)
Faculty of Law, Turku Institute of Advanced Studies, University of Turku, Turku, Finland
e-mail: dmcach@utu.fi

C. Keswani (ed.), *Bioeconomy for Sustainable Development*,
https://doi.org/10.1007/978-981-13-9431-7_15

on Trade-Related Aspects of Intellectual Property Rights (TRIPS Agreement) became of mandatory ratification and implementation by all countries wishing to take part and benefit from international trade within the context of the WTO. Hence, the disparities amongst the economies taking part in the negotiation of the TRIPS Agreement, and particularly the diverse public health needs, called for the revision and enactment of a new declaration by the WTO where public health concerns were addressed. The Doha Declaration on the TRIPS Agreement and public health was adopted by the WTO Ministerial Conference of 2001 in Doha on November 14, 2001. This declaration reaffirmed a series of flexibilities, also known as TRIPS flexibilities, that WTO member states could use to circumvent patent rights for better access to essential medicines. Nonetheless, TRIPS-plus agreements have proliferated and subsequently strengthening IP protection around the globe.

Pharmaceutical innovations as foreseen at the time of negotiating TRIPS did not necessarily envisage the development of biosimilars, the challenges to the IP framework due to disruptive innovation, nor that the incentives based on the patent system would require reassessment (Taubman and Gafale 2007). Generic competition, although desired in term of access to medicines, has not always been possible due to the expanding protection through IPRs, test data protection, and supplementary protection certificates (SPCs). Nevertheless, the pharmaceutical industry's current innovation landscape and society's increasing needs call to revisit the IP framework through a more development-oriented lens instead of the traditional right holder-based one.

In terms of innovation, the EU has pointed out the importance in finding alternative mechanisms to promote high-value innovations while at the same time ensuring that "innovation that matters is actually produced, patients have access to this and that health systems are financially sustainable" (EXPH – Expert Panel on Effective Ways to Investing in Health 2018). "Sustainability" is a concept ever more present in relation to access to health, drug discovery, and financing of R&D, and as such it will be examined. Drug discovery is an expensive and elaborated project, hence, justifying the market exclusivity and elevated prices. The average cost of developing a new drug and placing this on the market has been estimated at USD 2.6 billion (Mullard 2014). Data on R&D expenditure significantly varies, from USD 802 million in 2002 (Di Massi et al. 2003) to USD 240 billion in 2009 spent in health R&D globally, which 90% was in high-income countries (OECD 2017). Admittedly, patents continue to be a reward to the inventor for its contribution to society with the invention, and at the same time society is rewarded with knowledge unknown to them until the patent disclosure takes place (Capp 2003).

Part of the cost associated to medicines, or the rationale for the pricing level, is attributed to the cost related to regulatory approval. Thus, the European Medicines Agency (EMA) in 2016 introduced a scheme named PRIority MEdicines (PRIME), which is a voluntary scheme envisioned to support the development of medicines targeting unmet needs.[1] In brief, this scheme seems to foster collaboration between

[1] See PRIME: priority medicines, available at:<https://www.ema.europa.eu/en/human-regulatory/research-development/prime-priority-medicines> (accessed, 15 Feb 2019).

regulatory agencies, i.e. EMA, and drug developers to offset the lengthy regulatory processes delaying patients from obtaining needed medicines while, at the same time, optimising clinical trial design. The fact that PRIME is not contrary to the current patent system and other regulatory incentives denotes a move toward alternative "compliance" mechanisms employed by the "regulators"[2] based on society's health needs instead of the right holder's interests as a stand-alone. This book chapter explores how and if PRIME approximates to meta-regulation and CSR, thus bridging the gap between CSR and IP law.

This book chapter looks into corporate social responsibility (CSR) and meta-regulation as legislative technique as a way to implement regulatory incentives of sorts aiming to foster sustainable drug development. The chapter draws particular attention to supplementary protection certificates, the recent manufacturing waiver in the EU, and PRIME; by reviewing these regulatory incentives, the chapter reflects on the need to include CSR principles in future legislative reforms, which should not focus exclusively in recouping investments but instead in achieving a balance amongst stakeholders' and patients' interests, thus proposing meta-regulation as a means to achieve it.

15.2 Patents, the Classic Go-to Incentive

The TRIPS Agreement has been referred to as an "agent of change in the health sector" (Cullet 2003) by establishing protection for inventions related to pharmaceutical products and processes in Art. 27. The agreement, besides extending patent protection to all fields of technology and stating patentability requirements, also allows, in Art. 7 and 8, the use of other necessary measures to balance rights and obligations.

The wording in Art. 7 is clear in the fact that IP protection and enforcement need to be "conducive to social and economic welfare and to a balance of rights and obligations".[3] Additionally, Art. 8 reiterates the promotion of "public interest in sectors of vital importance to their socio-economic and technological development, provided that such measures are consistent with the provisions of this Agreement".[4]

[2] The term "legislator" is purposefully avoided, since the initiative although based in Law is not law.

[3] Marrakesh Agreement establishing the World Trade Organization, Geneva, 15 April 1994, entered into force 1 January 1995, 1867 United Nations Treaty Series 3, annex 1C – Agreement on Trade-Related Aspects of Intellectual Property Rights (1994), 'Art. 7: The protection and enforcement of intellectual property rights should contribute to the promotion of technological innovation and to the transfer and dissemination of technology, to the mutual advantage of producers and users of technological knowledge and in a manner conducive to social and economic welfare, and to a balance of rights and obligations'.

[4] Idem Art. 8(1): 'Members may, in formulating or amending their laws and regulations, adopt measures necessary to protect public health and nutrition, and to promote the public interest in sectors of vital importance to their socio-economic and technological development, provided that such measures are consistent with the provisions of this Agreement.

Traditionally, these two articles have given regulatory space for countries to make use of the well-known TRIPS flexibilities, e.g. compulsory licences. Nonetheless, the link between Arts. 7 and 8 TRIPS with CSR and even meta-regulation cannot be excluded. In the quest toward aligning or balancing rights, scholars claim that defining inventions by establishing patentability requirements but not defining what these means, except for the "inventive step", may in itself "open some room for flexibility at the national level"[5](Correa 1998).

Patents, as understood traditionally, grant exclusive rights to prevent others from making, using, or selling the new invention for a period of 20 years counted from the filling date.[6] The term of protection is heavily relied upon by the pharmaceutical industry, since they allow companies to recoup R&D costs. Thus, the patent system is said to contribute to the legal certainty for investors/ companies to engage in further R&D. Pharmaceutical products are rather particular, in a sense, since filling for patent protection does not ensure the product or process is apt for commercialisation. Thereafter, efficacy and safety need to be proven by engaging in costly clinical trials together with complying with another set of quality requisites before reaching the market. Obtaining marketing approval (MA) can take up to 12 years.

The patent system as the sole incentive, and model to finance R&D in the pharmaceutical sector, has been questioned over the years; in fact scholars began questioning the IP system to be the right outlet to prompt and finance innovation in all fields of technology (Love and Hubbard 2007). Without a doubt the pharmaceutical industry is highly dependent on the exclusive rights provided by patent protection and further entry market incentives to recoup their investment, e.g. SPCs. The success of the patent system is being challenged in terms of its ability to foster innovation but also by the development of new technologies seeking protection under this regulatory framework. Governmental health expenditure and the possibility to obtain larger profits are said to be the main drivers for innovation (Kremer 2002).

Additionally, the validity of the current regulatory framework related to new health technologies has also been challenged by the proliferation in the use of defensive patent strategies[7] which do not only delay market entry of competitors but

---

(2). Appropriate measures, provided that they are consistent with the provisions of this Agreement, may be needed to prevent the abuse of intellectual property rights by right holders or the resort to practices which unreasonably restrain trade or adversely affect the international transfer of technology'.

[5] Or at the regional level, case in point the EU. In other words, allowing for the insertion of voluntary incentives to foster collaboration or even reduce R&D costs by assisting in clinical trials or waiving fees as foreseen within PRIME. See further section 2.3.

[6] See Art. 33 TRIPS Agreement. At the EU level, patent protection is established at the Convention on the Grant of European Patents of 5 October 1973 as amended by the act revising Article 63 EPC of 17 December 1991 and by decisions of the Administrative Council of the European Patent Organisation of 21 December 1978, 13 December 1994, 20 October 1995, 5 December 1996, 10 December 1998, and 27 October 2005. See Art. 63.

[7] These patent strategies focus on excluding competitors without pursuing innovative efforts. See Commission, 'Pharmaceutical Sector Inquiry Final Report, Part I' (8 July, 2009); and 'Communication from the Commission – Executive Summary of the Pharmaceutical Sector Inquiry Report' (8 July, 2009).

potentially discourage innovation of new health technologies (Drexel 2013). Policy incoherencies and overlaps have emerged in the regulatory and incentive context; thus the High-Level Panel on Access to Medicines (HLP) stressed the need to find "solutions to remedying the policy incoherence between the justifiable rights of inventors, international human rights law, trade rules and public health in the context of health technologies that is impeding access and the right to health for millions" (Richard Elliott 2016).

Part of the economic challenges in creating incentives to prompt R&D relates to the aggregation of capital and the posterior need to delink the total R&D costs from the final price of medicines (Abbott 2017). Now when, for instance, traditional antibiotics are struggling to cope with new strains of old illnesses, "drug-resistant viruses, bacteria, parasites and fungi could cause 10 million deaths a year worldwide by 2050" (United Nations Secretary-General's High-Level Panel on Access to Medicine 2016). Therefore, creating new and sustainable incentives for R&D continues to be quintessential for innovation, while ensuring access to new drugs by delinking the costs of innovation from prices is more important than ever[8] (Cadillo Chandler 2016; United Nations Secretary-General's High-Level Panel on Access to Medicine 2016).

At the EU level, the discussion over incentives has focused on SPCs,[9] the recent manufacturing SPC waiver of export purposes,[10] and PRIME. Although SPCs are not an intellectual property right stemming from the TRIPS, these are sui generis rights existing in co-dependency with patent rights. Hence, a regulatory incentive is providing an additional layer of protection to pharmaceutical inventions. SPCs have been introduced in the EU as a way to compensate the right holder for the lengthy process while seeking for the necessary MA before placing the products on the market. SPCs may grant a maximum of 5 years of protection from the time the basic patent expires, thereafter extending the term of protection.

The pharmaceutical regulatory framework, as conceived in its beginning, aimed at protecting originators. Admittedly, the utilitarian patent approach contraposed to the trade-offs to society from the use of new health technologies has been the dominant approach justifying patent protection (Merges and Posner 1990; Posner 2005). However, globalisation seemed to have tipped off the balance in favour of generic and biosimilar manufacturers, whom will be able to manufacture these products – although only – for export purposes without infringing patent rights in the context of EU patent and SPC regulatory framework. This manufacturing waiver, which will eventually lead to access to cheaper drugs in the EU while also allowing the EU to remain as an

[8] World Health Organization, Sixty-Eight World Health Assembly, Decision A68/DIV./3 (5 June 2015)

[9] Regulation (Ec) No 469/2009 Of The European Parliament And Of The Council Of 6 May 2009 Concerning The Supplementary Protection Certificate For Medicinal Products (Codified Version)

[10] Impact Assessment Accompanying the Document Proposal for A Regulation Of The European Parliament And Of The Council Amending Regulation (Ec) No 469/2009 Concerning The Supplementary Protection Certificate For Medicinal Products.

innovation hub,[11] together with PRIME reveal a tendency within the EU to find alternative models of incentives creating synergies amongst interested parties.

As stated above, the protection and enforcement of IP rights was intended to prompt further transfer of technology and know-how and to achieve innovation conductive to social development.[12] However, by strengthening IP rights, innovation has not necessarily increased as hoped for. PRIME may be the first of many voluntary strategies to be implemented in synergy with the IP regulatory framework and additional layers of protection to increase innovation and sustainability in developing this.

15.3 Reviewing the EU's SPCs and Manufacturing Waiver

15.3.1 Supplementary Protection Certificates in Europe

Supplementary protection certificates were implemented in the EU by the European Economic Council (EEA) in 1992 under Council Regulation No. 1768/92, later on codified in the Regulation of the European Parliament and of the Council No. 469/2009 (SPC regulation).[13] This regulation aimed at compensating pharmaceutical companies for the time spent in complying with regulatory procedures, hence, creating an incentive to carry out R&D while allowing manufacturers to recoup the costs of their activity by extending the effective patent life (Mejer 2017). SPCs cannot be obtained in isolation from MA; in fact, it can only become into force once the basic patent expires and should not last for more than 5 years from the date in which it takes effect (Cook 2016). Even when these additional 5 years of protection may not be sufficient to recoup the costs, they are nevertheless of immense economic value for the pharmaceutical industry since the market peak for the commercialisation of medicinal products seems to be reached toward the end of the term of patent protection (Manley and Vickers 2015).

Initially, SPC regulation intended to harmonise protection and incentive for the pharmaceutical industry across the EU. However, the implementation amongst EU member states of the SPC regulation and its scope has been somewhat fragmented. For instance, expiry dates across member states is not homogenous (Mejer 2017), and also some discrepancies in terms of examination outcomes seem to persist. SPCs are a complex regulatory system, over which, the CJEU remains to clarify a few issues; however, these are not meant to be discussed within the book chapter. The purpose of reviewing SPC protection in the EU and the manufacturing waiver in connection to PRIME is to show the reader how alternative models may be beneficial to achieve what the IP system as a stand-alone has not been able to.

[11] Idem

[12] See Art. 7 and 8 TRIPS Agreement.

[13] See Regulation (EC) No 469/2009 Of The European Parliament And Of The Council Of 6 May 2009, and amendment full quote.

For SPCs, as established within the Regulation (EC) 469/2004, pharmaceutical research plays an important role within the continuous improvement of public health.[14] Thus, it is necessary to secure a term of at least 15 years of marketing exclusivity, particularly when medicinal products will not be able to reach the market without a marketing approval. SPCs have been pointed out to be, although national rights, sui generis right with one foot in patent system and the other in the regulatory system (Manley and Vickers 2015). While MA and data exclusivity fall outside the scope of this paper, it is important to address their relevance to the SPCs system, since the purpose of one is not excluded from the other. Data exclusivity "is intended to stimulate and protect the investment in expensive and time-consuming clinical trials" (Curley and Horst 2012), *and at the same time SPC* "compensates for the time lost during the 20-year patent term protection, whilst clinical trials are undertaken" (Curley and Horst 2012). To a certain extent, both rationales meet the utilitarian approach toward patent protection, where protection is granted as a reward to the inventor for the creation and dissemination of knowledge that otherwise would not be accessible to the public. Even when dissemination is required, protection on the other hand to make that dissemination worthwhile is equally important. Without getting into further details related to the rationales for the existence of three well-established systems, let's proceed to analyse the conditions to grant SPC protection within the EU, to thereafter examine the proposal for amendment, and PRIME.

Article 3 from the SPC regulation requires for the product to be protected by a basic patent and to have a valid authorisation to place the said product on the market as a medicinal product granted in accordance with EU rules; the product has not already been subject to the certificate and that the marketing authorisation referred to is the first authorisation to place the product on the market as a medicinal product. The requirements in themselves create the first link between both patents and MA, and the validity or effect of such MA either can be at the national level of an EU member state or may have effects within the single market as a whole, depending on the procedure followed to obtain it. In a similar fashion to MA, SPCs need to be applied for within each EU member state where the extension of the protection is to be sought, unless the upcoming UPC system incorporates a Unitary SPC of sorts; however, these too fall outside the scope of the present chapter.

SPC regulation entered into force on January 1993, later on in 2004, 2007, and 2013 in countries entering the EU during those years. Nevertheless, the regulation calls for the application for an SPC to be logged before the national patent office where protection is sought either within 6 months from the date of the first MA in the EU member state to commercialise the product was granted or within 6 months from the date in which the basic patent was granted if this was granted before the MA did.

---

[14] See Preamble from Regulation (EC) No 469/2009 Of The European Parliament And Of The Council Of 6 May 2009.

15.3.2 EU's SPC Manufacturing Waiver

The manufacturing waiver, as proposed by the EU Commission, brought to the spotlight the importance of creating further flexibilities for European small and medium enterprises (SMEs) in terms of manufacturing generics and biosimilars in the EU, to thereafter export them into countries where patent protection is not available, expired, or never granted. The waiver is essential to avoid patent infringement in the EU while at the same time emphasise the importance of SMEs for EU's economy. Additionally, EU based -SMEs- firms will continue to have the possibility to compete in the European market by having manufacturing capability to produce generics to be introduced onto the market in a date as close as possible to day 1 after patent and SPC expiration. This waiver seems to be a step in the right direction toward the use of TRIPS flexibilities within the EU.

The manufacturing waiver follows a lengthy review to the SPC regulatory framework as requested by the EU Commission during June 2016.[15]At that time the commission, requested all interested stakeholders to provide feedback on the challenges presented by SPCs. Although the review to the whole system is not complete, one of the most important outcomes is the manufacturing waiver. Several cases remain to be addressed by the CJEU, and admittedly the proposal is not covering these.

One of the main criticisms to the initial waiver proposal is the fact that public health or the rising cost of health within the EU will not be completely sorted out by the waiver in itself, if anything is somewhat a secondary interest as portrayed by the impact assessment.[16] One of the main arguments or rationales for the waiver seems to be the eminent delocalisation of EU-based SMEs into third-world countries to manufacture generics and/or biosimilars, hence, resulting in significant losses for the EU in relation to jobs, taxes, and skills just to name a few.

Public health, even when addressed within the proposal, is not necessarily central or pivotal for this shift, even when originators within the feedback/review process adamantly opposed to the manufacturing waiver. The impact assessment also seems to suggest the creation for a "Unitary SPC", which may grant more securities for right holders to ensure they won't face competition within the EU before SPC protection expires. Nonetheless, until this issue is addressed in the upcoming Unitary Patent Court (UPC),[17] nothing is for certain. Even when "Unitary SPCs" are a fascinating topic to address, this falls out of the scope of the present book chapter. The implementation of a manufacturing waiver for exports without the possibility to

---

[15] See Council Conclusions on strengthening the balance in the pharmaceutical systems in the EU and its member states. European Union: European Commission, Commission Staff Working Document Impact Assessment: Accompanying documents for Proposal for a Regulation of the European Parliament and of the Council amending Regulation (EC) No 469/2009 concerning the supplementary protection certificate for medicinal products, Brussels May 2018, SWD(2018) 240 Final.

[16] Idem.

[17] See Plomer, A. "A Unitary Patent for a (Dis)United Europe: The Long Shadow of History" in46 IIC – International Review of Intellectual Property and Competition Law 5, August 2015, pp. 508–533.

legally stockpile aiming for after SPC day-1 entry into the market only addresses the potential delocalisation of EU-based companies but not the creation of competition within the EU single market.

Amongst the arguments presented by the EU Commission in the impact assessment report,[18] the manufacturing waiver shall ensure Europe's position as a hub for pharmaceutical R&D, hence, protecting its competitive advantage in certain high-tech industries. Following the TRIPS Agreement, the implementation of higher standards of protection aimed at precisely protecting the competitive advantage of certain industries and preventing the narrowing gap that would diminish the said advantage (Muzaka 2011). The manufacturing waiver intends to address the potential competitive advantage loss not through higher standards, but on the opposite by allowing more flexibility in the manufacturing of generics and/or biosimilars with the purpose of competing in foreign markets.

Another aspect missing from the manufacturing waiver may be the inclusion of provisions addressing or securing subsequent innovation to pursue some sort of sustainable drug development. Sustainability, as highlighted within this book chapter's background, is a term with ever more importance when looking into innovation and regulation. Part of the challenge in the current innovation scheme is allegedly due to the regulatory hurdles and the exhaustion of easy innovation targets, thereafter demanding from the pharmaceutical and life science industry to find products that are not only novel but also effective enough in comparison to the existing ones (Mittra et al. 2011). The complexities of the current regulatory framework, i.e. patents, SPCs, and data exclusivity, mainly enable big players to compete in market. The EU manufacturing waiver, in this sense, is a step forward in looking after SME's potential to change the innovation landscape. Already in 2011, scholars pointed out the need to implement "smart regulation"[19] with the purpose of stimulating companies' structural changes instead of developing innovation strategies on the basis of the existing regulatory incentives (Mittra et al. 2011).

Incentives such as the EU Orphan Drug Scheme may need reformulation, since its success has also led to evaluate the possible abuse of the system by the lack of innovation in the targeted field (OECD 2017). In terms of the EU Orphan Drug Scheme,[20] its importance for this research is based on its success in achieving certain level of innovation within the EU. Just like SPCs, the classification of an orphan drug is not imposed by TRIPS; thus, it is an incentive outside the patent system. Orphan drugs have been regulated in the USA long before they were in the EU

[18] See European Union: European Commission, Commission Staff Working Document Impact Assessment: Accompanying documents for Proposal for a Regulation of the European Parliament and of the Council amending Regulation (EC) No 469/2009 concerning the supplementary protection certificate for medicinal products, Brussels May 2018, SWD(2018) 240 Final.

[19] Smart legislation is a terminology used by Mittra et al. (2011) requiring from regulators to design legislation implementation that is sensitive to opportunities emerging from life sciences and is not exclusively addressed to sustain multinational dominance.

[20] Regulation (EC) No 141/2000 of the European Parliament and of the Council of 16 December 1999 on orphan medicinal products (OJ L 18, 22.1.2000, p.1), last amended by Regulation (EC) No 596/2009 (OJ L 188, 18.07.2009, p. 14).

(1999); nevertheless, its regulation has allowed the pharmaceutical industry to foster innovation in a field where only a very low percent of the population is treated. The EC Orphan Drug Directive defines what shall fall within this category, consequently achieving the status that represents an additional set of benefits for the pharma company developing the drug. For instance, 10 years of market exclusivity,[21] faster procedure to obtain the MA, additional aid to support drug development, and exemption of fees related to protocol assistance are in a general fashion some of the benefits portrayed by the system. This designation is not exempt from abuses either, allegedly the pharmaceutical industry has used orphan drug designations strategically to gain position in the market both in the EU and USA (Nakov and Al 2016) (Prescrire International N° 171 2016). The other regulatory incentive addressing innovation for a targeted group is PRIME, which, as mentioned earlier, is of voluntary nature, and as such it is believed to have further links to CSR and meta-regulation.

15.4 Finding Alternatives: CSR and Meta-regulation

15.4.1 Corporate Social Responsibility in IP

Corporate social responsibility (CSR) is a concept generally used and analysed within the context of corporate governance or private governance. The emergence of questions related, for instance, to ethical responsibility and legal liability requires further analysis. The link between intellectual property rights, with innovation and CSR, does not seem to be widely explored. However, the existence of a link between innovation, e.g. in agriculture, and the use of genetically modified organisms (GMOs) to address sustainability is not unknown as a way to create social value. Corporate disclosure (Jackson 2014) and the creation of best practice guides are just a few of the mechanisms pointed out as means to increase accountability. Regardless of the context, there are questions emerging from the current use of CSR policies and their implications for IP holders. CSR may have positive connotations for the pharmaceutical and biotechnological industry if infused within the incentives designed in a complementary system or in law.

Admittedly the concept of CSR has evolved from being related strictly to the relationship between management and shareholders (Siegel 2001) into a measuring yardstick between society expectations and company values. Thus, translating into "a key ingredient for organisation's licence to operate" (Simon 2002). The European Commission has identified several concerns related to plausible damage caused by economic activity to the environment, consumers, and public authorities. Citizen's expectations over companies' performance seem to be driving forces behind the recent trend toward CSR (European Commission 2001).

[21] See SPC Dir granting two additional years of protection against unfair competition, note the term of protection is generally 8.

The adoption of CSR in a company does not necessarily correlate to "better behaviour" on their behalf. On the opposite, studies have shown that companies engaging in stricter CSR policies tend to also engage in further irresponsible behaviour (Jackson 2014). Striking a balance amongst different stakeholders' interests and rights, i.e. industry, society, and governments, is a crucial aspect to promote legislative reforms seeking to benefit all from inventive activities. In this respect, CSR may play an important role in achieving such a balance.

For CSR to be effective, there are several elements to take into consideration, i.e. mandatory disclosure. Auditing mechanisms need to be readily available to attest on the veracity of company reports. Defining "how-good is good" will or should improve a company's accountability for the information disclosed. "Mandatory disclosure will prompt normative pressure to undertake actions" (Jackson 2014). Disclosure is particularly sensitive for intellectual property rights, i.e. patent, trade secrets, and even more in relation to MA and SPCs. Thus, distinguishing amongst the information for public disclosure may be key when suggesting CSR as a mechanism to increase companies' accountability. In this regard, PRIME, by providing scientific assessment at clinical trial design early stages, may be bridging some of the gaps.

Much has been said about the role of intellectual property rights to foster innovation. However, recent health crises and the need to provide sustainable means or food sources for a growing world population have called for further attention to corporate social responsibility issues. On the one hand, the pharmaceutical industry heavily relies on the patent system to protect their inventions, but on the other hand, the current discourse has led to raise the question on whether or not this industry should also share or have co-responsibility with the fulfilment of the human right to health. From the bio-agricultural industry perspective, the patent system also protects their inventions; however striving use of GMOs as a sustainable means to source foods leads to ponder not only on environmental consequences but also on health consequences in the long run due to the use of these inventions in everyday life. GMOs although mentioned are not part of the analysis within this chapter; nevertheless it is worth showing the reader the various connections between technological developments and CSR.

Another example where CSR may have infused regulation and compliance is the Kimberley Process. This process was set up as an international scheme of certification requirement for rough-diamond trade amongst diamond traders, aiming to avoid taking part in the commercialisation of conflict diamonds.[22]

The Kimberly Process Certification Scheme (KPCS) came as a form of company self-governance seeking to increase transparency in diamond trading activities. The ethical component is expressed by excluding conflict diamonds, which at the same time works as an incentive since a country not complying with the KPCS will be expelled from the scheme (Aaken 2008). This certification scheme even in a way seems to present elements of CSR, since it is not strictly related to the relationship

[22] *See* Kimberley Process Certification Scheme (KPCS) available at: <https://www.kimberleyprocess.com>.

between management and shareholders (Siegel 2001), but seems to fulfil a function of a measuring yardstick between society expectations – no trading with conflict diamonds – and company values, only trading with diamonds following a particular standard. Thus, translating into "a key ingredient for organisation's licence to operate" (Simon 2002).

The draft report from the EU Options for Improving Access to Medicines[23] calls on the commission to not only promote R&D driven by patient's needs, while fostering social responsibility by setting up an EU public platform for R&D funded by contributions from profit made by the pharmaceutical industry through sales to public health systems, but also call to increase transparency on the costs of R&D. Thus, further collaboration between stakeholders may lead to increasing both transparency and accountability and subsequently finding ways to reduce cost of R&D for major public health needs or priority medicines.

In the past CSR was considered a form of philanthropy; however there seems to be a shift in the discourse calling for CSR practices to increase accountability and transparency. It has been argued that CSR as a form of compliance will not be plausible unless there are incentives and a regulating entity, i.e. the KPCS. The gaps between CSR and meta-regulation and IP are narrowing as policy implementation develops, particularly in the case of sustainability. In this regard, sustainability as a tool to achieve balance can be foreseen within EU Economic Partnership Agreements, thereafter stressing the imminent link between exclusivity and societal progress (Grosse Ruse-Khan 2010). As stated above, meta-regulation has been used in other fields and commonly through corporate social responsibility.

In terms of IP, the wording in Art. 7 is clear in the fact that IP protection and enforcement needs to be "conducive to social and economic welfare, and to a balance of rights and obligations".[24] Additionally, Art. 8 reiterates the promotion of "public interest in sectors of vital importance to their socio-economic and technological development, provided that such measures are consistent with the provisions of this Agreement".[25] Traditionally, these two articles have given regulatory space for countries to make use of the well-known TRIPS flexibilities. Nonetheless, the link between Arts. 7 and 8 TRIPS with CSR or even meta-regulation seems to be present.

15.4.2 Meta-regulation Reaching Out to IP-Protected Inventions: Did PRIority MEdicines (PRIME) by EMA Achieved This?

During 2015, the European Medicines Agency (EMA) and its scientific committees worked on a number of initiatives that, as highlighted in by the EU Medicines Regulatory Network in their Strategy 2020,[26] aimed to find ways that support the

[23] European Parliament, Options for improving access to medicines, European Parliament resolution of 2 March 2017 on EU options for improving access to medicines (2016/2057(INI)).

[24] Art. 7 TRIPS Agreement.

[25] Art. 8(1) TRIPS.

[26] EU Medicines Agencies Network Strategy to 2020, Working together to improve health (EMA/MB/151414/2015, 17 December 2015).

development of medicines with the intent of accelerating patients' access to medicines that address unmet medical needs or major public health needs. Thereafter became necessary to develop a scheme "to reinforce early dialogue and regulatory support to stimulate innovation, optimise development and enable accelerated assessment of PRIority MEdicines (referred to as PRIME),"[27] Early dialogue to facilitate early assessment of medicines has been highlighted by the OECD as a positive step forward in allowing for medicines to reach patients faster and also allows developers to identify negative effects earlier on to improve the targeted drug therapy (OECD 2017). Early assessment is also based on Recital 33 and Article 14 (9) of Regulation (EC) No 726/2004. This regulation provided that "an applicant may request an accelerated assessment procedure in order to meet, in particular the legitimate expectations of patients and to take account of the increasingly rapid progress of science and therapies, for medicinal products of major interest from the point of view of public health and in particular from the viewpoint of therapeutic innovation".[28]

Identifying negative effects to improve the drug therapy earlier in the process may reduce economic hurdles, as pointed out by EMA in the first review of the scheme in 2018. Tailoring the scheme to the developer's needs at the different stages up to the MA application is amongst the benefits portrayed by the system. Additionally, the incentives or benefits are delineated at two levels, first "in early stages of development, following demonstrated proof of principle", and second "at clinical stages of the development, following demonstrated proof of concept". The first level, to label it somehow, focuses on SMEs and the academic sector, which in essence reflects on EU's policy focus in strengthening these industry sectors as the SPC manufacturing waiver did.

The benefits have been summarised as follows:

- *Raising awareness of regulatory requirements early in the development, by providing scientific and regulatory advice on the overall development plan and at major development milestones, with the possibility to involve multiple stakeholders* (e.g. *health technology assessment (HTA) bodies, patients).*
- *Eligibility to PRIME may help these applicants to overcome financial hurdles[2] to progress through later stages of the development.*
- *Upon request, SMEs and applicants from the academic sector may also be eligible for fee reductions on their scientific advice requests, upon case-by-case decisions.*

[27] EMA, Enhanced early dialogue to facilitate accelerated assessment of PRIority Medicines (PRIME) EMA/CHMP/57760/2015.

[28] European Medicines Agency, Committee for Medicinal Products for Human Use, Enhanced early dialogue to facilitate accelerated assessment of PRIority Medicines (PRIME), 7 May 2018, EMA/CHMP/57760/2015, Rev. 1.

In clinical stages of the development, the following demonstrated proof of concept:

- *Early appointment of CHMP Rapporteur[4] (in line with current process, objective criteria and methodology).*
- *An initial kick-off meeting with multidisciplinary participation from the EU network (SAWP, CAT, COMP, PDCO, PRAC, and experts, as relevant), including the CHMP Rapporteur[4], to discuss the proposed development programme, gives preliminary guidance on requirements for MAA and to develop a schedule for giving regulatory and scientific advice and for submissions of applications to fulfil legislative requirements (e.g. paediatric investigation plan) and scientific advice on key decision points/issues for the preparation of MAA with the possibility to involve multiple stakeholders (e.g. health technology assessment (HTA) bodies, patients), when relevant. This may also include scientific advice on risk management plan and post-authorisation activities.*[29]

PRIME is conceptualised within the so-called adaptive pathways that entail strategies aimed at granting "approval in stages, beginning with a restricted patient population then expanding to wider patient populations; or confirming the benefit-risk balance of a product, following a conditional approval based on early data (using surrogate endpoints) considered predictive of important clinical outcomes" (OECD 2017). Initially, PRIME is not a law as conceived within the context of the IP regulatory framework, nor other regulatory incentives as the SPC system, but instead is part of the alternative models to steer faster access to medicines.

As a voluntary scheme, PRIME furthers synergies between regulators, developers, and public health interests. This combination of elements could be the first connecting point to CSR principles and meta-regulation. Meta-regulation is a technique or approach likely to point out at nontraditional governance mechanisms to be recognised as law in an extended sense, leading to accountability and transparency (Parker 2007), and in terms of CSR – although primarily voluntary – has been conceptualised as policies aiming to do social good, providing early access to patients, beyond the interests of the firm and that is required by law.

15.5 Concluding Remarks

By addressing the importance of intellectual property rights, patents case in point, at the beginning of the chapter, emphasis was made on the role of the TRIPS Agreement for innovation. On the one hand, IP protection was extended to all fields of technology and, on the other hand, regulating and providing incentives for new pharmaceutical innovation. Even when the provisions within TRIPS intended to

[29] European Medicines Agency, Committee for Medicinal Products for Human Use, Enhanced early dialogue to facilitate accelerated assessment of PRIority Medicines (PRIME), 7 May 2018, EMA/CHMP/57760/2015, Rev. 1.

provide a balance, in practice, IP protection has been strengthened, and additional regulatory incentives have been necessary to prompt R&D. Interestingly enough, the wording in Art. 7 and 8 TRIPS seems to provide links to look for alternative models that were not thought of before in the context of IP.

Even when sustainability, as such, is not a concept exhaustively addressed in the book chapter, its mention has served as a connector to CSR and meta-regulation. Some suggest that the creative use of government intervention may be strategically more promising than following the traditional incentives (Ashford and Hall 2011). The gaps between CSR and meta-regulation and IP are narrowing as policy implementation develops, particularly in the case of sustainability. PRIME may be the first of many voluntary strategies to be implemented in synergy with the IP regulatory framework and additional layers of protection to increase innovation and sustainability in developing this.

SPC regulation in the EU extends the term of patent protection by allowing the originator to retain exclusive rights longer, as a means to compensate for the lengthy procedures to obtain MA. This sui generis system, perhaps as envisaged at the beginning, was more lenient toward right holders than public health concerns. However, dismissing the role of IP protection and other incentives for the development of innovation aiming to alleviate health concerns would not be adequate either. Globalisation seemed to have tipped off the balance in favour of generic and biosimilar manufacturers, particularly SMEs, whom will be able to manufacture these products – although only – for export purposes without infringing patent rights in the context of EU patent and SPC regulatory framework. This waiver will eventually lead to access to cheaper drugs in the EU while also allowing the EU to remain as an innovation hub. PRIME reveals the need to find alternative models of incentives creating synergies amongst interested parties, which may lead to a reduction in R&D costs which ultimately benefit patients. Whether meta-regulation or voluntary schemes will continue to infuse incentives targeting pharmaceutical innovation remains to be seen. Reviewing PRIME effectiveness in the future will shed light on the usefulness of implementing alternative regulation mechanisms to achieve further innovation and sustainable health technology development.

References

Aaken AV (2008) Effectuating public international law through market mechanisms? Comparative Research in Law & Economy, Research Paper 34/2008, 04(07)

Abbott F (2017) Excessive pharmaceutical prices and competition law: doctrinal development to protect public health. UC Irvine Law Rev 6(3)

Ashford N, Hall R (2011) The importance of regulation-induced innovation for sustainable development. Sustainability:270–292

Cadillo Chandler D (2016) The never ending story of access to medicines. WIPO J: Anal Intellect Prop 8(1):42–51

Capp DA (2003) A Propiedade Intelectual Na Constituição. In: Fabris S (ed) Límites Jurídicos da Regulação e Defensa da Concorrência (p. 52). Porto Alegre

Cook T (2016) Pharmaceuticals in biotechnology and the law. Lexis Nexis

Correa C (1998) Implementing the TRIPS agreement in the patents field: options for developing countries. J World Intellect Prop 1(1):75–99
Cullet P (2003) Patents and medicines: the relationship between TRIPS and the human right to health. Int Aff 79(1):139–160
Curley D, Horst MH (2012) Patents and regulatory data exclusivity for medicinal products. In: Wilkof N, Basheer S (eds) Overlapping intellectual property rights. Oxford University Press, Oxford, pp 119–136
Di Massi JA et al (2003) The price of innovation: new estimates of drug development costs. J Health Econ 22(2):151–185
Drexel J (2013) AstraZeneca and the EU Sector inquiry: when do patent filings violate competition law? Max Planck Inst Intellect Prop Compet Law Res Pap 12(2). https://doi.org/10.2139/ssrn.2009276
European Commission. (2001). Promoting a European framework for corporate social responsibility. COM
EXPH – Expert Panel on Effective Ways to Investing in Helath (2018) Opinion on Innivative payment models for high-cost innovative medicines. Publications Office of the European Union, Luxemburg
Grosse Ruse-Khan H (2010) Sustainable development in international intellectual property law – new approaches from EU economic partnership agreements? ICTSD programme on IPRs and sustainable development (Paper N 29)
Jackson G (2014) A socio-political perspective on corporate social responsibility: understanding regulatory substitution and the persistence of irresponsibility. In: Hilty RM-B (ed) Corporate social responsibility: Verbindliche standards des Wettbewerbsrechts? Spinger, Berlin, pp 19–31
Kremer M (2002) Pharmaceuticals and the developing world. J Econ Perspect 16(4):67–90
Love J, Hubbard T (2007) The big idea: prizes to stimulate R&D for new medicines. Chicago Kent Law Rev 82(3):1519–1554
Manley M i, Vickers M (2015) Supplementary protection certificates. In: Manley M i, Vickers M (eds) Navigating European pharmaceutical law. Oxford University Press, Oxford, pp 277–299
Mejer M (2017) 25 years of SPC protection for medicinal products in Europe: insights and challenges
Merges R, Posner R (1990) On the complex economics of the patent scope. Colum L Rev 90:839
Mittra J, Tait J, Wield D (2011) The furute of pharmaceutical innovation: new challenges and opportunities. Innov Pharm Technol 36:32–34
Mullard A (2014) New drugs cost US$2.6 billion to develop. Nat Rev Drug Discov 13:877
Muzaka V (2011) The politics of intellectual property and access to medicines. Palgrave Macmillan, Hampshire
Nakov Z, Al E (2016) Orphan diseases, orphan drugs and orphan regulation in USA and EU. IOSR J Pharm 6(11):5–10
OECD (2017) New health technologies: managing access, value and sustainability. OECD Publishing, Paris
Parker C (2007) Meta-regulation: legal accountability for corporate social responsibility. In: McBarnet D, Voiculescu A, Campbell T (eds) The new corporate accountability: corporate social responsibility and the law. Cambridge University Press, Cambridge
Posner R (2005) Intellectual property: the law and economics approach. J Econ Perspect 19(2):57–73
Prescrire International N° 171 (2016) New drugs and indications 2015, May, 25
Richard Elliott wt-L (2016) United Nations Secretary-General's high-level panel on access to medicines, background paper: international legal norms: the right to health and the justifiable rights of inventors
Siegel AM (2001) Corporate social responsibility: a theory of the firm perspective. Acad Manag Rev 26(1):117–127

Simon A (2002) Corporate social responsibility and biotechnology: identifying social aspects for European biotechnology companies. International Institute for Industrial Environmental Economics, Lund

Taubman A, Gafale R (2007) Public sector IP management in the life sciences: reconciling practice and policy-perspectives from WIPO. In: Krattinge A, Mohoney R, Nelsen L et al (eds) Intellectual property management in health and agricultural innovation: a handbook of best practices. MIHR, Oxford, pp 229–246

United Nations Secretary-General's High-Level Panel on Access to Medicine (2016) Promoting innovation and access to health technologies. United Nations, New York

Function of the Medicinal Plants of the Mangroves in a Society of High Marginalization in Tabasco, Mexico

16

Ángel Sol Sánchez, Gloria Isela Hernández Melchor, and Juan Manuel Zaldívar Crúz

Abstract

Medicinal plants were studied in 4 communities from the coast of Tabasco; it registered 63 species, which are used to treat 37 local diseases. Those plants have different origins; 81.5% were located in family gardens, and people daily coexist with them; two species were cultivated for consumption, but a secondary use is medicinal. Also 2 species were registered in the fences, and 4 come from mangroves and 2 from paddocks, and 14 are wild. Here we found six biological forms: tree, shrubs, herbs, vines, epiphytes, parasites, and thick vines. Because the soil is very saline, only a few species grow as trees in the orchards. The other species, mainly grasses and vines, are planted in discarded dishes and hung to the trees or arranged in the branches of the same. For these plants, substrate from other agricultural areas of the region is used. It is important to mention that only tree or shrub species are grown in the gardens of family orchards, such as the nance (*Byrsonima crassifolia* (L.) Kunth.) and the cinnamomum (*Cinnamomum zeylanicum* Breyn). The rest of the species are usually planted in waste containers such as buckets or frets that no longer fulfill their main function, but the substrate is not from the area; this is because in coastal communities the soil is too saline, due to which many species do not grow in these conditions, so over time the ladies have brought from communities far from the coast fertile soil where watermelon, corn, or sorghum is produced and have used it as a substrate to grow their medicinal plants. The main uses were for treatment of pain, spells, treatment of nerves, expulsion of intestinal parasites, inflammations rash, hair treatment, constipation, emmenagogue, diarrhea, diabetes, and rheumatism, the other treated with two species each.

Á. Sol Sánchez · J. M. Zaldívar Crúz (✉)
Colegio de Postgraduados en Ciencias Agrícolas, Cárdenas, Tabasco, México
e-mail: zaldivar@colpos.mx

G. I. H. Melchor
Sociedad Iberoamericana de Bioeconomia y Cambio Climático. Asociación Civíl, Cárdenas, Tabasco, México

C. Keswani (ed.), *Bioeconomy for Sustainable Development*,
https://doi.org/10.1007/978-981-13-9431-7_16

Keywords
Mangroves · Medicinal plants · Traditional knowledge

16.1 Introduction

Since ancient times medicinal plants have played an important role in the health and economy of rural villages, rural communities that are located outside the areas of economic development and generally remain in a state of poverty, in some cases extreme poverty. This situation leads them to depend on the resources of the area for their survival such as food, medicine, clothing, footwear, and housing (Keswani et al. 2017; Singh et al. 2016).

Of these satisfiers the two most important are food and medicine, since food is daily and medicine is constant; for this reason the collection of medicinal plants is one of the main sources of income in communities that have access to a market.

Countries such as Albania generate a large part of their trade from *Salvia* (*Salvia officinalis*), which, although produced in large volume, is processed in importing countries. Currently Europe is experiencing a growing market in the consumption of medicinal plants that had gone to the second place by the use of synthetic drugs; however the consumption of medicinal plants is in the process of growth since its consumption is safer and without collateral effects.

The use of plants is not restricted to medicinal plants but to a wide range of uses such as colorants, homeopaths, cosmetics, and pharmaceuticals, among many others (Ocampo 2002). In Latinamérica the use and trade of medicinal plants is very broad, and some species characterize some countries, such as guaraná (*Paullinia cupana)* in South America, cat's claw (*Uncaria tomentosa*) in Peru, ipecacuanha (*Psychotria ipecacuanha*) in Costa Rica and Nicaragua, and cocolmeca (*Smilax domingensis*) and calaguala (*Phlebodium aureum* L.) in Guatemala, while Mexico, with little more than 3100 species of medicinal plants, commercializes, besides the matter, essential oils of 70 species that generate income of several million dollars annually.

In México, one of the main vending markets for medicinal plants is the Sonora market where it is possible to obtain species with different uses. It is estimated that the market for medicinal plants in Mexico covers 5000 species, of which 250 are clinically validated. From this total, the Mexican population generally employs between 3500 and 4000 species, of which 250 species are often used naturally. A high percentage of these species (3600) are wild and come from collection, and only 370 come from family gardens. In general, there are records that at least 35 species used as medicinal plants are in threatened status (Belluci and Paola 2002).

In the case of the mangrove, this has been cited as a medicinal species; there are references that in the archipelago of Fiji the use of the red mangrove root is of daily and is commonly used to cure fever, cough, and asthma among others. This is possibly due to the anti-inflammatory, antioxidant, and antiviral substances that it possesses. Also, there are conventional medicines made with the root of this species. It has also been used in burn treatments (Fondén et al. 2015). This means a possibility for the exploitation of the roots of this species through a management plan, since these sites generate welfare not by the sale of medicinal plants, but because they generate medicinal species that avoid economic expenses in medicines or medical (Alongi 2002).

In this sense, the restoration and conservation of mangroves should be a priority, in which the government and local communities should be integrated. If this possibility is not considered, there is a risk that the mangrove area will decrease given that many nuclei of human population are settled on the coasts where there is a wide use of mangrove areas for shrimp culture, fish culture, wood extraction, and others (Romañach et al. 2018).

In the rural communities of Mexico, the use of medicinal plants is very broad, large volumes of unprocessed plants are used to alleviate common or frequent diseases given the impossibility of traveling to the cities or paying a consultation. Bautista-García et al. (2016) cite the presence of 203 species of useful plants in the ejido La Encrucijada of Cárdenas, Tabasco, of which 42 correspond to medicinal plants, representing 16% of the total registered flora. The diversity of medicinal flora is generally collected in different areas, and a certain percentage is grown in family gardens, since they do not grow wild. The richness of medicinal species differs from the geographical area and environments. In the tropics, secondary environments have a great diversity of medicinal plants in comparison with other places such as jungles or orchards (Sol-Sánchez 1993).

This diversity of medicinal plants could be affected by the impacts of climate change, which strongly affects mangroves in different parts of the world and which alters the main functions of these ecosystems, and there could be a general increase in atmospheric Co_2, increase in temperature, storms, sea level rise, marine currents, and alteration of the precipitation regimen, among many others (Ward et al. 2016; Gilman et al. 2008).

16.2 Therapeutic Properties of Medicinal Plants (Pengelly 1996; Bruneton 2001)

The active principles are substances located in different parts or organs of plants and that alter or modify the functioning of the organs and systems of the human and animal body when they are ingested. In traditional medicine they are used as a drug or medication.

The active ingredients are the substances used as drugs or medicines. These active principles are classified into glycosides, polyphenols, terpenoids, alkaloids, and other active ingredients.

16.2.1 Heterosides or Glucosides

They are compounds that are formed in two parts; a sugar (glucose) and a nonsugar (glucone, aglycone, or genin). They are classified according to the structural properties of the non-sugar part which generates its pharmacological function (Table 16.1).

Table 16.1 Classification of the actives ingredients of the plants

Type	Properties or uses
Anthraquinones	Purgatives
Cardiotonic	Diuretic, cardiac tonic
Cyanogenic	Anesthetics antispasmodics, hypotensive
Cumarinics	Antibacterial anticoagulant, sunscreen
Phenolics	Febrifuges and antipyretics
Flavonoids	Hair fragility, vitamin C
Ranunculosides	Irritating to the skin
Saponosides	Hemolysis emollients, dermatitis
Sulfur	Antibiotics

Anthraquinones are used as laxatives or purgatives. Some of the species with these properties are sabila or *Aloe* and the rhubarb (*Rheum* spp.). The active ingredients that act on the cardiac muscle belong to the group of cardiotonic and exert therapeutic action in it, generating alterations in the rhythm. Depending on the locality is the existence and use of one or another species used, such as *Digitalis purpurea* or *Nerium oleander* that contains those types of substances.

Within the cyanogenic plants are located those that release hydrocyanic acid, which are capable of causing poisoning and that could be fatal. Some botanical families with these properties are Rosaceae, Euphorbiaceae, Fabaceae, and Poaceae. In relation to coumarins, these are flavorings that protect capillary fragility, and some act as sedatives or hypnotics. Some may act as appetite stimulants, antirheumatics, and antiarthritic diuretics. In relation to polyphenols, these are substances with a benzenic nucleus with hydroxyl group ion; they bind to sugars to form heterosides, but they are also in free form. Examples are lignin and tannins. The most important groups are phenolic acids, flavonoid coumarins, lignans, tannins, and quinones.

Phenolic acids have diverse applications as antioxidants, analgesics, and choleretics. A well known example is eugenol, an antiseptic and analgesic that is used in almost all parts of the world in dental treatments. Some phenols are estragole, myristicin, apiol, and atenol.

As for the flavonoids, these are the yellow pigments derived from the phenylbenzo-ɣ-pyrone or phenyl chromone. The main types are chalcones, flavones, flavonols, anthocyanidins, condensed tannins, xanthines, and aurones. This group presents a great pharmacological importance. Some species with these flavonoids are passiflora (*Passiflora* sp.) and licorice and ginkgo among many others.

The coumarins are benzo-α-pyrones. It is a very broad group of phenolic active principles, present in some medicinal plants. The pharmacological interest is not very great, but it has several uses due to its effects on the vascular and venous systems and its use in the treatment of psoriasis. Some of the coumarins are present in plants such as the *Melilotus*, the esculosido of *Aesculus* sp., the visnadina of the visnaga, and the dicumarol.

The lignans result from the union of two units of phenyl propane (C6-C3) and are abundant in plants.

Tannins are complex polyphenolic substances. They are found in roots, barks, and sometimes in leaves. Its properties are very broad such as antibacterial, astringent, and antiseptic. It is present in families such as the Salicaceae, Rosaceae, or Fabaceae, among others.

Quinones are the result of the oxidation of phenols, and there are several types: para-benzoquinones, naphthoquinones, anthracyclines, anthraquinones, and phenanthraquinones.

The terpenoids include essential oils, lactones, and saponins.

Iridoids are monoterpenic compounds. It is usually found in vegetables such as the Gentianaceae and Valerianaceae botanical family. An example is the valerian.

This group also includes very common lactones in the family of Asteraceae, Lauraceae, and Magnoliaceae and saponins, whose power is foaming; some are hemolytic and toxic.

16.2.2 Alkaloids

These are the largest group of interest in pharmacology. This knowledge begins with morphine isolated in 1805 and strychnine in 1819; in 1930 more than 300 were isolated; in 1950 more than 100 were isolated and in 1973 between 5000 and 6000 of the known drugs.

There are several types depending on the molecule from which they derive such as tropane, quinoline, and isoquinoline. For the first case, some produce delirium, vertigo, or death at high concentrations. Examples of these drugs are cocaine and lysergic acid. Caffeine, atropine, daturine, and hyoscyamine are used in the treatment of Parkinson's disease.

Of the group derived from quinoline, the family Rubiaceae has several species of the genus *Cinchona*. Used against malaria, due to the quinine that is obtained from them.

From the group derived from isoquinoline, opium is obtained from the scratching of the immature capsules of *Papaver somniferum* to produce morphine very useful for intense pain, and which acts as a sedative but generates secondary effects that are difficult to counteract.

Other active ingredients found in plants are the mucilage and gums, considered as heterogeneous polysaccharides formed by sugars, which have various uses as emollients for external use in the form of poultices, but also internally.

16.3 Ways to Prepare Plants with Medicinal Properties for Its Use (Penner and Martínez 2010)

16.3.1 Infusions

It consists of placing some part of the plant in hot water, letting it rest, and then drinking it as many times as necessary.

16.3.2 Cooking

It consists of placing some part of the plant in water and putting it to the fire to cook; when it has already boiled for at least 5 min, put out the fire, let it stand, and take the resulting product either hot or as daily water. If what is boiled is the stem or roots, the water should boil at least 15 min to release the compounds of the plant.

16.3.3 Extracts

It consists of macerating on a mortar with pistils some part of the fresh plant of interest; if you do not have a mortar, this can be done on any kitchen utensil. The macerate part is the employee. This maceration can be done accompanied with a little alcohol.

16.3.4 Tinctures

It is the product obtained by the extraction of fresh natural inks from some plants? Alcohol and water are generally used as an ally for extraction. It is obtained after macerating the most useful part of the plant for several days in alcohol and water. After several days in maceration, strained, mixed, and obtained the active principle of plant in its pure state.

16.3.5 Syrup

It consists of liquefying the parts of the required plants, adding honey or sugar, putting it to boil over low heat, and stirring constantly, and when it has thickened, remove it from the claw, and drink it by spoonfuls. The most common are syrups for colds.

16.3.6 Powder

They consist of drying the required parts of the plants and then grinding them to pulverize them in manual mills.

16.4 Drops

They come from infusions, orally. Some plants, such as the purple maguey (*Rhoeo discolor*), are placed on the medium-roasted fire, and, if the leaf is soft, it is squeezed, and drops are obtained from the plant that are placed directly on the part to be cured.

16.4.1 Juice

Depending on the part of the fruit to use, it is necessary to squeeze until the desired juice is obtained. In vegetative parts that cannot be obtained, they are macerated until the vegetative part is obtained in aqueous form.

16.4.2 Maceration

It consists of crumbling and crushing leaves, flowers, fruits, and seeds, in water, alcohol, wine, oil, and/or brandy and then using it.

16.4.3 Bath

It consists of preparing a bath with water and medicinal plants to cure diseases of the skin. Examples of plants are the cundeamor (*Momordica charantia*) and the bitter and sour orange.

They are prepared with a concentrated cooking or infusion of plants, which are mixed with the warm water of the tub.

16.4.4 Cataplasm

It consists of preparing the vegetative part that you want to use and then applying it to the affected part. Example majagua leaves are placed in the sun, and when they are almost withered, they are applied on grains that match on the face or body macerate.

16.4.5 Compresses

It consists of preparing the plants to be used, either in oils or only heated and placed on a cloth and applied on the affected part. Generally, you should stay there for several hours.

Inhalation is the aspiration of the vapors that are obtained in a decoction of the plants.

Under this knowledge, medicinal plants were studied in some communities of the coast of Tabasco with the objective of identifying those medicinal flora in the mangroves and surrounding communities and their applications.

The methodology applied was developed in two periods: the first from 2007 to 2011 and the second from 2014 to 2017 where the useful flora was inventoried in the communities studied. The work consisted of three stages of which only four communities completed the three stages considered in the project. These communities were ejidos: Las Coloradas, Golpe, Mingo, and Sinaloa first section.

The methodology consisted of informal visits to the communities to identify possible sources of primary information, application of a semi-structured interview, and field trips to identify medicinal plants and uses.

16.5 Results and Discussion

In general, the diversity of the medicinal flora was 63 species, which are used to treat 37 diseases. But it should be considered that a high percentage is not species-specific to the area, but has been brought from other locations in the state of Tabasco, depending on the places visited by the inhabitants. In the same area, Gómez (2011) records 41 medicinal species, and for the ejido La Encrucijada, a community close to the study area, Bautista-García et al. (2016) reported 42 medicinal species.

The treated diseases are diverse, but the spells and rashes are the pads that are treated with a greater number of plants, with five species each, elimination of parasites and treatment of the nerves four species each, and three for pains in general. The rest of the ailments are only treated with one or two species (Fig. 16.1).

Regarding the origin of the plants, the largest proportion was located in the family gardens; for this reason 81.5% are part of the diversity of plants that coexist daily in the gardens of the communities: 2 of them are species cultivated mainly for consumption and secondarily as medicine, 2 more were located in live fences, 4 come from the mangroves, 2 come from paddocks, and 14 are wild, being able to be

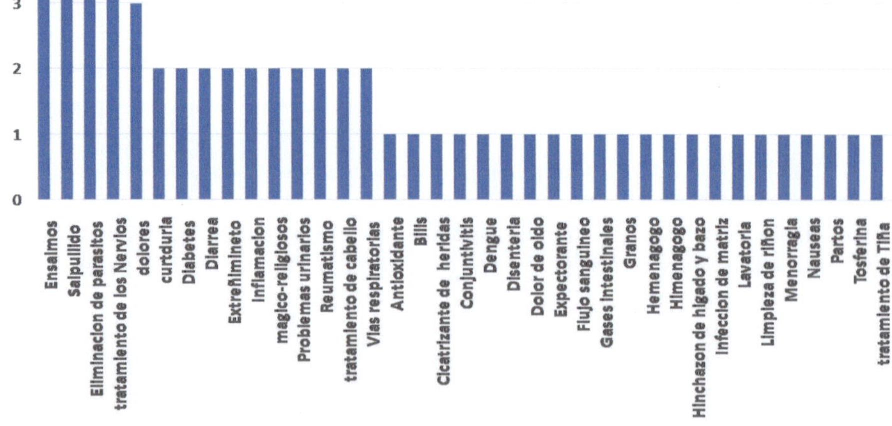

Fig. 16.1 Diseases and number of species to treat them

located on roadsides or abandoned environments. It is important to mention that only tree or shrub species are grown in the gardens of family orchards, such as the nance (*Byrsonima crassiolia* (L.) Kunth.) and the cinnamomum (*Cinnamomum zeylanicum* Breyn). The rest of the species are usually planted in waste containers such as buckets or frets that no longer fulfill their main function, but the substrate is not from the area; this is because in coastal communities the soil is too saline, due to which many species do not grow in these conditions, so over time the ladies have brought from communities far from the coast fertile soil where watermelon, corn, or sorghum is produced and have used it as a substrate to grow their medicinal plants

For this reason, it is common to find plants hanging from trees in containers of different shapes and sizes. It is important to note that the number of plants per species is minimal, almost always one plant per species; occasionally there could be more, but in very few orchards.

The diversity of species in the orchard registers greater security in the cure of diseases. The number of conditions that can be cured with medicinal plants in the communities worked is 37, being the most common plants used to heal nerves are 5 species, 4 to expel parasites, 2 species for rheumatism, 2 for inflammation and for constipation, 2 for diarrhea, 2 to regulate diabetes, and 2 for hair treatment; others showed only 1 use.

It is important to mention that the use cited here corresponds to the one that the population considered as the main one, but species that had more than one use were obtained, such as the *Sambucus mexicana* that presented the greatest diversity of medicinal uses. Also, the use of the plants is done fresh, and usually the form of preparation is tea, poultices, or prepared in alcohol and is usually prepared at the time of use.

An important aspect is the fact that the population uses plants that are not found in the locality, and they have to go out and bring them from the outside like avocado (*Persea americana* L.) and majagua (*Hampea nutricia* Fryxell). The avocado seed is cut into small pieces, put in alcohol, and left to stand for 2 weeks and then used to cure rheumatism and arthritis, while majagua is used to cure mumps.

Also, since corn is not produced in the locality, corn hair used to cure urinary tract problems is obtained with those who plant corn outside the area; they are usually given as a gift, but at other times, they exchange medicinal plants.

Some species are very difficult to obtain such as the calaguala (*Phlebodium aureum* L.), but they are very important for women during their period of ovulation, due to it being the remedy to regulate bleeding. Another species is the momo (*Piper auritum* H.B.K.) which, although it is not found in the community, is used frequently as a regulator of contractions at the time of birth and to deflate the womb after birth.

The use of medicinal plants is always monitored by an adult or older person who knows very well the use and proportions. Escamilla and Moreno (2015), in addition to listing the species, makes a brief description of the way of use in two communities from Veracruz, México.

This use is carried out throughout the year, its use being more frequent in the rainy season or cold days. When the rainy season is short, respiratory diseases usually do not occur, but when this cold or cool period lasts for up to 6 or 7 days, the population becomes ill from the throat, and there is usually an exchange of plants.

For example, summer rains are not as important as autumn and winter since the conditions in the first case are scarce. According to field information, the neediest communities from the point of view of health are those that have more medicinal species, as a response to a primary need whose expenses cannot be met due to the high cost of medicines. Hence, there is a direct dependence on medicinal plants to which they are given daily care, providing them with water, organic fertilizer, support in some cases, and elimination of pests and diseases.

The most diverse community in medicinal plants was Sinaloa first section; there were registered trees, shrubs, vines, lianas, epiphytes, and parasites. The least diverse was the ejido Las Coloradas, where a complete abandonment of traditional medicine was observed and where the inhabitants attribute such abandonment to the fact that the soil is not conducive to cultivating plants that are not typical of saline environments.

In this community the population has entered a stage of self-medication with medicines acquired in the city of Cárdenas, where the patient does not attend the consultation but rather orders the medicine with someone who travels to the city, which causes in many cases the disease to complicate.

In both cases there are particularities, for example, in Sinaloa, the population is self-employed in the capture of fish, oysters, and crab, but they do not stay long time in the mud, while in Las Coloradas, although the activities are almost the same, they almost always work in the mud, and they stay wet throughout the day, so there are more cases of respiratory diseases. However, the richness of medicinal species is very low. The communities of Golpe and Mingo recorded a richness intermediate of medicinal plants.

In relation to the origin of the medicinal species, the vast majority is exchange or gift-giving except for some such as the *Kalanchoe flamma* and the *Tagetes erecta*; these are bought in the city, and generally these are not given, as the vegetative parts do not always ensure a new plant.

In the four communities, the number of people who continue with the use of medicinal plants is scarce, concentrating on one lady or family nucleus per community, and the rest of the population gets what they need in them. Annex 1 cites the list of registered plants in the communities

16.6 Conclusion

It is concluded that there is a high dependence on medicinal plants in the region, but its acquisition and conservation is difficult, given the saline conditions of the place. It is necessary to deepen in doses of application given that usually speaking of immeasurable quantities such as a pisca, three fingers, or a fist. Traditional medicine is a very important activity in the region since it allows not to spend on commercial medicines.

At a specific level, the treatments cured with the 63 registered species are 31, of which 8 are used for respiratory problems. For treatment of pain, for spells, and for the treatment of nerves, five species are used in each case. Four species are used for the expulsion of intestinal parasites. For inflammations three species are used; for rash, hair treatment, constipation, emmenagogue, diarrhea, diabetes, and rheumatism, treatment is with two species each, and the other species are used for a condition each.

Appendix 1: Medicinal Plants Registered in the Communities Studied

Allium sativum L.	*Mentha sativa* L.
Aloe vera (L.) Burm. f.	*Momordica charantia* L.
Anacardium occidentale L.	*Murraya paniculata* (L.) Jacq.
Aristolochia pentandra Jacq.	*Ocimum basilicum* L.
Artemisia mexicana Willd. ex Spreng.	*Pedilanthus tithymaloides* (L.) Poit.
Averrhoa carambola L.	*Petiveria alliacea*
Avicennia germinans L.	*Piper auritum* H.B.K.
Bouganvillea glabra Choisy	*Phlebodium aureum*
Bursera graveolens Tr. et Planch	*Plectranthus amboinicus* (Lour.) Spreng.
Bursera simaruba (L.) Sarg.	*Plectranthus tomentosus* Benth. ex E. Mey.
Byrsonima crassifolia (L.) Kunth	*Porophyllum ruderale* Jacq. Cass
Capraria biflora L.	*Psidium guajava* L.
Carica papaya L.	*Rhizophora mangle* L.
Catharanthus roseus (L.) G. Don	*Ricinus communis* L.
Chenopodium ambrosioides L.	*Ruta chalepensis* L.
Cinnamomum zeylanicum Breyn	*Ruta graveolens* L.
Citrus aurantium L.	*Sambucus mexicana* C. Presl ex DC.
Citrus limon (L.) Burm.	*Sansevieria zeylanica* Willd.
Cucurbita pepo L.	*Senna alata* L.
Cydista aequinoctialis (L.) Miers	*Strutantus cassytoides*
Cymbopogon citratus Stapf.	*Tagetes erecta* L.
Eryngium foetidum L.	*Tamarindus indica* L.
Gliricidia sepium (Jacq.) Steud	*Tithonia diversifolia* (Hemsl.) Gray
Justicia spicigera Schltdl.	*Tradescantia spathacea* Sw.
Kalanchoe flammea stapf.	*Tradescantia zebrina* Purpusii
Kalanchoe gastonis-bonnieri Raym-H.	*Turnera ulmifolia* L.
Laguncularia racemosa L.	*Vitex aff. negundo* L.
Manilkara zapota (L) V. Royen	*Zea mays*
Melissa officinalis L.	*Zebrina pendula* Schnizl.
Mentha piperita L.	*Zingiber officinale* Roscoe

References

Alongi DM (2002) Present state and future of the world's mangrove forests. Environ Conserv Found Environ Conserv 29(3):331–349

Bautista-García G, Sol-Sánchez A, Velázquez-Martínez A, Llanderal-Ocampo T (2016) Composición florística e importancia socioeconómica de los huertos familiares del Ejido La Encrucijada, Cárdenas, Tabasco. Rev Mex Cien Agrícolas Pub Esp Núm 14:2725–2740

Belluci S, Paola A (2002) La herbolaria en los mercados tradicionales. Rev Cent Investig. Universidad La Salle [en línea] 5 (junio-julio): [Fecha de consulta: 28 de enero de 2019] Disponible en: http://www.redalyc.org/articulo.oa?id=34251806. ISSN:1405-6690

Bruneton J (2001) Farmacognosia: Fitoquímica de las plantas medicinales. "da edición, Editorial Acribia S.A. 1074 p

Escamilla P. B.E. y Moreno C. P. 2015. Plantas medicinales municipio de la Matamba y el Piñonal. Jamapa. Instituto de Ecología A. C. ISBN 978-607-7579-44-1. 108 p

Fondén RVK, Delmás FA, Herrero PC, Torres NA (2015) Eficacia de la crema de Rhizophora mangle L. al 50% en el tratamiento local de la quemaduras dérmicas. AB. Multimed Rev. Med. Gramma Multimed 2015: 19 (2). version On-line ISSN1028-4818. pp 1–16.

Gilman EL, Ellison J, Duke NC, Field C (2008) Threats to mangroves from climate change and adaptation options: a review. Aquat Bot 89:237–250. https://doi.org/10.1016/j.aquabot.2007.12.009

Gómez GE (2011) Etnobotánica del Ejido Sinaloa 1ª Sección, Cárdenas, Tabasco, México. Tesis de maestría. Colegio de Postgraduados, Postgrado en Producción Agroalimentaria en el Trópico. 82 p

Keswani C, Bisen K, Singh SP, Singh HB (2017) Traditional knowledge and medicinal plants of India in intellectual property landscape. Med Plants 9(1):1–11

Ocampo RA (2002) Situación actual del comercio de plantas medicinales en América latina Vol. 1 (4). Boletín Latinoamericano y del caribe de plantas Medicinales y Aromáticas. ISSN:0717-7917

Pengelly A (1996) The constituents of medicinal plants: an introduction to the chemistry & therapeutics of herbal medicines. Second edition. Sunflowers herbals. 109 p

Penner R, Martínez M (2010) Plantas medicinales y aromáticas: una alternativa de producción comercial. USAID PARAGUAY 60 p

Romañach SS, DeAngelisb DL, Koh HL, Li Y, Teh SY, Raja BR s, Zhai L (2018) Conservation and restoration of mangroves: global status, perspectives, and prognosis. Ocean Coast Manag 154(2018):72–82

Singh HB, Jha A, Keswani C (eds) (2016) Intellectual property issues in biotechnology. CABI, Wallingford. 304 pages, ISBN-13:9781780646534

Sol–Sánchez A (1993) Utilización de los recursos vegetales por los habitantes del Ejido Linda Vista, Palenque, Chiapas. Tesis profesional, División Académica de Ciencias Biológicas. Universidad Juárez Autónoma de Tabasco, Tabasco, México 86 pp

Ward RD, Fries DA, Day RH, MacKenzie RA (2016) Impacts of climate change on mangrove ecosystems: a region by region overview. Ecosystem Health and sustainability 2(4). Special feature Wetlands and Global Climate and Land-use Change 25 p

The Global Economic Impact of Neurodegenerative Diseases: Opportunities and Challenges

17

Walia Zahra, Sachchida Nand Rai, Hareram Birla, Saumitra Sen Singh, Hagera Dilnashin, Aaina Singh Rathore, and Surya Pratap Singh

Abstract

The main challenge in today's world to the healthcare system is the elevated occurrence of the neurodegenerative disorders. Progress in the field of bioinformatics and biomedical research has allowed us to understand the pathobiology of the neurodegenerative disorders in a detailed manner. The threat of these diseases increases with aging, and Parkinson's disease, Alzheimer's disease, amyotrophic lateral sclerosis, and Huntington's disease are the major ones affecting the public health and posing the higher economic burden. The research centers, pharmaceutical companies, and academic institutions are conducting research work in collaborations these days to enhance the development of new therapeutic strategies and develop novel drugs in a sustainable way. This can help in the development of safer therapies with reduced risk and can help in developing the authentic and evident biomarkers so that the disease can be diagnosed at early stages and treated accordingly. The advancement can therefore help in improving the quality of life of patients suffering from these debilitating neurodegenerative disorders and can also help in providing the job opportunities to the students interested in drug development program.

Keywords

Neurodegenerative disorders · Parkinson's disease · Alzheimer's disease · Amyotrophic lateral sclerosis · Huntington's disease · Drug development · Novel therapeutics

W. Zahra · S. N. Rai · H. Birla (✉) · S. S. Singh · H. Dilnashin · A. S. Rathore · S. P. Singh
Department of Biochemistry, Institute of Science, Banaras Hindu University, Varanasi, India

C. Keswani (ed.), *Bioeconomy for Sustainable Development*,
https://doi.org/10.1007/978-981-13-9431-7_17

17.1 Introduction

The age-dependent damage and loss of neurons leads to the diseases referred to as neurodegenerative diseases. The progressive loss of neurons from the regions of the brain and/or spinal cord is the pathologies exhibited by neurodegenerative disorders (Yacoubian 2017). Degeneration of the neurons may lead to cognitive impairment, motor dysfunction, loss of decision making, etc. Since the mechanisms behind the pathophysiologies of these diseases are not fully investigated, hence, the proper treatment of the diseases does not exist (Agnati et al. 1992; Durães et al. 2018). The treatment for these diseases have to be identified at urgent basis as the World Health Organization (WHO) has suggested that the neurodegenerative disorders that mainly affect the motor functions can be the second-most prevalent cause of death after cardiovascular diseases (Gammon 2014; Gitler et al. 2017). Diseases such as Parkinson's disease (PD), Alzheimer's disease (AD), Huntington's disease (HD), and amyotrophic lateral sclerosis (ALS) are the mostly occurring neurodegenerative diseases which have the greater threats to the public health and heavier economic burden. Genetic factors are also involved in the progression of neurodegeneration, but deposition and aggregation of toxic proteins in the brain and mitochondrial dysfunction play a major role in the disease progression. The cure of these diseases is still unknown, but drugs are present to manage the symptoms of neurodegeneration such as L-Dopa for PD, riluzole for ALS, and memantine and donepezil for AD (Parayath et al. 2017). Progress in the field of biomedical research and informatics has improved the knowledge about the impact of gene, epigenetic modification, drugs, aging, and microbes on the neurodegenerative diseases. The rapidly aging population has increased the risk of incidence of neurodegenerative diseases which leads to the higher rate of illness and mortality and hence higher costs for treatments. For example, in Europe, there are approximately 16% of people who are above 65 years in age, and by 2030, the percentage is expected to rise up to 25%. Dementias are the most prevalent neurodegenerative disorders affecting about seven million people and are estimated to double by 2040. These diseases usually prevail for a period of 2–10 years and require special care and treatments for the concerned patients. About €130 billion is spent to treat the patients suffering from neurodegenerative diseases per year (Yacoubian 2017).

Therefore, the neurodegenerative diseases can be deeply investigated to attain an elaborated overview, and the treatment strategies can be built for providing better care to the struggling patients. Up till now, AD and PD are the two mostly investigated diseases as suggested by research programs and databases such as MDSgene, PDGene, JPND research program, and AlzGene (Delamarre and Meissner 2017; Yacoubian 2017).

Thus, the collaborative networks consisting of research institutes, medical centers, and skilled specialists should be constructed for bench to bedside research which can help in providing the safer therapies, help in identifying the disease at their earlier stages, and help in providing the improved prognosis to enhance the patient's quality of life (Keswani et al. 2017; Singh et al. 2016a, b).

17.2 Overview of PD

PD is characterized by the progressive loss of dopaminergic neurons from the substantia nigra pars compacta and formation of Lewy bodies in the midbrain (Yacoubian 2017). More than ten million people with variable prevalence based on gender, age and geographical regions are affected with PD. The cost for treating the patients of PD in the USA is about USD 14.4 billion/year, and it is suggested that the number will double by 2040, while in Italy about € 8.340 are spent per patient (von Campenhausen et al. 2011). A report by the Global Forecast to 2022 also suggests that the PD treatment market was valued at USD 3.99 billion in 2016 and is expected to reach USD 5.69 billion in 2022 (www.marketsandmarkets.com). Moreover on February 25, 2019, the Michael J. Fox Foundation for Parkinson's Research has announced 127 new grant awards of about USD 24 million which are based on developing the research strategy to define, measure, and treat PD. A total of USD 5,354,712 was given to support 32 new grants to understand causes and progression of PD. To assess the effectiveness of treatments for PD, about 63 new grants were supported, and a total of USD 8,674,094 was funded by the Michael J. Fox Foundation, while 21 new grants were supported with the amount of USD 9,393,888 to develop treatments for slowing the progression of PD (The Michael J. Fox Foundation for Parkinson's disease 2019).

The therapy for the treatment of PD is still unknown. Still specific drugs are available to control different symptoms of PD such as rigidity, resting tremor, and bradykinesia despite having pathological and clinical heterogeneity of the disease. Levodopa (L-Dopa) has come up as the most promising drug for the treatment of PD as it is effective in enhancing the neurotransmission of dopaminergic neurons and effectively managing the motor symptoms in the early and later stages of PD (Schumacher-Schuh et al. 2014; Payami 2017; Domingo and Klein 2018). Other than L-Dopa, two other classes of drugs, i.e., "dopaminergic" and "nondopaminergic" drugs, are used as the current therapies of PD (Lin et al. 2017). Dopaminergic drugs include apomorphine, entacapone, and selegiline and are recommended to reduce the motor fluctuations of PD. Amantadine, on the other hand, is the example of nondopaminergic drugs and is effective in reducing the L-Dopa-induced dyskinesia. Other nondopaminergic drugs are still under preclinical and clinical trials (Lin et al. 2017). Other epigenetic modulators destined to target HDACs (histone deacetylase), DNMTs (DNA methyltransferase), and HATs (histone acetyltransferase) are also under preclinical and clinical trials thought to be as novel therapeutic drugs for PD (Hwang et al. 2017).

Research centers work in collaboration to reduce the risk of PD and understand the pathobiogenesis behind it for effective therapeutics and drug development. For example, in June 2018, Parkinson's Institute and Clinical Center (PICC) and Axial Biotherapeutics have announced their collaboration for investigating the role of gastrointestinal metabolites that may cause PD. The team believes in conducting the advanced clinical research and considers that the origin of PD occurs in the gastrointestinal system (Parkinson's Institute and Clinical Center 2017, 2018). Similarly, the Parkinson's Institute and Clinical Center and Denali Therapeutics had also announced their partnership to explore novel clinical endpoints for understanding the progression of PD (Business Wire 2018). Also, the Parkinson's Institute and Clinical Center has also collaborated with Merck to study the effects of LRRK2 kinase inhibitors on the nonmotor symptoms in a preclinical model of gastrointestinal dysfunction in PD (www.thepi.org). These centers also get awarded for their contribution in PD research, such as the Parkinson's Institute and Clinical Center was awarded with USD 1.9 million from California Institute for Regenerative Medicine to study the effects of downregulation of key protein (α-synuclein) in PD (www.thepi.org).

17.3 Overview of ALS

The chronically lethal neurodegenerative disorder, ALS, was first discovered by Jean-Martin Charcot, a French neurologist (Mulder et al. 1986; Rowland 2001; Harris 2014). The symptoms of ALS are usually irreversible, and deterioration of upper and lower motor neurons occurs in the brainstem, motor cortex, and spinal cord (Gordon et al. 2006). About 1–2 persons can be affected with ALS per 100,000 individuals (Neymotin et al. 2009), and its occurrence rate is about 6–8 per 100,000 people/year. Ninety to 95% patients are of sporadic origin, while 5–10% patients have the inherited form of the disease (Pandya et al. 2013). About 60% mutations are found to be responsible for the familial form of ALS (Gordon 2013). The symptoms of ALS reflected by the patients include spasm, twitching, speech problems, and weakness (Perry et al. 2012). Generally, the average survival time for ALS patients is about 19 months from diagnosis and 30 months from onset (Boillée et al. 2006; Logroscino et al. 2008). The pathology of ALS is still unknown, but it has been studied to include oxidative stress, glutamate toxicity, endoplasmic reticulum (ER) stress, mitochondrial dysfunction, protein misfolding and aggregation, etc. (Mátyus et al. 2012). Thus, the therapy for the treatment of ALS is generally based on targeting these dysfunctions of the cell (Lu et al. 2016).

Till date, riluzole is the only drug approved by the US Food and Drug Administration (FDA) and is known to have potential against glutamate toxicity. The drug has been found effective in delaying the start of respiratory problems, and hence riluzole has been demonstrated to slightly delay the onset of respiratory dysfunction and prolonging the survival by 2–3 months and increasing the survival rate by 9% in the 1st year (Deng et al. 2009). The drug does not prevent the firing in human myotubes, but it is effective in blocking the Na^+ currents and affecting the action potentials (Deflorio et al. 2014). Moreover, severe patients of ALS are also not much affected by riluzole treatment as it is unable to enhance the motor functions or myodynamia (Borras-Blasco et al. 1998). Thus, newer drugs for the patients of ALS are being proposed, and some are present in the advanced stages of clinical trials. In this context, the ALS Association has also announced to support about 58 new research grants by providing the fund of USD 11,621,638 so as to find the effective treatments and a cure for ALS in July 2015 (ALS Association 2015) (Table 17.1).

Table 17.1 Some ALS drugs and their status in the USA (www.alsresearchforum.org)

Drug name	Company	Status in the USA
123C4	Iron Horse Therapeutics	Preclinical
AMX0035	Amylyx Pharmaceuticals	Phase II
AT-1501	Anelixis Therapeutics	Phase I
Anavex 2-73	Anavex Life Science Corporation	Preclinical
Arimoclomol	Orphazyme	Phase III
CAT-4001	Catabasis Pharmaceuticals	Preclinical
CN S10-NPC-GDNF	Svendsen La	Phase I
DNL747	Denali Therapeutics	Phase I
Dimethyl fumarate	Biogen	Phase II
Ibudilast	MediciNova	Phase II
Masitinib	AB Science	Phase III
ODM-109	Orion Pharma	Phase III
Riluzole	Covis Pharma	FDA approved
Edaravone	Mitsubishi Tanabe Pharma Corporation	FDA approved

17.4 Overview on AD

It is the most common neurodegenerative disorder in the elderly and responsible for about 80% of dementia cases. The patients suffer from loss of memory, inability to learn, deteriorating behavioral function, etc. The pathology of AD is marked by the deposition of amyloid plaques in the brain that ultimately leads to the loss of neurons and synapses (Kumar and Singh 2015; Wang et al. 2017; Hurtado-Puerto et al. 2018).The frequency of AD increases as the world's population ages and hence is a major threat to the health of the people. After the age of 65, the frequency of AD is found to be doubled every 5 years, and in the USA, it is suggested that the frequency will reach up to 14 million from 5.5 million by the year 2050 (Brookmeyer et al. 2007; Alzheimer's Association 2018). On the other hand, about 135 million people will be suffering with AD by the year 2050 which means 100 million people will be added in the list when compared to today's AD cases (35 million) (Prince et al. 2016). Thus, one can imagine the cost that will be spent in managing the aggravating symptoms of the disease in addition with the human sufferings if the possible majors have not been taken (Brookmeyer et al. 2007; Brodaty et al. 2013; Roberts and Knopman 2013). Presently the cure of AD is still unknown, and the drugs that are being used are only capable of treating cognitive abnormalities that too at the early stages of AD (Kumar and Singh 2015). Galantamine is the only drug that is being widely marketed for the treatment of AD. Because of the presence of tertiary ammonium base in its structure, galantamine can easily cross the blood-brain barrier and inhibit brain acetylcholinesterase (Heinrich 2010; Mucke 2015). The drug was being investigated for its therapeutic effects since the 1980s, and in 2000 it was introduced as an anti-AD drug. Till date, galantamine acts as the most effective drug in preventing the severe symptoms in AD patients (Heinrich 2010).

The US Government has targeted to prevent the occurrence of AD by 2025 which has also been supported by other countries (Vradenburg 2015; Aisen et al. 2017). The goal is to prevent or delay the onset, slow the advancement of disease, and improve the cognitive and behavioral abnormalities. Though the failure rate of new drug development and disease-modifying therapies are 99% and 100%, respectively, still there is a need to increase our knowledge and discover the advanced drug therapy in order to reduce the socioeconomic burden on the society (Cummings et al. 2014) (Fig. 17.1).

The AD drug development programs cost for about USD 5.6 billion, and it takes about 13 years from clinical studies for the drug to be approved by the US Food and Drug Administration (FDA) (Scott et al. 2014). Thus, a pharmaceutical industry, eager to get an approval for the new drug, might have to spend about USD 2.8 billion (DiMasi et al. 2016) (Fig. 17.2).

Figures 17.1 and 17.2 show the duration and approximate cost of drug development for AD and include both cost of capital and cost of failure sustained by the companies working for AD drug development (Scott et al. 2014). From the data available, it might be suggested that pharmaceutical companies have to pay more for the Phase III trials as compared to other stages.

The US National Institutes of Health (NIH) is the principle funder in health research and spends almost USD 34 billion annually. The cost spent on the disease and the cost devoted to the research are, however, mismatched. For example, the US society spends more than USD 216 billion on AD, while the NH budget for the disease is USD 1.8 billion. Thus, it can be said that for every USD 1 spent on AD, less than 1% is dedicated to research (Martorell et al. 2013; Deb et al. 2017). Though AD has the greater impact on the economy of the USA as compared to cancer and cardiovascular diseases, it shares the lesser budget than any of these two disorders, i.e., for cancer USD 6.0 billion and for cardiovascular disease USD 2.2 billion (www.nih.gov).

NIH has a funding agency named as National Institute on Aging (NIA) dedicated to funding of AD research. A trial coordinating center known as the Alzheimer's

Fig. 17.1 Time duration for drug approval in moths at difference stages for drug development of AD. (Based on data from Scott et al. 2014)

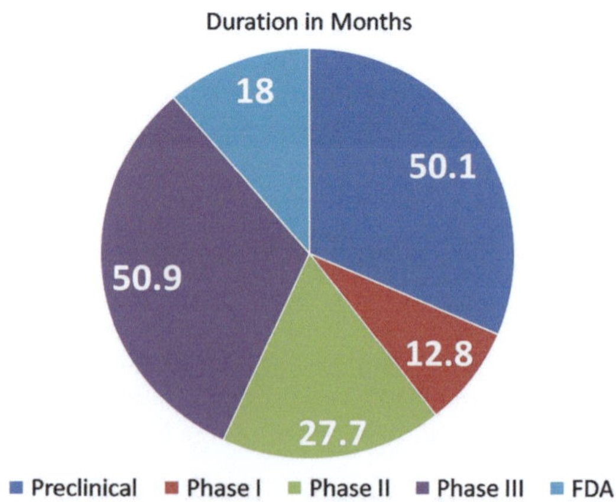

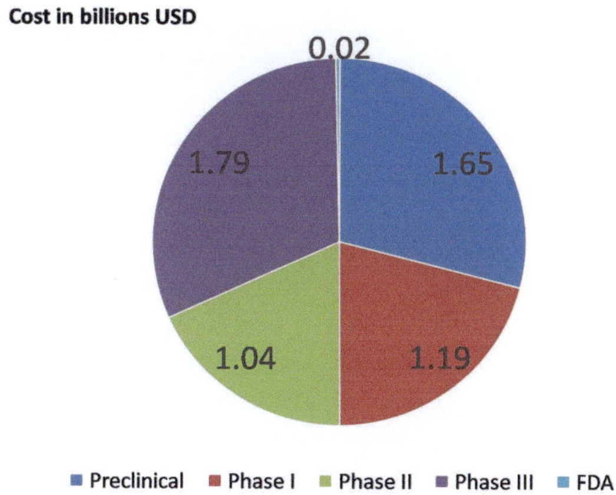

Fig. 17.2 Cost of drug development in billions (USD) at different stages of drug development for AD. (Based on data from Scott et al. 2014)

Clinical Trial Consortium is funded by the NIA to conduct the clinical trials on AD and its related disorders and develop the newer therapies for AD and related dementias (Cummings et al. 2018). NIA also has a public-private partnership with Alzheimer's Disease Neuroimaging Initiative (ADNI) which aims to put on the clinical trials and collect the data related to trial planning. In the USA the states also fund for AD centers and its related research projects. California, Texas, New York, and Nevada are the ones which are involved in the major funding for AD research.

Some NIA-supported programs for AD drug development:

- Alzheimer's Clinical Trial Consortium
- Alzheimer's Disease Neuroimaging Initiative
- Trial-Ready Cohort for Preclinical and Prodromal AD
- AD Genetics Consortium
- National Cell Repository for AD
- Dominantly Inherited AD Network
- DIAN-Treatment Unit (DIAN-TU)
- Alzheimer's Prevention Initiative
- Alzheimer's Disease Centers
- National Alzheimer's Coordinating Center
- Alzheimer's Drug Development Program
- Pilot Clinical Trials for the Spectrum of Alzheimer's Disease and Age-related Cognitive Decline (PAR-18-175)
- Phase III Clinical Trials for the Spectrum of Alzheimer's Disease and Age-related Cognitive Decline (PAR-18-028)
- AD Sequencing Project
- Molecular Mechanisms of the Vascular Etiology of Alzheimer's Disease Consortium
- Alzheimer's Preclinical Efficacy Database (AlzPED) Accelerating Medicines Partnership-Alzheimer's Disease Target Discovery and Preclinical Validation Project (Cummings et al. 2018)

The biotechnology companies are the venture-backed drug development centers which employ the use of technology on living organisms and biological systems (Lawrence 2017). High return is often produced by the development of AD drugs (Fleming 2015). United Kingdom-based Dementia Discovery Fund and Dolby Family Ventures are the specifically involved in supporting the cost of drugs for AD treatment. About USD 50,000,000 has been donated to the Dementia Discovery Fund by Bill Gates of the Gates Foundation, and the same amount is provided to the additional venture capital for the drug development in the biotechnology sector (Cummings et al. 2018).

The largest noncorporate funder for the research in AD is the Alzheimer's Association. It has spent almost USD 90 million in research which comprises USD 25 million invested in new projects and multiyear commitments in 2016 (Cummings et al. 2018).

The pharmaceutical industry, Biopharma, is involved in the funding of 60% of US research and developmental activities which cost about USD 75 billion in 2016. About 70% of the total clinical trials for AD are sponsored or cosponsored by the pharmaceutical industries (Cummings et al. 2017).

The high expense for AD research disappoints the companies working in this area and the funding agencies to fund for the development of the drugs, thus creating a hurdle in the development of new therapies for AD. Thus, it is suggested that an estimate amount of USD 38.4 billion is required to make a strong pipeline for AD therapy (Lo et al. 2014). Public-private partnerships have played an important role in progressing research by managing the distribution of cost in between the private and federal sources (Widdus 2005; Muller and Weigelt 2010; Murphy et al. 2014; Portilla and Rohrbaugh 2014). This is particularly operational in the precompetitive arenas like development of biomarkers and can be especially effective in precompetitive arenas such as biomarker development, advancement of the therapeutics, and disease modeling. Alzheimer's Disease Neuroimaging Initiative (ADNI) is one of the examples of the research program that is funded by both NIH and numerous pharmaceutical companies (Sidders et al. 2014).

Some advocacy groups have involved the venture funding methods in order to fund the drug development directly (Ramsey et al. 2017). Another innovative funding approach, known as crowdfunding, is dependent on web-based means for raising funds and is quite helpful in producing the small amount of funds for inaugurating the new drug development programs (Dragojlovic and Lynd 2014; Carter et al. 2017). On the other hand, the collaboration between two or more pharmaceutical companies helps in distributing the financial risk in the way of drug development for AD and hence seems to be a popular strategy. Examples of such collaborative developments include the development of inhibitor for β-site amyloid precursor protein-cleaving enzyme by Eli Lilly and AstraZeneca and codevelopment of inhibitor for β-site amyloid precursor protein-cleaving enzyme and an anti-amyloid antibody by Eisai and Biogen. Also, the collaboration between Banner Alzheimer Institute (a private institute), NIH, and two pharmaceutical companies forms the basis Alzheimer Prevention Initiative (Reiman et al. 2011). Similarly, the Oxford Biomedical Research Centre represents those research centers which are involved in advancing the treatment strategies and represents the interface of academic universities and healthcare systems (Greenhalgh et al. 2017). The organizations such as NIH, Alzheimer's Association, and others are awarded with funds on the basis of their rank or score in a competitive manner. It has been observed that the resources as well as skills from different laboratories are required in order to address an identified problem which occurs only when the participants willingly agree on collaborating and share the data with each other. Thus, all collaborators are funded accordingly once these requirements are fulfilled. Also for a sustainable AD drug development, innovation in the field is required, and the pipeline for producing the promising and effective drug for AD is needed to be more structured. It has been seen that sometimes the flawed agents get funding while the promising drug candidate are not supported (Cummings et al. 2018). Hence, to limit the burgeoning numbers of AD patients, the current funding system must work in a comprehensive manner and more wisely to raise the advanced therapeutic approaches and promising drugs for AD.

17.5 Overview on HD

The neurodegenerative disorder HD is inherited in an autosomal dominant manner and is diagnosed on the basis of symptoms such as adult-onset of motor dysfunctions, intellectual deterioration, and psychiatric disturbances (McColgan and Tabrizi 2018). Other symptoms such as sleep and circadian rhythm disturbances or dysfunction of autonomic nervous system are also the debilitating ones but are lesser observed in the patients of HD (Bano et al. 2011). The neuropathological features of HD are mainly detected in the striatum, but other brain areas such as thalamus, cerebellar cortex, and cerebellum are also affected as suggested by the postmortem of the brain of HD patients (MacDonald et al. 1993). The onset of the neurodegeneration is generally seen in between 30 and 50 years of age and lasts for up to 17–20 years. The patients gets worsen, and they become more dependent on others for their daily life until death. Hence, newer strategies for the neuroprotective intervention of HD can be identified and further explored by understanding the detrimental processes responsible for cellular anomalies in the patients of HD (Nguyen and Weydt 2018).

According to the analysis and research done by GlobalData, it is suggested that the disease market will have a rise from USD 252.6 million in 2014 to about USD 2.6 billion in 2024, which represents a high value of compound annual growth rate (CAGR), i.e., 25.6% mainly affecting the countries like Germany, Spain, the USA, the UK, France, Italy, and Japan (www.rdmag.com).

17.6 Conclusion

The research and treatment development for neurodegenerative disorders costs an enormous amount. Thus, funding from different organizations is being promoted to fund the research centers, academic institutions, and pharmaceutical companies involved in finding the advanced therapeutic strategies and effective drugs for the treatment of different neurodegenerative disorders. These organizations are involved in developing the treatment to prevent the disease onset, delay, or slow the progress and improve the symptoms of the neurodegenerative disorders. The investment by funding agencies in this arena can decrease the cost of these drugs making them cost-effective thereby improving the quality of life of the patients suffering from the neurodegenerative disorders such as PD, AD, ALS, and HD. The collaborations in between different research arenas can also help in the development of promising biomarkers and can open the career options for students in drug discovery and development.

References

Agnati LF, Zoli M, Biagini G, Fuxe K (1992) Neuronal plasticity and ageing processes in the frame of the 'red queen theory'. Acta Physiol Scand 145(4):301–309

Aisen PS, Cummings J, Jack CR, Morris JC, Sperling R, Frölich L, Jones RW, Dowsett SA, Matthews BR, Raskin J, Scheltens P (2017) On the path to 2025: understanding the Alzheimer's disease continuum. Alzheimers Res Ther 9(1):60

ALS Association (2015) Available on URL: http://www.alsa.org/news/media/press-releases/new-research-grants-2015.html. Accessed 06/03/2019

Alzheimer's Association (2018) Alzheimer's disease facts and figures. Alzheimers Dement 14(3):367–429

Bano D, Zanetti F, Mende Y, Nicotera P (2011) Neurodegenerative processes in Huntington's disease. Cell Death Dis 2(11):e228

Boillée S, Yamanaka K, Lobsiger CS, Copeland NG, Jenkins NA, Kassiotis G, Kollias G, Cleveland DW (2006) Onset and progression in inherited ALS determined by motor neurons and microglia. Science 312(5778):1389–1392

Borras-Blasco J, Plaza-Macías I, Navarro-Ruiz A, Peris-Marti J, Anton-Cano A (1998) Riluzole as a treatment for amyotrophic lateral sclerosis. Rev Neurol (160):1021–1027

Brodaty H, Heffernan M, Kochan NA, Draper B, Trollor JN, Reppermund S, Slavin MJ, Sachdev PS (2013) Mild cognitive impairment in a community sample: the Sydney memory and ageing study. Alzheimer's Dement. 1 9(3):310–317

Brookmeyer R, Johnson E, Ziegler-Graham K, Arrighi HM (2007) Forecasting the global burden of Alzheimer's disease. Alzheimer's Dement. 1 3(3):186–191

Business Wire (2018) Available on URL: https://www.businesswire.com/news/home/20180305005266/en/Parkinson%E2%80%99s-Institute-Clinical-Center-Collaborates-Denali-Therapeutics. Accessed 04/03/2019

Carter AJ, Donner A, Lee WH, Bountra C (2017) Establishing a reliable framework for harnessing the creative power of the scientific crowd. PLoS Biol 15(2):e2001387

Cummings JL, Morstorf T, Zhong K (2014) Alzheimer's disease drug-development pipeline: few candidates, frequent failures. Alzheimers Res Ther 6(4):37

Cummings J, Lee G, Mortsdorf T, Ritter A, Zhong K (2017) Alzheimer's disease drug development pipeline: 2017. Alzheimer's Dement: Transl Res Clin Interv 3(3):367–384

Cummings J, Reiber C, Kumar P (2018) The price of progress: funding and financing Alzheimer's disease drug development. Alzheimer's Dement: Trans Res Clin Interv 4:330–343

Deb A, Thornton JD, Sambamoorthi U, Innes K (2017) Direct and indirect cost of managing alzheimer's disease and related dementias in the United States. Expert Rev Pharmacoecon Outcomes Res 17(2):189–202

Deflorio C, Onesti E, Lauro C, Tartaglia G, Giovannelli A, Limatola C, Inghilleri M, Grassi F (2014) Partial block by riluzole of muscle sodium channels in myotubes from amyotrophic lateral sclerosis patients. Neurol Res Int 2014:946073

Delamarre A, Meissner WG (2017) Epidemiology, environmental risk factors and genetics of Parkinson's disease. La Presse Med 46(2):175–181
Deng Y, Xu Z, Xu B, Tian Y, Xin X, Deng X, Gao J (2009) The protective effect of riluzole on manganese caused disruption of glutamate–glutamine cycle in rats. Brain Res 1289:106–117
DiMasi JA, Grabowski HG, Hansen RW (2016) Innovation in the pharmaceutical industry: new estimates of R&D costs. J Health Econ 1(47):20–33
Domingo A, Klein C (2018) Genetics of Parkinson disease. Handb Clin Neurol 147:211–227 Elsevier
Dragojlovic N, Lynd LD (2014) Crowdfunding drug development: the state of play in oncology and rare diseases. Drug Discov Today 19(11):1775–1780
Durães F, Pinto M, Sousa E (2018) Old drugs as new treatments for neurodegenerative diseases. Pharmaceuticals 11(2):44
Fleming JJ (2015) The decline of venture capital investment in early-stage life sciences poses a challenge to continued innovation. Health Aff 34(2):271–276
Gammon K (2014) Neurodegenerative disease: brain windfall. Nature 515(7526):299–300
Gitler AD, Dhillon P, Shorter J (2017) Neurodegenerative disease: models, mechanisms, and a new hope. Dis Model Mech 10:499–502
Gordon PH (2013) Amyotrophic lateral sclerosis: an update for 2013 clinical features, pathophysiology, management and therapeutic trials. Aging Dis 4(5):295
Gordon PH, Cheng B, Katz IB, Pinto M, Hays AP, Mitsumoto H, Rowland LP (2006) The natural history of primary lateral sclerosis. Neurology 66(5):647–653
Greenhalgh T, Ovseiko PV, Fahy N, Shaw S, Kerr P, Rushforth AD, Channon KM, Kiparoglou V (2017) Maximising value from a United Kingdom biomedical research Centre: study protocol. Health Res Policy Syst 15(1):70
Harris BT (2014) Amyotrophic lateral sclerosis. In: Pathobiology of human disease: a dynamic encyclopedia of disease mechanisms. Elsevier, Waltham
Heinrich M (2010) Galanthamine from Galanthus and other Amaryllidaceae–chemistry and biology based on traditional use. Alkaloids Chem Biol 68:157–165. Academic Press
Hurtado-Puerto AM, Russo C, Fregni F (2018) Alzheimer's disease. In: Neuromethods. Humana Press, New York
Hwang JY, Aromolaran KA, Zukin RS (2017) The emerging field of epigenetics in neurodegeneration and neuroprotection. Nat Rev Neurosci 18(6):347
Keswani C, Bisen K, Singh SP, Singh HB (2017) Traditional knowledge and medicinal plants of India in intellectual property landscape. Med Plants 9(1):1–11
Kumar A, Singh A (2015) A review on Alzheimer's disease pathophysiology and its management: an update. Pharmacol Rep 67(2):195–203
Lawrence S (2017) Biotech's wellspring—a survey of the health of the private sector in 2016. Nat Biotechnol 35(5):413
Lin JY, Xie CL, Zhang SF, Yuan W, Liu ZG (2017) Current experimental studies of gene therapy in Parkinson's disease. Front Aging Neurosci 9:126
Lo AW, Ho C, Cummings J, Kosik KS (2014) Parallel discovery of Alzheimer's therapeutics. Sci Transl Med 6(241):241cm5
Logroscino G, Traynor BJ, Hardiman O, Couratier P, Mitchell JD, Swingler RJ, Beghi E (2008) Descriptive epidemiology of amyotrophic lateral sclerosis: new evidence and unsolved issues. J Neurol Neurosurg Psychiatry 79(1):6–11
Lu H, Le WD, Xie YY, Wang XP (2016) Current therapy of drugs in amyotrophic lateral sclerosis. Curr Neuropharmacol 14(4):314–321
MacDonald ME, Ambrose CM, Duyao MP, Myers RH, Lin C, Srinidhi L, Barnes G, Taylor SA, James M, Groot N, MacFarlane H (1993) A novel gene containing a trinucleotide repeat that is expanded and unstable on Huntington's disease chromosomes. Cell 72(6):971–983
Martorell P, Hurd MD, Delavande A, Mullen KJ, Langa KM (2013) Monetary costs of dementia in the United States. N Engl J Med 368(14):1326–1334
Matyus P, Dunkel P, Chai CL, Sperlagh B, Huleatt PB (2012) Clinical utility of neuroprotective agents in neurodegenerative diseases: current status of drug development for Alzheimer's,

Parkinson's and Huntington's diseases, and amyotrophic lateral sclerosis. Expert Opin Investig Drugs 21(9):1267–1308
McColgan P, Tabrizi SJ (2018) Huntington's disease: a clinical review. Eur J Neurol 25(1):24–34
Mucke HA (2015) The case of galantamine: repurposing and late blooming of a cholinergic drug. Future Sci OA 1(4):FSO73
Mulder DW, Kurland LT, Offord KP, Beard CM (1986) Familial adult motor neuron disease: amyotrophic lateral sclerosis. Neurology 36(4):511–517
Muller S, Weigelt J (2010) Open-access public-private partnerships to enable drug discovery – new approaches. IDrugs: Invest Drugs J 13(3):175–180
Murphy DG, Goldman M, Loth E, Spooren W (2014) Public-private partnership: a new engine for translational research in neurosciences. Neuron 84(3):533–536
Neymotin A, Petri S, Calingasan NY, Wille E, Schafer P, Stewart C, Hensley K, Beal MF, Kiaei M (2009) Lenalidomide (Revlimid®) administration at symptom onset is neuroprotective in a mouse model of amyotrophic lateral sclerosis. Exp Neurol 220(1):191–197
Nguyen HHP, Weydt P (2018) Huntington disease. Med Genet 30(2):246–251
Pandya RS, Zhu H, Li W, Bowser R, Friedlander RM, Wang X (2013) Therapeutic neuroprotective agents for amyotrophic lateral sclerosis. Cell Mol Life Sci 70(24):4729–4745
Parayath NN, Pawar G, Avachat C, Miyake MM, Bleier B, Amiji MM (2017) Neurodegenerative disease. In: Nanomedicine for inflammatory diseases. CRC Press, Boca Raton/London, pp 289–318
Parkinson's Institute and Clinical Center (2017) Available on URL: http://www.thepi.org/clientuploads/directory/News/PI_CIRM_PR%201_20_2017.pdf. Accessed 05/03/2019
Parkinson's Institute and Clinical Center (2018) Available on URL: http://www.thepi.org/news/2018/02/05/news-articles/exciting-partnership-to-advance-precision-medicine-for-lrrk2/. Accessed 05/03/2019
Payami H (2017) The emerging science of precision medicine and pharmacogenomics for Parkinson's disease. Mov Disord 32(8):1139–1146
Perry JJ, Pratt AJ, Getzoff ED (2012) Amyotrophic lateral sclerosis: update and new developments. Degenerative Neurol Neuromuscul Dis 2012(2):1
Portilla L, L Rohrbaugh M (2014) Leveraging public private partnerships to innovate under challenging budget times. Curr Top Med Chem 14(3):326–329
Prince M, Comas-Herrera A, Knapp M, et al (2016) World Alzheimer report 2016. In: https://www.alz.co.uk
Ramsey BW, Nepom GT, Lonial S (2017) Academic, foundation, and industry collaboration in finding new therapies. N Engl J Med 376(18):1762–1769
Reiman EM, Langbaum J, Fleisher AS, Caselli RJ, Chen K, Ayutyanont N, Quiroz YT, Kosik KS, Lopera F, Tariot PN (2011) Alzheimer's prevention initiative: a plan to accelerate the evaluation of presymptomatic treatments. J Alzheimers Dis 26(s3):321–329
Roberts R, Knopman DS (2013) Classification and epidemiology of MCI. Clin Geriatr Med 29(4):753–772
Rowland LP (2001) How amyotrophic lateral sclerosis got its name: the clinical-pathologic genius of Jean-Martin Charcot. Arch Neurol 58(3):512–515
Schumacher-Schuh AF, Rieder CR, Hutz MH (2014) Parkinson's disease pharmacogenomics: new findings and perspectives. Pharmacogenomics 15(9):1253–1271
Scott TJ, O'connor AC, Link AN, Beaulieu TJ (2014) Economic analysis of opportunities to accelerate Alzheimer's disease research and development. Ann N Y Acad Sci 1313(1):17–34
Sidders B, Brockel C, Gutteridge A, Harland L, Jansen PG, McEwen R, Michalovich D, Seidel H, Weiss B, Williams-Jones B, Woodwark M (2014) Precompetitive activity to address the biological data needs of drug discovery. Nat Rev Drug Discov 13(2):83
Singh HB, Jha A, Keswani C (eds) (2016a) Intellectual property issues in biotechnology. CABI, Wallingford. 304 pages, ISBN-13: 9781780646534
Singh HB, Jha A, Keswani C (2016b) Biotechnology in agriculture, medicine and industry: an overview. In: Singh HB, Jha A, Keswani C (eds) Intellectual property issues in biotechnology. CABI, Wallingford, pp 1–4

The Michael J. Fox Foundation for Parkinson's disease (2019) Available from URL: https://www.michaeljfox.org/foundation/news-detail.php?what-we-fund-24-million-in-new-grants-for-parkinson-research. Accessed 06/03/2019

von Campenhausen S, Winter Y, e Silva AR, Sampaio C, Ruzicka E, Barone P, Poewe W, Guekht A, Mateus C, Pfeiffer KP, Berger K (2011) Costs of illness and care in Parkinson's disease: an evaluation in six countries. Eur Neuropsychopharmacol 21(2):180–191

Vradenburg G (2015) A pivotal moment in Alzheimer's disease and dementia: how global unity of purpose and action can beat the disease by 2025. Expert Rev Neurother 15(1):73–82

Wang J, Gu BJ, Masters CL, Wang YJ (2017) A systemic view of Alzheimer disease—insights from amyloid-β metabolism beyond the brain. Nat Rev Neurol 13(10):612

Widdus R (2005) Public-private partnerships: an overview. Trans R Soc Trop Med Hyg 99(Supplement_1):S1–S8

Yacoubian TA (2017) Drug discovery approaches for the treatment of neurodegenerative disorders. Academic Press/Elsevier, Amsterdam

Conjugated Recombinant Proteins as Emerging New Drugs

18

Nasir Mahmood, Sarah Bushra Nasir,
and Kathleen Hefferon

Abstract

Conjugation combines protein domains to achieve new functionalities in therapeutic drugs by utilizing their functional and structural versatility creating highly diverse macromolecules. Several human recombinant protein drugs are being used for the treatment of different diseases. However to improve these recombinant protein drugs' bioavailability and to incorporate different new features in the single-protein drug, sortase-mediated conjugation is a promising technology. The sortase A enzyme from *Staphylococcus aureus* can link any two proteins if one contains N-terminal triglycine (GGG) motif and second contains C-terminal LPXTG motif. In this new protein recombinant drugs can be developed. It is estimated that in the future, conventional recombinant protein drugs will be modified with the help of conjugation technology, and new recombinant drugs will be available with further enhancement of recombinant drugs business worldwide.

Keywords

Conjugated proteins · Sortase · Pegylation · Recombinant proteins · Drugs

N. Mahmood
Department of Biochemistry, University of Health Sciences, Lahore, Pakistan

Department of Cell and Systems Biology, University of Toronto, Toronto, Canada

S. B. Nasir
Abdus Salam School of Sciences, Nusrat Jahan College, Rabwah, Chiniot, Pakistan

K. Hefferon (✉)
Department of Cell and Systems Biology, University of Toronto, Toronto, Canada
e-mail: kathleen.hefferon@alumni.utoronto.ca

C. Keswani (ed.), *Bioeconomy for Sustainable Development*,
https://doi.org/10.1007/978-981-13-9431-7_18

18.1 Introduction

Proteins are linear polymers and macromolecules made up of amino acid monomers, capable of versatile activities in living systems including transport and storage of oxygen molecules, immune protection, mechanical support, growth, differentiation, catalysts, and others (Berg et al. 2002). Several human recombinant proteins are being used as drugs, and their chemical modifications are being performed to enhance their qualities. However these chemical modifications have several limitations; therefore, sortase-mediated conjugation is a good alternative and promising technology to introduce several features of different proteins into single recombinant protein drug.

18.2 Protein Conjugation

There is an ever-growing trend of bioconjugates in the therapeutics world today (Aggarwal 2011) due to the valuable advantages they offer including optimized chemical/physical properties, controlled drug release, reduced toxic effect, and better disease targeting (Li and Mahato 2017). Therapeutic bioconjugates have also become prominent in the fight against cancer (Krug et al. 2004), infectious bacteria (Verez-Bencomo et al. 2004), and malaria (Seeberger and Werz 2007).

Proteins can be modified by introducing or altering an amino acid at the post-translational level or by isolating the particular protein (Van Vught et al. 2014). Proteins yield biohybrid molecules called bioconjugates when covalently conjugated with polymers. The bioconjugate is a macromolecule that exhibits the properties of both the protein and the polymer (Borchmann et al. 2014). It has three main components including carrier molecules, therapeutic agents, and linkers (Li and Mahato 2017). Protein therapeutics mostly utilizes endogenous proteins thereby minimizing the chances of an immunogenic reaction (Dozier and Distefano 2015). They are large sized and have a specific conformation and so are specialized for binding activity with a reduced risk of cross-reactivity.

18.3 Pegylation of Proteins

Pegylation is one of the most common approaches toward protein drug therapeutics (Mero et al. 2011). Protein modification was first studied by Abuckowski et al. (1977) using polyethylene glycol (PEG) and bovine albumin. These studies paved the way for protein-polymer conjugation to impart stability to the bioconjugate in vivo. Pegylation provided desirable outcomes to the pharmaceutical world such as an increased in vivo half-life, reduced immunogenic response, and aggregation. Pegylation works equally well with liposomes (Mero et al. 2011). It also imparts greater solubility and increased residence time (Tegnér 2015). Pegylation works by hiding the protein surface and increasing the molecular weight of the protein resulting in lesser renal ultrafiltration and thereby lesser exposure to the antibodies and antigen-presenting cells and less destruction by proteolytic enzymes (Veronese 2001).

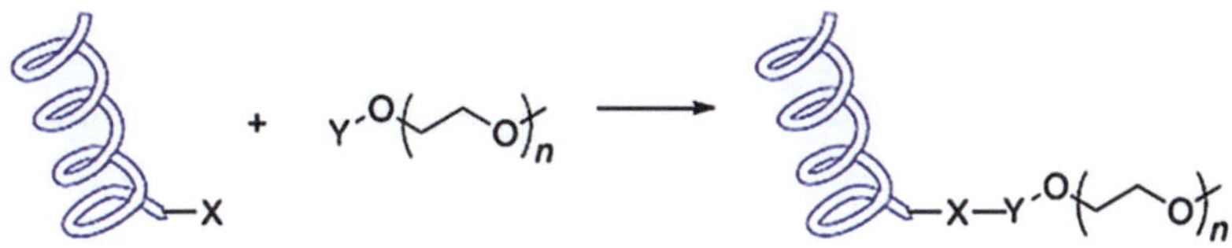

Fig. 18.1 Representation of a generalized protein pegylation reaction. The functional group (X) of the protein reacts with the complementary group (Y) of the PEG polymer yielding the PEG-protein conjugate. (Dozier and Distefano 2015)

Pegylation is generally performed in a batch reactor which yields 10% multipegylated and 60% monopegylated proteins. The SEC column (size exclusion chromatography column) or IEC column (ion exchange chromatography column) separates the products from the residues. SERC (size exclusion reaction chromatography) which is underway its development process is expected to deliver better separation of pegylated products (Tegnér 2015).

Pegylation involves covalent bonding between the protein and polyethylene glycol (PEG) molecule to impart enhanced pharmaceutical value to the protein. Functional groups such as imidazolyl formate, succinimidyl succinate, and cyanuric chloride are initially attached to the terminal ends of the PEG molecule (Harris and Chess 2003). In the next step, the N-terminal of amino acid is covalently attached to PEG molecule. Amino acids including histidine, serine, aspartic acid, glutamic acid, arginine, tyrosine, cysteine, threonine, and ∞ or ε ends of lysine are possible choices for attachment to the PEG molecule (Roberts and Harris 1998). The N-terminal of the alpha amino group is preferred over lysine epsilon amino group due to a reduced pKa value (Kinstler et al. 2002) (Fig. 18.1).

The Food and Drug Administration (FDA) has approved over ten pegylation drugs currently (Dozier and Distefano 2015), while others are underway the development process (Alconcel et al. 2011). Some of these drugs include uricase against gout (Sherman et al. 2008), interferon-α2a and (INF-α2a) interferon-α2b (INF-α2b) against hepatitis B and C (McHutchison et al. 2009; Wursthorn et al. 2006; Lau et al. 2005), and L-asparaginase against leukemia (Graham 2003).

18.4 Limitation of Pegylated Protein Drugs

Despite the fact that PEG is considered as the best polymer for protein conjugation, its immunogenic behavior and nondegradability are the chief concerns to its use (Pelegri-O'Day et al. 2014; Knop et al. 2010). In addition, there are reports of hypersensitivity and vacuolation as a reaction to pegylated drugs (Zhang et al. 2014). Animal-based studies have also revealed that anti-PEG antibodies are produced as a result of these drugs (Armstrong et al. 2003; Sroda et al. 2005; Cheng et al. 2005; Shimizu et al. 2012). Anti-PEG antibodies against PEG-asparaginase (Armstrong et al. 2007) and PEG-uricase (Sherman et al. 2008) were detected in the patients' serum indicating the inefficiency of the drugs. Additionally, the nondegradability of PEG can damage liver function or support its accumulation in tissues (Veronese and Pasut 2005). A study by Garratty (2004) revealed an increase of 19.8% anti-PEG antibodies in a healthy population in comparison to only 0.2%, 20 years ago (Richter and Akerblom 1983) which is surprising. Early clearance of PEG-asparaginase (Armstrong et al. 2007) and PEG-uricase (Sundy et al. 2007) from the patients' blood was reported in two different studies pointing to the reduced therapeutic potential of the drug. Vacuolation in response to pegylated proteins was reported in macrophages and histiocytes of several organs and renal tubular cells of the animal system cells.

18.5 Conjugation Versus Pegylation of Proteins

Pegylation is a complex technique that does not have a standard protocol and varies from one project to the other (Mero et al. 2011). It was concluded that efficiency of a pegylated drug can vary from 7% to 98% according to the molecular weight of the PEG molecule and protocol used. Moreover, it is not possible to achieve pure monomethoxylated PEG due to the presence of 1–10% PEG diol (Veronese 2001). Nonspecific pegylation produces several isoforms of the product despite sufficient purification efforts. Branched pegylation is helpful in reducing the immunogenic behavior of the PEGylated product but limits its bioactivity (Monfardini et al. 1995). Site-specific pegylation have been reported to support aggregation of the conjugated product (Veronese et al. 2007). Altered viscosity as a result of using PEG has also been reported by Kerwin et al. (2002) in a mixture of TNF receptor 1 with PEG which was shown to be five times more than the separate constituents. The PEG-related moieties display polydispersity which is another complication along with stearic hindrance with large proteins due to excessive pegylation resulting in poor binding affinity and/or activity. The above data points to the need for improved PEG conjugates; however, the large number of patents associated discourages the development of new conjugates.

Many bioconjugation techniques prevalent today lack efficiency and compatibility and are nonstoichiometric and non-site-specific. In addition, the conjugated product is difficult to separate from the unconjugated one. Sortase has helped overcome this problem by combining the ligand purification and site-specific C-terminal conjugation together by the STEPL (sortase-tag expressed protein ligation) technique. STEPL was found to generate tumor-related affinity ligands with subsequent azide modifications at high levels of purity with no residual unconjugated products (Warden-Rothman et al. 2013).

Although techniques like native chemical ligation (Dawson and Kent 2000), enzyme-based methods, intein (Jackson et al. 1994; Blaschke et al. 2000), and several others have been utilized for protein conjugation, however, they present practical complications, a need for additional protein and lack of site-specificity. The efficiency of sortase A to catalyze versatile protein pairings in conjugation reactions comprising enzymes, antibodies, fluorescent, cell surface, and blood proteins in diverse hosts comprising *E.coli*, yeast, and mammals was shown to produce a yield of 30–85%. The main fusions catalyzed by sortase A were antibody fusions and diverse LPETG motifs bearing triglycine reactants establishing the fusion efficiency of sortase A regarding intricate disulfide protein domains. The results clearly showed that sortase A is efficiently capable of fusion between intricate disulfide domains of protein as well as other protein combinations as compared to the inefficient and unproductive genetic fusions (Levary et al. 2011).

18.6 Sortase as a Tool for Conjugated Proteins Production

Sortase is an enzyme of the transpeptidases family associated with the membrane of the gram-positive bacteria. It helps the bacterial surface virulence proteins in the covalent attachment with its cell wall (Mazmanian et al. 2001; Ton-That et al. 2004; Desvaux et al. 2006; Marraffini et al. 2006). Sortase has four subfamilies based on primary sequence (Fig. 18.2): sortase A, sortase B, sortase C, and sortase D (Comfort and Clubb 2004; Dramsi et al. 2005). Sortase was first identified in *Staphylococcus aureus*' surface protein and named Srt A after the protein, sorting A (Mazmanian 1999) (Fig. 18.2).

Srt A was established as a tool in biotechnology after it was shown to locate peptides containing the LPXTG motif without the hydrophobic domain or tail. The sortase A substrate protein can be recognized by a sorting signal located on its C-terminal. The sorting signal consists of a conserved LPXTG motif, hydrophobic domain, and a positively charged tail. The Srt A functions by first attaching to its

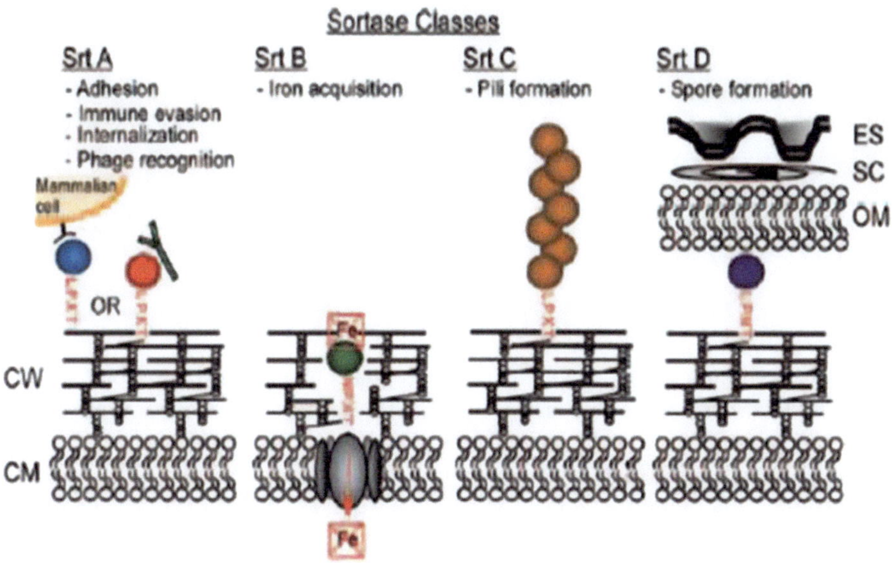

Fig. 18.2 Sortases are grouped into four different classes on the basis of sequence homology and diverse functions in the gram-positive bacteria. (Maresso and Schneewind 2008)

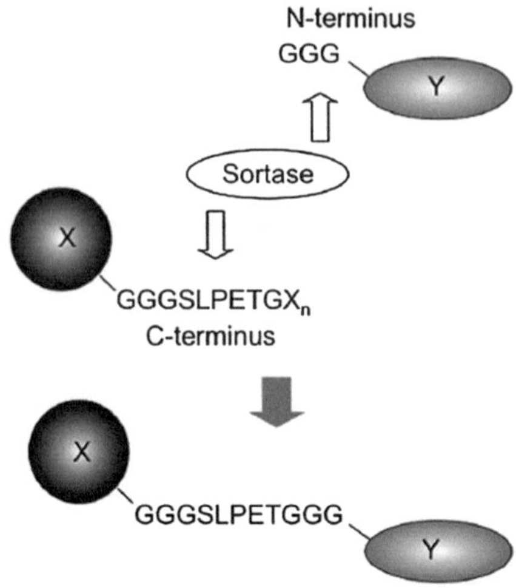

Fig. 18.3 Schematic demonstration of sortase A-mediated reaction displaying fusion of two protein terminals

substrate and then cleaving the threonine-glycine bond in the LPXTG motif thereby catalyzing the creation of threonine-pentaglycine amide bond in the cell wall of bacteria (Zong et al. 2004) (Fig. 18.3).

The sortase A enzyme from *Staphylococcus aureus* (SaSrtA) is widely researched since the year 1999 when it was discovered (Mazmanian 1999). Its function was highlighted by studies revealing that Srt A lacking mutants of *Staphylococcus aureus* were not able to attach cell surface proteins promoting attenuated virulence (Clancy et al. 2010; Mazmanian et al. 2000).

Srt A has been particularly a highly effective agent for its ability to ligate biomolecules through covalent bonding (Jacobitz et al. 2017).

Sortase was classified as a molecular stapler by Parthasarathy et al. (2007) in a study that used sortase A to covalently link a fluorescent-labeled protein with chemically altered polystyrene beads. Srt A labeling of proteins can be successfully carried out on both protein terminals. Some applications of sortase-mediated conjugation include immobilization on solid support, protein cyclization, fluorescent tagging, and pegylation (Popp and Ploegh 2011; Popp et al. 2011). As a drug against cancer, sortase A works site specifically in the antibody drug conjugate (ADC) to link the cytotoxic agent and the antibody together in a bioconjugate (Schwartz 2004a; Gordon et al. 2015).

Chimeric proteins are valuable therapeutic agents that support N-C and C-N terminal fusions but not the unusual N-N and C-C terminal ones. Sortase A has the ability to provide the protein with the required molecular probes to achieve the desired N-N and C-C terminal chimeric proteins. Some examples of these proteins are ubiquitin, interleukin-2, interferon-α, and several antibodies (Witte et al. 2013). The peptide sequences for sortase-based fusions can be easily accessed by following the standard protocols for peptide synthesis (solid phase) for both the N and C protein terminals. The technique is diverse as it allows functional groups that are not genetically encoded to be attached to the peptide sequences (Antos et al. 2009).

18.7 World Economy of Non-conjugated and Conjugated Recombinant Protein Drugs

Many therapeutic protein drugs and peptides have developed since the twentieth century. According to a report by PHARMA 2010, biotech products account for over 35% of the 37 novel active substances introduced in 2001 (Tang et al. 2004). In the year 2007, the worldwide sale of biotech drugs doubled (12.5%) in comparison with the conventional molecule drugs (6.4%) with a total profit of USD 75 billion. In 2008, biotech drugs contributed to one fifth of the best seller drugs (Malik 2008).

Proteins are distinct as therapeutics since they offer specific action mechanisms and are highly efficient (Pisal et al. 2010b). They are a highly valuable class of drugs especially for patients who are in need of novel treatment options. Recombinant protein drugs formulated to treat a wide range of clinical implications include autoimmune disorders/inflammation, genetic disorders, infectious agent exposure, and cancers. The latest developments in protein-engineering skills have facilitated drug manufacturers to exploit and modify the best functional features of desired proteins while maintaining product efficacy and safety.

Extraction of protein from natural sources or chemically synthesizing them is associated with inherent constraints for a large-scale production. Therefore, recombinant DNA technology is a desirable option for large-scale production of therapeutic proteins. Till today, among the 650 protein pharmaceutical drugs approved worldwide, around 400 drugs have been produced by recombinant technologies (Global Data 2015). The number of other recombinant drugs under development is around 1300 (Sanchez-Garcia et al. 2016).

In order to maintain the technological advancement and lead in recombinant proteins production, active research efforts and development is necessary. European countries invest 2% of their GDP on research with the USA being the leader of innovative products having a record of issuing 81% biotechnology patents all over the world. On the other hand, nondeveloped countries can only manage to spend 0.5% of the GDP on research having modern pharmaceutical products inaccessible largely (Palomares et al. 2002).

18.8 Market Potential of the Conjugated Recombinant Protein Drugs

Recombinant protein drugs are being used to treat a wide range of diseases including autoimmune disorders, inflammation, genetic disorders, infectious diseases, and cancers. Worldwide to date 650 proteins drugs have been approved, and among them, 400 have been produced by recombinant technologies (Global Data 2015). The recombinant drugs under trial and synthesis phase are around 1300 (Sanchez-Garcia et al. 2016).

The worldwide sale of recombinant proteins drugs in year 2007 doubled (12.5%) in comparison with the conventional molecule drugs (6.4%) with a total profit of USD 75 billion, while in 2008 biotech drugs contributed to one fifth of the best seller drugs (Malik 2008). There is shift in the makeup of the total pharmaceutical market and share of biological protein drugs hold from 16% in 2006 to 25% to date (QI MIDAS 2016). The three largest recombinant protein drugs biological therapy areas include autoimmune, diabetes and oncology have collectively worth USD 110 billion, over half of all biologic revenue (MIDAS 2015). The latest developments in protein-engineering skills have facilitated drug manufacturers to exploit and modify the best functional features of desired proteins while maintaining product efficacy and safety. In this regard recombinant protein drug conjugation is very useful which involves combining or merger of protein domains of different protein drugs to achieve new functionalities in therapeutic protein drugs. Therefore it is estimated that in upcoming years, the modified versions of conventional recombinant protein drugs with the help of conjugation technology will be available which resulted in further enhancement of recombinant drugs business worldwide (Singh et al. 2016, 2019a, b).

18.9 Future Scope of the Conjugated Protein Drugs

In summary, the potential for recombinant protein drugs is growing in the presence of a steadily expanding platform for protein production. Despite the protein production potential of mammalian cell lines, the cost-effectiveness and versatility of microbial cells especially *E. coli* cannot be ignored. Recombinant protein drugs are closely related to the success of biotech industry within the last 30 years with some highly efficient pharmaceutical products. There are currently more than 30 proteins being utilized for several clinical applications with more than 300 proteins under clinical trials. Protein conjugation is playing an important role in the advancement of recombinant drugs to deliver more refined protein constructs as compared to the plain versions, which is remarkable progress for novel proteins delivery. Conjugation combines protein domains to achieve new functionalities in therapeutic drugs by utilizing their functional and structural versatility creating highly diverse macromolecules. It is anticipated that protein engineering will be carried out more extensively in the future with the expectation that new generation protein drugs will carry neo-sequences nonexistent in nature.

References

Abuchowski A, van Es T, Palczuk N, Davis F (1977) Alteration of immunological properties of bovine serum albumin by covalent attachment of polyethylene glycol. J Biol Chem 252:3578–3581

Aggarwal S (2011) What's fueling the biotech engine. 2010 to 2011. Nat Biotech 29:1083–1089

Alconcel SN, Baas AS, Maynard HD (2011) FDA-approved poly (ethylene glycol)–protein conjugate drugs. Polym Chem 2(7):1442–1448

Antos JM, Chew GL, Guimaraes CP, Yoder NC, Grotenbreg GM, Popp MW, Ploegh HL (2009) Site-specific N- and C-terminal labeling of a single polypeptide using sortases of different specificity. J Am Chem Soc 131(31):10800–10801

Armstrong JK, Wenby RB, Meiselman HJ, Fisher TC (2003) In vivo survival of poly(ethylene glycol)-coated red blood cells in the rabbit. Blood 102:94A
Armstrong JK, Hempel G, Koling S, Chan LS, Fisher T, Meiselman HJ, Garratty G (2007) Antibody against poly(ethylene glycol) adversely affects PEG-asparaginase therapy in acute lymphoblastic leukemia patients. Cancer 110:103–111. PubMed: 17516438
Berg JM, Tymoczko JL, Stryer L (2002) Biochemistry, 5th edn. W H Freeman, New York. Chapter 3, Protein Structure and Function. Available from: https://www.ncbi.nlm.nih.gov/books/NBK21177/
Blaschke UK, Silberstein J, Muir TW (2000) Protein engineering by expressed protein ligation. Methods Enzymol 328:478–496
Borchmann DE, Carberry TP, Weck M (2014) "Bio"-macromolecules: polymer-protein conjugates as emerging scaffolds for therapeutics. Macromol Rapid Commun 35(1):27–43
Cheng TL, Cheng CM, Chen BM, Tsao DA, Chuang KH, Hsiao SW, Lin YH, Roffler SR (2005) Monoclonal anti-body-based quantitation of poly(ethylene glycol)-derivatized proteins, liposomes, and nanoparticles. Bioconjug Chem 16:1225–1231
Clancy KW, Melvin JA, McCafferty DG (2010) Sortase transpeptidases: insights into mechanism, substrate specificity, and inhibition. Biopolymers 94:385–396
Comfort D, Clubb RT (2004 May 1) A comparative genome analysis identifies distinct sorting pathways in gram-positive bacteria. Infect Immun 72(5):2710–22
Dawson PE, Kent SB (2000) Synthesis of native proteins by chemical ligation. Annu Rev Biochem 69:923–960
Desvaux M, Dumas E, Chafsey I, He'braud M (2006) FEMS Microbiol Lett 256:1–15
Dozier JK, Distefano MD (2015) Site-specific PEGylation of therapeutic proteins. Int J Mol Sci 16(10):25831–25864
Dramsi S, Trieu-Cuot P, Bierne H (2005) Sorting sortases: a nomenclature proposal for the various sortases of gram-positive bacteria. Res Microbiol 156(3):289–297
Garratty G (2004) Progress in modulating the RBC membrane to produce transfusable universal/stealth donor RBCs. Transfus Med Rev 18:245–256
Global Data (2015). http://www.globaldata.com. 2015
Gordon MR et al (2015) Field guide to challenges and opportunities in antibody drug conjugates for chemists. Bioconjug Chem 26:2198–2215
Graham ML (2003) Pegaspargase: a review of clinical studies. Adv Drug Deliv Rev 55:1293–1302
Harris JM, Chess RB (2003) Effect of pegylation on pharmaceuticals. Nat Rev Drug Discov 2(3):214–221
Jackson DY, Burnier J, Quan C, Stanley M, Tom J et al (1994) A designed peptide ligase for total synthesis of ribonuclease A with unnatural catalytic residues. Science 266:243–247
Jacobitz AW, Kattke MD, Wereszczynski J, Clubb RT (2017) Sortase Transpeptidases: structural biology and catalytic mechanism. Adv Protein Chem Struct Biol 109:223–264. https://doi.org/10.1016/bs.apcsb.2017.04.008
Kerwin BA, Chang BS, Gegg CV, Gonnelli M, Li T, Strambini GB (2002) Interactions between PEG and type I soluble tumor necrosis factor receptor: modulation by pH and by PEGylation at the N terminus. Protein Sci 11:1825–1833
Kinstler O, Moulinex G, Treheit M, Ladd D, Gegg C (2002) Mono-N-terminal poly(ethylene glycol)-protein conjugates. Adv Drug Deliv Rev 54:477–485
Knop K, Hoogenboom R, Fischer D, Schubert US (2010) Poly(ethylene glycol) in drug delivery: pros and cons as well as potential alternatives. Angew Chem Int Ed 49:6288–6308
Krug LM, Ragupathi G, Ng KK, Hood C, Jennings HJ, Guo Z, Kris MG, Miller V, Pizzo B, Tyson L, Baez V, Livingston PO (2004) Clin Cancer Res 10:916–923
Lau GKK, Piratvisuth T, Luo KX, Marcellin P, Thongsawat S, Cooksley G, Gane E, Fried MW, Chow WC, Paik SW et al (2005) Peginterferon α-2a, lamivudine, and the combination for HBeAg-positive chronic hepatitis b. N Engl J Med 352:2682–2695
Levary DA, Parthasarathy R, Boder ET, Ackerman ME (2011) Protein-protein fusion catalyzed by sortase A. PLoS One 6(4):e18342

Li F, Mahato RI (2017) Bioconjugate therapeutics: current progress and future perspective. Mol Pharm 14(5):1321–1324

Malik NN (2008) Drug discovery: past, present and future. Drug Discov Today 13(21–22):909–912

Maresso AW, Schneewind O (2008) Sortase as a target of anti-infective therapy. Pharmacol Rev 60(1):128–141

Marraffini LA, Dedent AC, Schneewind O (2006) Microbiol Mol Biol Rev 70:192–221

Mazmanian SK (1999) Staphylococcus aureus Sortase, an enzyme that anchors surface proteins to the cell wall. Science 285:760–763

Mazmanian SK, Liu G, Jensen ER, Lenoy E, Schneewind O (2000) Staphylococcus aureus sortase mutants defective in the display of surface proteins and in the pathogenesis of animal infections. Proc Natl Acad Sci U S A 97:5510–5515

Mazmanian SK, Ton-That H, Schneewind O (2001) Mol Microbiol 40:1049–1057

McHutchison JG, Lawitz EJ, Shiffman ML, Muir AJ, Galler GW, McCone J, Nyberg LM, Lee WM, Ghalib RH, Schiff ER et al (2009) Peginterferon α-2b or α-2a with ribavirin for treatment of hepatitis c infection. N Engl J Med 361:580–593

Mero A, Clementi C, Veronese FM, Pasut G (2011) Covalent conjugation of poly (ethylene glycol) to proteins and peptides: strategies and methods. In: Bioconjugation protocols. Humana Press, Totowa, pp 95–129

Monfardini C, Schiavon O, Caliceti P et al (1995) A branched monomethoxypoly(ethylene glycol) for protein modification. Bioconjug Chem 6:62–69

Palomares LA, Kuri-Breña F, Ramírez OT (2002) Industrial recombinant protein production. In: The Encyclopedia of Life Support Systems, vol 6. EOLSS Publishers, Oxford. (3.8)

Parthasarathy R, Subramanian S, Boder ET (2007) Sortase a as a novel molecular "stapler" for sequence-specific protein conjugation. Bioconjug Chem 18(2):469–476

Pelegri-O'Day EM, Lin E-W, Maynard HD (2014) Therapeutic protein–polymer conjugates: advancing beyond PEGylation. J Am Chem Soc 136(41):14323–14332. PubMed: 25216406

Pisal DS, Kosloski MP, Balu-Iyer SV (2010a) Delivery of therapeutic proteins. J Pharm Sci 99(6):2557–2575

Pisal DS, Kosloski MP, Balu-Iyer SV (2010b) Delivery of therapeutic proteins. J Pharm Sci 99(6):2557–2575

Popp MWL, Ploegh HL (2011) Making and breaking peptide bonds: protein engineering using sortase. Angew Chem Int Ed 50:5024–5032

Popp MW, Dougan SK, Chuang TY, Spooner E, Ploegh HL (2011) Sortase-catalyzed transformations that improve the properties of cytokines. Proc Natl Acad Sci U S A 108:3169–3174

QI MIDAS MAT Q2 (2016) LCUS$ used for growth figures

QI MIDAS MAT Q4 (2015) ex-manufacturer level pricing, excludes discounts and rebates

Richter AW, Åkerblom E (1983) Antibodies against polyethylene glycol produced in animals by immunization with monomethoxy polyethylene glycol modified proteins. Int Arch Allergy Appl Immunol 70:124–131

Roberts MJ, Harris JM (1998) Attachment of degradable poly(ethylene glycol) to proteins has the potential to increase therapeutic efficacy. J Pharm Sci 87(11):1440–1445

Sanchez-Garcia L, Martín L, Mangues R, Ferrer-Miralles N, Vázquez E, Villaverde A (2016) Recombinant pharmaceuticals from microbial cells: a 2015 update. Microb Cell Factories 15(1):33

Schwartz RS (2004a) Paul Ehrlich's magic bullets. New Engl J Med 350:1079–1080

Schwartz RS (2004b) Paul Ehrlich's magic bullets. N Engl J Med 350(11):1079–1080

Seeberger PH, Werz DB (2007) Synthesis and medical applications of oligosaccharides. Nature 446(7139):1046

Sherman MR, Saifer MG, Perez-Ruiz F (2008) PEG-Uricase in the management of treatment-resistant gout and hyperuricemia. Adv Drug Deliv Rev 60:59–68

Shimizu T, Ichihara M, Yoshioka Y, Ishida T, Nakagawa S, Kiwada H (2012) Intravenous administration of polyethylene glycol-coated (PEGylated) proteins and PEGylated adenovirus elicits an anti-PEG immunoglobulin M response. Biol Pharm Bull 35:1336–1342

Singh HB, Jha A, Keswani C (eds) (2016) Intellectual property issues in biotechnology. CABI, Wallingford. 304 pages, ISBN-13: 9781780646534

Singh HB, Keswani C, Singh SP (eds) (2019a) Intellectual property issues in microbiology. Springer-Nature, Singapore. 425 pages, ISBN- 9789811374654

Singh HB, Keswani C, Reddy MS, Royano ES, García-Estrada C (2019b) Secondary metabolites of plant growth promoting Rhizomicroorganisms: discovery and applications. Springer-Nature, Singapore. 392 pages, ISBN- 978-981-13-5861-6

Sroda K, Rydlewski J, Langner M, Kozubek A, Grzybek M, Sikorski AF (2005) Repeated injections of PEG-PE liposomes generate anti- PEG antibodies. Cell Mol Biol Lett 10:37–47

Sundy JS, Ganson NJ, Scarlett E, Rehrig CD, Huang W, Hershfield MS (2007) Pharmacokinetics and pharmacodynamics of intravenous PEGylated recombinant mammalian urate oxidase in patients with refractory gout. Arthritis Rheum 56:1021–1028. PubMed: 17328081

Tang L, Persky AM, Hochhaus G, Meibohm B (2004) Pharmacokinetic aspects of biotechnology products. J Pharm Sci 93(9):2184–2204

Tegnér F (2015) Optimization of a PEGylation process A combined reaction and separation with size exclusion reaction chromatography. Department of Chemical Engineering Lund University May 2015

Ton-That H, Marraffini LA, Schneewind O (2004) Biochim Biophys Acta 1694:269–278

Van Vught R, Pieters RJ, Breukink E (2014) Site-specific functionalization of proteins and their applications to therapeutic antibodies. Comput Struct Biotechnol J 9(14):e201402001

Verez-Bencomo V, Fernandez-Santana V, Hardy E, Toledo ME, Rodríguez MC, Heynngnezz L, Rodriguez A, Baly A, Herrera L, Izquierdo M, Villar A (2004) A synthetic conjugate polysaccharide vaccine against Haemophilus influenzae type B. Science 305(5683):522–525

Veronese FM (2001) Peptide and protein PEGylation: a review of problems and solutions. Biomaterials 22:405–417

Veronese FM, Pasut G (2005) PEGylation, successful approach to drug delivery. Drug Discov Today 10:1451–1458. PubMed: 16243265

Veronese FM, Mero A, Caboi F, Sergi M, Marongiu C, Pasut G (2007) Bioconjug Chem 18(6):1824–1830

Warden-Rothman R, Caturegli I, Popik V, Tsourkas A (2013) Sortase-Tag Expressed Protein Ligation (STEPL): combining protein purification and site-specific bioconjugation into a single step. Anal Chem 85(22):11090. https://doi.org/10.1021/ac402871k

Witte MD, Theile CS, Wu T, Guimaraes CP, Blom AEM, Ploegh HL (2013) Production of unnaturally linked chimeric proteins using a combination of sortase-catalyzed transpeptidation and click chemistry. Nat Protoc 8:1808–1819

Wursthorn K, Lutgehetmann M, Dandri M, Volz T, Buggisch P, Zollner B, Longerich T, Schirmacher P, Metzler F, Zankel M et al (2006) Peginterferon α-2b plus adefovir induce strong cccDNA decline and HBsAg reduction in patients with chronic hepatitis b. Hepatology 44:675–684

Zhang F, Liu M-r, Wana H-t (2014) Discussion about several potential drawbacks of PEGylated therapeutic proteins. Biol Pharm Bull 37(3):335–339

Zong Y, Bice TW, Ton-That H, Schneewind O, Narayana SV (2004) Crystal structures of Staphylococcus aureus sortase A and its substrate complex. J Biol Chem 279:31383–31389

Economic Importance of Medicinal Plants in Asian Countries

19

Walia Zahra, Sachchida Nand Rai, Hareram Birla, Saumitra Sen Singh, Aaina Singh Rathore, Hagera Dilnashin, Chetan Keswani, and Surya Pratap Singh

Abstract

Due to their characteristic benefits and medicinal value, medicinal plants are gaining importance worldwide. It is becoming famous, and people are using herbal therapy as an alternative medicinal therapy. Because of their increased traditional use and cultural acceptability, these medicinal plants are greatly admired and also have minimal side effects and thus are gaining global importance. Herbal drugs which are cost-effective than synthetic drugs in many cases are being promoted by most of the developing countries, and they have started discovering and filing patents on the medicinal plants and their derivatives. Many drugs still have not undergone the process of drug approval and are not yet validated for their safety and efficacy. These medicinal plant-derived drugs can be formulated by medicine-based industries. The international trade of medicinal plants and their products was estimated to be USD 60 billion in 2010, and by 2050, it is expected to reach USD 5 trillion. Asian countries are very rich in medicinal plant species and are the major exporters of these plants and their products. These medicinal plants can be popularized and used to improve the economy of low-income countries of Asia and create livelihoods for its people. Moreover, overexploitation of these medicinal plants should be limited, the valuable species of high marketing value should be conserved, and their cultivation should be promoted for future use.

Keywords

Medicinal plants · Herbal drugs · Traditional knowledge · Herbal drugs · Asian countries · Global consumption

W. Zahra · S. N. Rai · H. Birla · S. S. Singh · A. S. Rathore · H. Dilnashin · C. Keswani
S. P. Singh (✉)
Department of Biochemistry, Institute of Science, Banaras Hindu University, Varanasi, India

C. Keswani (ed.), *Bioeconomy for Sustainable Development*,
https://doi.org/10.1007/978-981-13-9431-7_19

19.1 Introduction

In both traditional and modern medicinal systems, the interest in using medicinal plant-derived drugs has increased. Due to this, the increment on the demand for medicinal plants has also occurred on global basis resulting in the collection of these plants from their native place. These plants play an important role in maintaining the economy of the low-income countries as the medicines derived from them are quite essential in keeping the population healthy which in turn is valuable for sustaining the economy of the countries. Cultivation of medicinal plants within these countries offers employment to the poor people, hence helping in their livelihood. Valuable medicinal plants are reported to grow in South Asian countries, mostly in fragile ecosystems which are particularly inhabited by rural peoples and native communities. Humans are familiar with these plants since olden times, and the ancient literature suggests the use of these plants in religious ceremonies and for curing diseases in the countries like India. The use of medicinal plants in low-income countries should be promoted to make them self-reliable on their pharmaceutical resources which would also be quite effective in managing the endemic diseases condition. This can also be helpful in alleviating the challenges of poverty, hunger, ill-health, and illiteracy which are being largely faced by these Asian countries. These plants and their derivative drugs are prescribed in both developing and developed countries, and the global trade based on these drugs was estimated to be USD 32–43 billion. This should be considered as priority for other developing countries too in order to enhance their self-reliability on their own pharmaceutics which will strengthen their economy. To accomplish this, the developing countries of Asia must realize the value of their inhabitant medicinal plants and their economic importance. The advertisement of their valuable plants can then attract the investors in the field of the pharmaceutical production and help in the production of their own medicinal products. This strategy will end up in boosting the economy of the developing countries by offering jobs to the poor and improving the healthcare delivery system which will allow the population to purchase the medicine at affordable prices. Thus, the use of their inhabitant treasure, i.e., medicinal plants, will not only address the increased income of the country but will also allow access to the medicines by the population enhancing the country's health services. Some of the valuable medicinal plants found in Asia-Pacific region include the species of *Cassia*, *Atropa*, *Podophyllum*, *Rauvolfia*, *Psoralea*, *Catharanthus*, *Hyoscyamus*, and *Papaver.*

19.2 The Growing Need to Spend Money on Pharmaceutical Development

Kaplan and Mathers (2011) have suggested that countries throughout the world are nowadays threatened with the shift of the diseases from acute to chronic which directly set the allegations toward the supply and demand of the pharmaceutical products. The growing population of the developing countries is seen mostly to undergo the abovementioned situation. The unavailability of essential medicines is observed, and the distribution of the medicinal products for both communicable and noncommunicable diseases does not meet to the need of large number of people in the population of developing countries (Cameron et al. 2009; Kaplan and Mathers 2011).

Although countries have spent large amount of money on the development of pharma products, a little impact on the citizens' health has been observed. The experience was considered to be harsh, and the balanced approach toward the manufacture, delivery, and consumption of medicine was thought to be done with cost-effective methodology. This approach has possibly improved the situation of the developing countries and has considerably reduced their expenses of the medicine production. Still, some countries especially from the Southeast Asia were unable to implement the approach and failed in reducing the high pharmaceutical expenditures. Accordingly, India (18.8%), Burma or Myanmar (24.5%), Nepal (44.3%), Thailand (30.5%), and Bangladesh (63%) were observed to have high percentage of pharmaceutical expenses (Bukar et al. 2016). Due to this unpleasant experience, it was suggested that these developing nations should identify their resources to become self-reliable and effectively implement this to raise their economy. This can be acknowledged as most people in developing counties rely on medicinal plants to be used as medicines irrespective of their reasons of cultivation, cost, and accessibility (Bukar et al. 2016). Nigeria has set the example of such developing countries in which medicinal plants are widely used (Okoli et al. 2007; Dahlberg and Trygger 2009). These developing countries are wealthier in the sense that they have valuable resources in the form of medicinal plants that can be potentially subjected for medicine production (Messiaen and Rouamba 2004; Okoli et al. 2007; Bukar et al. 2016; Rai et al. 2017; Singh et al. 2018; Birla et al. 2019).

This can allow citizens of developing nations to afford the cost-effective pharmaceutics and have the liberty to choose them. The production of these products in developing countries might also allow them to increase their global pharmaceutical production which was observed to be higher in the high-income countries, i.e., 89.1% in 1985 and 92.9% in 1999, as compared to the combined shared value of 10.9% and 7.1% during the same time in low-income countries (Bukar et al. 2016). Hence, the dependency of the developing nations on their counterpart developed countries can be reduced as the import of the pharmaceutical products would not be much needed.

19.3 Global Consumption of Pharmaceutical Products from Developing Nations

"High-value minor crops" is the phrase given to herbs, spices, and medicinal and aromatic plants that are generally the small contributors to the agricultural output of a country (Sher et al. 2014). Resurgence of the traditional medicine system at the global level has led to the increased rate of marketing of herbal drugs. Accordingly by 2050, the world herbal trade is expected to reach USD 7 trillion. Reports have widely suggested the inability to fulfill the need of essential medicines due to either unavailability and higher importation cost or their inadequate distribution in developing countries (Bukar et al. 2016; Keswani et al. 2017). Though being the small individual contributors, these medicinal plants have helped in reaching their global trade to be USD 60 billion in 2006 (Sher et al. 2014). From Asia and Africa, Europe has been found to import about USD 1 billion in medicinal and aromatic plants (Sher and Hussain 2009; Ghimire et al. 2004). This trade is assumed to be expanded significantly by the year 2050 as herbal medicines are attaining high popularity these days (Kuipers 1997; Al-Quran 2008; Khan et al. 2011). Despite contributing a smaller portion of the agricultural output, these medicinal plants have the highest value per weight among the traded plants. These are referred to as the pharmaceutical cash crops that have high potential ability for their respective countries based on their agricultural practices and contributing to the regional economy (Dubey et al. 2004; Kuniyal et al. 2015). The south and western countries of Asia have been addressed for collecting, cultivating, and trading the medicinal and aromatic plants from centuries (Ali-Shtayeh et al. 2000; Lev and Amar 2002; Ghorbani 2005; Al-Quran 2008; Mati and De Boer 2011).The cinnamon leaves from Meghalaya, an Indian state, have been reported to be about 2800 tons which is closer to about a million dollar as reported by a study (Karki et al. 2003). Nepal has also been found to export medicinal raw materials of an estimated USD 18–20 million to India and other countries (Schippmann et al. 2006). China is said to be rich in about 4941 medicinal plant species out of a total 26,092 species, while India is found to be rich in 3000 medicinal plants out of 15,000 plant species. Malaysia, Indonesia, and Nepal have comparatively lesser number of plant species than India and China. In the fast growing global trade, India only shares about 1.6% in the field of herbal medicine (Wakdikar 2004). The Indian share of world export though has grown by 4.95%, whereas the growth of China was found to be 7.38% between the years 1991 and 2002 (Verma and Singh 2008; Keswani et al. 2017). AYUSH has suggested that in 2009, India ranks as the second largest exporter after China (Keswani et al. 2017). The Indian herbs exported in the year 2007–2008 was about USD 96 million, while in 2006–2007 it was about USD 76.9 million (Schippmann et al. 2003).

Malaysian industry of herbs has the business of USD 315 million per annum and is reported to be growing steadily at the rate of 20%. In 1994, the estimated market value was estimated to be USD10 million in 1994 which has grown up to USD 15.8 million in 1996. The products mostly demanded from Malaysia are garlic, evening primrose oil, and *Ginkgo*. About 1000 manufacturers are involved in the production of herbal medicines, and the global market for phytomedicines is about USD 20 billion which has an average growth rate of 15–20% in the year 2012 (Nirmal et al. 2013).

After Brazil, Indonesia is known as the second largest biodiversity center. Generally, about 40% of the population of Indonesia consumes herbal drugs for their healthcare requirements. The herbal industry has developed rapidly in Indonesia in the recent years, and 468 registered industries have grown to the number of 807 in the year 2000 from 1992 which reflects the increased consumption of herbal medicines in the country (Nirmal et al. 2013).

Cambodian people have a tradition of using traditional medicine to fulfill their healthcare needs, and the nation is capable of providing the raw material obtained from more than 500 medicinal plants (Nirmal et al. 2013).

Thailand is wealthy in the case of medicinal and aromatic plants. About 1400 species of medicinal and aromatic plants are present in about 10,000 species of plants. 248 manufacturing units in Bangkok and 451 in rural areas have contributed in the production of traditional drugs in Thailand in the year 2000. The market value is being targeted by Thailand's Public Health Ministry to rise the marketing of herbal products from USD 9.69 million to USD 16.15 million (Nirmal et al. 2013).

The national health and development is very much dependent on the medicinal plants and herbal drugs in Vietnam. Almost 10,000 tons of medicinal herbs are harvested, and 40,000 tons are being imported annually. Around 3850 species of medicinal plants are present in Vietnam. Vietnam has earned millions of dollars by exporting the valuable medicinal herbs (Nirmal et al. 2013).

In Sri Lanka, also the medicinal plants are widely used in traditional medicine systems. India has exported about USD 10 million worth of medicinal plants in Department of Ayurveda, while about 60 species of medicinal plants of annual value of USD 1.27 million are imported in Sri Lanka from India and other countries in 1996. The herbal market value in 2007 was about USD 20 billion, and it is expected to reach USD 5 trillion in the coming future (Nirmal et al. 2013).

19.4 Research and Economic Priority of Medicinal Plants

A report given by the WHO in 2003 has indicated the sale of about 30% of the pharmaceutics throughout the world, which were derived from medicinal plants. The global sale of herbal products was of about USD 600 million with 80% of population of developing countries using the products for fulfilling their healthcare needs in 2002. Therefore, interest of the low-income countries could be safeguarded by commercializing medicinal plants strategically (Kala 2005b). By bio-hunting of the new drugs derived from these plants and due to the more costly prescribed drugs for personal health, interest in medicinal plants has been aroused as witnessed by the World Bank (Lambert et al. 1997) and Hoareau and DaSilva (1999). According to their opinion, they have suggested that medicinal plants will continue to play an important role in the health aid system based on the previous research. Also, development in information technology and upgrowing interest in medicinal plants have fueled the need for the electronic information about the medicinal plant and their emergence as promising health aid. This has resulted in arranging the screening program to identify the bioactive components from these medicinal plants and developing newer drugs from them (Hoareau and DaSilva 1999; Meena et al. 2009). The pharma products derived from the plants also show promising aspects for their commercial development as they are advantageous in terms of production scale and easy to store and also allow to deliver cost-effective drugs to the low-income countries (Ma et al. 2005; Pandey et al. 2013). The conditions and trend related to medicinal plants have been reported by FAO (Bukar et al. 2016), which talks about their production and market value and informs the world about both the possibilities and difficulties associated with the trade of medicinal plants.

Also, without incorporating the medicinal plants, it is impossible to implement the health policies prevailing in developing countries that mostly deal with health problems in which the infectious diseases play a prominent role (Farnsworth et al. 1985; Liu 1995). Thus, the blending of the policy practice with that of the cultural way to cure a disease is much relevant to the context (Reynolds and Sofowora 2007). China has thus set an example in this regard as medicinal plants are widely used in about 40% of the cases when the primary healthcare is needed (Sofowora et al. 2013). This fact seems to be quite convincing as over 80% of the world population is dependent on the plants and herbs of medicinal values as primary medicinal system. About 5 billion people worldwide address plant-derived products as the remedy of both acute and chronic diseases. Thus, medicines originated from medicinal plants should be implemented in modern medicinal system and should be practiced clinically. Thus, when the disease pattern in the geographical area will be investigated, cooperation between the policy makers and the scientists will be there and overdependence on the imported pharmaceuticals will be reduced, and when there is a strong political will, the deteriorating health of the population of the low-income countries will be alleviated, and the economy of the countries will be strengthened (Bukar et al. 2016) (Table 19.1).

19.5 Importance of Medicinal Plants

19.5.1 The Social Point of View

In many Asian countries, medicinal plants are being used for centuries to fulfill the nutritional needs and provide the primary healthcare as part of their culture and tradition (Farnsworth and Soejarto 1991). Thus, the heritage is quite familiar with the native people and they are very much accustomed about their cultivation, harvesting, processing them into the products they need and fulfill their requirements was according to the household or medicinal necessity. The cultivation and management of medicinal plants easily go with the fact of making the rural women employed. Women are usually acknowledged of successfully administrating the chores required as the activities related to medicinal plant production fit with the daily requirements and work schedules of women. They are involved from collecting the raw materials to drying it and transporting it to the market. They also get training from the herbal drug industry for administering the tasks. Thus, medicinal plants provide the livelihood in the rural areas and help in fulfilling the basic healthcare need. Arya Vaidya Sala (AVS), Kottakkal, Kerala, makes the example of such employment generator which allows doing business and providing traditional medicine service simultaneously. These industries are thus helpful not only in strengthening the social network but also in preserving the traditional knowledge of

Table 19.1 Some plants of medicinal values (Balasankar et al. 2013; Ghosh and Patil 2010; Khushbu et al. 2011; Gawade and Fegade 2012; Umadevi et al. 2012; Dastagir and Rizvi 2016; Lim 2016)

S. no.	Common name	Scientific name	Family	Features	Native countries	Treatment
1.	Periwinkle	*Catharanthus roseus*	Apocynaceae	A 40–90 cm tall, everblooming, drought-resistant, perennial crop with white or purple flowers	Native: Malagasy	Leaves: cancer
					Grown in India, Malagasy, Israel, USA	Roots: high blood pressure
2.	Senna	*Cassia senna*	Fabaceae	60–15 cm tall, drought-resistant, perennial crop	Native: Somalia, India, and Arab	Leaves and pods: constipation
					Grown in India, Arabia, Egypt, Sudan	
3.	Winter cherry or Ashwagandha	*Withania somnifera*	Solanaceae	30–150 cm tall, drought-resistant, annual crop	Native: India and Pakistan	Roots: used as stimulant or tonic
					Grown in India	
4.	Indian long pepper	*Piper longum*	Piperaceae	90–120 cm tall, perennial, climber	Native: India, Nepal, Indonesia	Roots: respiratory disorder
					Grown in Indonesia and India	Fruits used as spices
5.	Serpentine root	*Rauvolfia serpentina*	Apocynaceae	30–100 cm tall, perennial crop	Native: Bangladesh, Indonesia, Myanmar, India, Thailand, Sri Lanka	Roots: mental disorder, insomnia, high blood pressure
6.	Medicinal yam	*Dioscorea alata*	Dioscoreaceae	4–5 m tall, perennial climbers	Native: Africa, Asia, America	Tubers: for manufacturing steroidal drugs
					Grown in Mexico, India, China	
7.	Liquorice	*Glycyrrhiza glabra*	Fabaceae	A 60–200 cm tall, drought-resistant, perennial crop	Afghanistan, China, Italy	Rhizomes and roots: as flavoring agent and for treatment of cough, ulcers

medicines and in providing opportunities to the native youth to get employed. In this way, they not only inherit the knowledge from their elders but also are capable of earning their livelihood and providing strength to their society (Karki et al. 2003).

19.5.2 Traditional Knowledge Protection

The traditional knowledge regarding the use of medicinal plants is vanishing day by day, and it is very much important to take the measures to preserve it for its implementation. The mountains of Himalayas are the greatest source of traditional medicines and form the basis of Ayurveda. The native people from Himalayas are aware of their valuable healthy traditions, and the traditional heritage has been explored and practiced for years to provide proper medication to the inhabitants (Hamilton 2004; Karki et al. 2003). Thus, if proper measures are taken and if the traditional knowledge of the medicines can be investigated, a large number of employments can be produced within the rural areas. If the enterprise is incorporated, thousands of jobs can be produced to harness the traditional medicinal knowledge with economic opportunities. Thus, medicinal plants are helpful in contributing to the economic growth in the areas which have limited resources, educational options, lack of infrastructure, and underdeveloped commercial activities (Singh et al. 2016a, b, 2019a, b). Himalayas being rich in the biological and cultural diversity which is a result of the millions of years of evolution should be protected for their indigenous cultural and medicinal values (Ramakrishnan 1992).

19.5.3 The Environmental Point of View

The production of good-quality certified drugs of herbal origin has become a necessity these days as population is most concern about using the nonchemical-based products. Medicinal plants appear to be the most eco-friendly alternative in terms of production of the drug and providing health products useful in households and can be industrialized (Balunas and Kinghorn 2005; Karki et al. 2003). The shrubs, trees, and grasses of medicinal origin are found abundantly in South Asia. The tropical and subtropical forests of the area hold important medicinal value and can be harvested to yield the environment-friendly drugs of botanical origin. By promoting the community-based conservations, these forests can be preserved for its entry in the world drug market. People are also greatly encouraged to lend their participation in conserving the forest ecosystem by developing the medicinal plant-based job opportunities.

19.5.4 Commercialization of Medicinal Plants

The commercialization of medicinal plants in South Asia holds high economic importance as it allows the local traders to sell their traditionally renowned products at higher prices and open a national or global market for the products that are newly identified for their medical importance. If the partnership develops in between the local traders and the private sectors owing to produce the drugs in the industries, both the partners get equally benefitted. By growing the desirable crops, with their associated species and the intercrops alongside, the cultivators ensure to complement the medically important trees and conserve the forests simultaneously. Many medicinal plant roots help in stabilizing the soil of their areas and inhibit the soil erosion from hilly slopes. Medicinal plants have the quality to grow in adverse environmental conditions such as in poor soil fertility and low rainfall and hence are naturally regenerated. The mix plantation is allowed on enormous land areas including the shifting fallow land. Different configurations of crop geometry can be adapted to grow the plants of medicinal value as mostly the species are shade tolerant and others include trees, shrubs, herbs, and climbers (Rao et al. 2004; Chapman and Chomchalow 2003; Karki et al. 2003).

Nowadays, the demand for medicinal plants is increased by major herbal drug industries as an essential raw material. Thus, there is a high opportunity to create jobs for the unemployed people as the collection; processing transportation of these plant products requires high labor input. Thus, cash earnings can be increased for local people by enhancement of traditional processing by the industries (Karki et al. 2003)

In certain states of South Asia, medicinal plants can be cultivated simultaneously with traditional farming systems, i.e., through mixed farming. This helps in the growth of medicinal plants in the existing cropping system enabling them to grow in different eco-physical conditions when the selected species are cultivated with mixed or companion crops. The improvement of the soil quality by crop rotation enhances the livelihood in South Asia (Schippmann et al. 2006; Meena et al. 2009) (Table 19.2).

19.6 Conservation of Medicinal Plants

Medicinal plants are generally collected from the forests in large amount which ranges in between 70% and 99% in most of the countries (Parrotta and Agnoletti 2007). This has led to depletion in the number of plants as the demand for herbal drugs has become high. As a result, most of the medicinal plant species face the threat of extinction. According to the International Union for Conservation of Nature and Natural Resources (IUCN), about 300 medicinal plant species are at the threat of extinction (Gurib-Fakim 2006).

There are numerous reasons to conserve the medicinal plants:

- The harvesting practices for collection of medicinal plants from the wild have become rather destructive. Hence important measures have to be taken to save the species by nondestructive, high-quality collection methods.
- A steady decline in the use of medicinal plants by local communities has been observed which is directly related to their loss of interest in the plant conservation.

Table 19.2 Recent patents on medicinal and aromatic plants (Keswani et al. 2017)

Patent number	Title	References
USPP12030P2	Hybrid mint plant named "Neerkalka"	Kumar et al. (2001)
USPP12426P2	Mint plant named "Kosi"	Kumar et al. (2002)
USPP13336P2	High-yielding and stable plant of *Cymbopogon flexuosus* called "Chirharit"	Patra et al. (2002a)
USPP12791P2	Novel, high-yielding stable *Mentha arvensis* plant named "Damroo"	Patra et al. (2002b)
USPP13279P2	Mint plant named "Saksham"	Khanuja et al. (2002)
US PP13110 P2	*Lippia alba* plant named "Bhurakshak"	Kumar et al. (2002)
US6534696B1	"Rakshit," a poppy plant	Dhawan et al. (2003)
US6548746B1	"Dhawal," a high alkaloid-producing periwinkle plant	Kulkarni et al. (2003)
USPP14090P3	Peppermint plant named "Pranjal"	Dwivedi et al. (2003)
US6831214B2	"Vaishnavi," a high-yielding self -pollinated *Cymbopogon martinii*	Patra et al. (2004)
USPP14538P2	Mint plant named "Sambhav"	Khanuja et al. (2004)
USPP16566P3	Mint plant "Kushal" for late transplanting	Khanuja et al. (2006b)
US20050150027P1	Mint plant named "CIM-Indus"	Khanuja et al. (2006a)
USPP16712P3	Citral-rich high-yielding lemongrass plant "Nima" of *Cymbopogon flexuosus*	Lal et al. (2006)
US20090191292	High essential oil- and eugenol-yielding cultivar of *Ocimum sanctum* "CIM-Ayu"	Lal et al. (2007a)
USPP17505P3	*Plantago ovata* plant named "Mayuri"	Lal et al. (2007b)
US7442854B2	High-yielding multiple disease-resistant/stable variety "Madakini" of opium poppy	Shukla et al. (2008)
US7435877B2	Distinct type cultivar of *Ocimum basilicum* "CIM-Saumya"	Khanuja et al. (2008a)
US7375260B2	High artemisinin-yielding artemisia plant named "CIM-Arogya"	Khanuja et al. (2008b)
US20050050593A1	High herb-, phyllanthin-, and hypophyllanthin-yielding cultivar of *Phyllanthus amarus* "CIM-Jeevan"	Gupta et al. (2008)

(continued)

Table 19.2 (continued)

Patent number	Title	References
US20150056255 A1	Product comprising a plant for medicinal, cosmetic, coloring, or dermatologic use	Ragot et al.(2014)
US PP24545 P3	*Heuchera* plant named "Ginger Snap"	Egger and Terra Nova Nurseries, Inc (2014)
US 20140259228 A1	*Cannabis* plant named "Avidekel"	Cohen (2014)
US20140245494 A1	*Cannabis* plant named "Erez"	Cohen (2015)
USPP26474 P3	Autotetraploid *Vetiveria zizanioides* plant useful for carbon sequestration and soil conservation named "CIMAPKH40"	Lavania et al. (2016)

The rural people should be encouraged and should be provided job opportunities in the conservation field so that the native species can be consereved.

- As the demand for medicinal plants is high for their commercial sale and the international demand for the products is high, hence actions should be taken to conserve the species and export them in limited amount.
- The collectors are often offered with minimal prices, and the difference from the market price is in between 50% and 255% in India, while in Mexico the collectors are only offered with about 6% of final market price (Duke 1990; Rao et al. 2004).

19.7 Contribution of Medicinal Plants to the Livelihood

Medicinal plants in a particular area play a key role in maintaining the health of the native communities. Also, they are subjected to produce income sources for the "localities." Thus, they provide the livelihood for a large number of people living in that area. Since large proportions of plants are being collected from the forests which is non-sustainable, the pressure on the resources has become high. Thus, cultivation of demanded species of medicinal plants can be subjected to provide the alternative livelihood source to the farmers of that area. Countries including India have made some policies to enhance the promotion of medicinal plants (Gurib-Fakim 2006). Almost 1.5 million traditional medicine practitioners prescribe herbal medicines for curing different diseases in India (Vaidya and Devasagayam 2007; Verma and Singh 2008).

Though only fewer number of medicinal plant species are being cultivated on larger scale, the cultivation in many cases includes those plants too that do not have high global demand (Balunas and Kinghorn 2005; Gurib-Fakim 2006). China being the major cultivator of medicinal plants only cultivates about 100–250 species at the larger scale, and 80% of the medicinal plants are obtained from the wild habitats (Schippmann et al. 2003), while Europe has been seen to cultivate only 130–140 species of medicinal plants out of 1200–1300 species of inhabitant native plants (Wurtele et al. 2012).

While cultivation of medicinal plants holds greater potential, it is also limited by different issues. Because farmers are often offered with lower prices, it becomes difficult for them to take the initiative in the field of medicinal plant cultivation, whereas underdeveloped cultivation technology, scarcity of planting material, longer growth periods, and lack of industries in the area are the other different reasons for inability of doing cultivation practices. Thus, by removing these barriers, cultivation practice to increase the number of medicinal plants can be seen as new approach to earn livelihood in the rural areas of low-income Asian countries. Farmers are generally challenged with the difficulties of cultivation as they have no experience in it. Therefore, there is a powerful need to develop the technologies which can aid the farmers in cultivation, harvesting, and storage of medicinal plant products. Many research institutes are also researching on the medicinal properties of these plants, but to farmers it has not been much helpful which is because of the following reasons:

1. As research is being done on larger number of plant species, the resources provided for the research are not evenly distributed. Thus the research done in this area is not much successful. So as to use the resources in an efficient way, the research has to be constricted on lesser number of species, and this can be achieved by omitting the lack of coordination in between the institutes. Moreover, when the coordination will be done by one agency with close collaboration with other institutes, the research would be done with greater effort.
2. Researchers are focused on the development of the cultivation technology. The problems that occur here are with the packaging, storage, and transport of the products with retention of its quality. Measures should be taken thus to overcome these problems in order to achieve high quality of products.
3. The association between the research institutes and medicinal plant farmers are not as strong as most of the research is being done in laboratories. Research involving farms is negligible, so the farmer's contribution is insignificant in the research. Due to this, the problems faced by the farmers on land are not properly addressed, and researchers are unable to benefit by the experience of the farmers. Thus, the gap between the researchers and farmers has to be fulfilled, and farmers should be trained by experienced researchers to enhance the cultivation technology of medicinal plants.
4. There is also a lack of collaborations in between the research institutes and industries. The industries are quite uncertain of collaborating with the research institutes as they are unsure about the commercial return.
5. Researchers have also suggested about the shortage of planting materials to be cultivated in larger areas. Farmers have reported that the material provided by the governmental agencies is of bad quality. Thus, the involvement of private sectors is being proposed for better quality production and marketing of planting materials.

6. Moreover, researchers are often focused on the development of cultivation technology for species that are facing threat. It is suggested to promote the cultivation of economically important species by the farmers supported by the research programs (Lubbe and Verpoorte 2011).

19.8 Contribution of Medicinal Plants of the Himalayan States of India

More than 8000 species of medicinal plants are reported to treat different health problems in India (Das et al. 2017).The regions of Himalayas are specially gifted with large number of medicinal plant species (Kala 2000). The diverse agroclimatic conditions ranging from rainforests found in the northeast to the dry deciduous and alpine meadows found in the northwest are responsible for the diversity. Though only 15% of the country's geographical region is occupied by the Himalayas, then there are about 30% widespread medicinal species found in India which are predominantly rich in the medicinal value (Dahanukar et al. 2000; Kala 2005a).

The states of India that are in vicinity of the Himalayas are rich in large variety of medicinal plants. For example, Himachal Pradesh is found to be rich in about 500 medicinal plants out of about 3000 plant species, while Uttaranchal, Meghalaya, and Arunachal Pradesh are also gifted with greater number of medicinal plants. Medicinal plants have contributed to the rural income particularly in the region of Himalayas. The Great Himalayan National Park in the Kullu Valley has set an outstanding example for this as around 11,000 people are habitant of 5 km-wide belt around the park. Because of the limitation of other employment in such area, the medicinal plant collection serves as an important source for the native people. In 1997, the medicinal plant in combination with guchhi has provided Rs. 10,000 per family to the villagers around the park (Khare 2008).

Also, most healers of the area use these medicinal plants to resolve the health issues and depend on their traditional knowledge. Bhotia people living at high altitude of the central Himalayas use about 150 different medicinal plants to treat different number of diseases (Maikhuri et al. 1998). Although India is quite diverse, there is still high pressure on prevailing resources. Thus, conserving the medicinal plants has become a necessity as they on one hand provide health benefits to the community, while on the other hand they are used to fulfill the demand of newer and safer drugs of natural origin throughout the world.

19.9 Other Important Recommendations

1. Strong policies should be made and implemented for the collection of medicinal plant species which should be transparent. Improvement of the performance by the government agencies such as forest department is strictly required as they are involved in the collection and transportation of medicinal plant species.

The unnecessary and illegal collection of plant species can only be prohibited when the agencies will perform their duties honestly.

2. For the conservation of the traditional knowledge, popularization of medicinal plants is necessary. State government should take measures to aware the population about the importance of these plants of medicinal value. The resources should also be increased to improve the healthcare-based systems. This will provide the motivation to the growth of the relevant sector.
3. The effective participation off the native rural people should be promoted.
4. The farmers involve in the cultivation of medicinal plants face many difficulties. Moreover, the amount of resources and the production of planting material should be increased.
5. The collection and cultivation points should be provided with processing facilities so that earning of the farmers and cultivators are increased, and the quality of the product is enhanced.
6. Farmers should be offered with good marketing facilities. The innovative marketing techniques should be implemented so that difficulties faced by the farmer is reduced and their income is improved.
7. Insurance scheme should be introduced to reduce the risk faced by the farmers.
8. The species selected to be cultivated should be based on the availability of the technology and its economic importance. It should not be based only on the criteria of conservation.
9. The information regarding the prices and the demand for various species is not reliable which should be corrected.
10. The threat of illegal patents is often faced by medicinal plant conservation and local communities. The local farmers must be aware by the respective government agencies.
11. As civil society plays a key role in establishing the medicinal plant sector, policies should be introduced that can facilitate its role.
12. Finally, the conservation and cultivation of these plants should be promoted, and the central government should provide technical, and final support for this to the respective state government of medicinal plants is a national issue (Cunningham 1998; Firenzuoli and Gori 2007; World Health Organization 2007).

19.10 Conclusion

Medicinal plants are valuable for different reasons. For the basic healthcare needs the rural population of Asian countries require high quality raw material for the production of drugs from medicinal plants. The large number of people in the rural areas can also be provided with income opportunities by incorporating them in the processing and collection of medicinal plants. Since enormous varieties of the plants species are found in the Himalayan region, thus they play evident role providing the livelihood to the native communities. The demand for the medicinal plants has grown faster in the recent years, and it has led to their unlimited collection, thereby extreme pressure on the wild resources has been observed. Due to this, many of the species are at the threat of extinction. Measures to limit the excessive exploitation of these medicinal plant species should be thus inherited, and the cultivation of economically important plant must be proposed. Although, the government is taking newer initiatives, but the implementation of the policies remains to be poor. Overexploitation can only be stopped when these policies will be implemented and hence significant improvement of the product from cultivation to marketing will be viewed.

References

Ali-Shtayeh MS, Yaniv Z, Mahajna J (2000) Ethnobotanical survey in the Palestinian area: a classification of the healing potential of medicinal plants. J Ethnopharmacol 73(1–2):221–232

Al-Quran S (2008) Taxonomical and pharmacological survey of therapeutic plants in Jordan. J Nat Prod 1(1):10–26

Balasankar D, Vanilarasu K, Preetha PS, Rajeswari S, Umadevi M, Bhowmik D (2013) *Senna*–A medical miracle plant. J Med Plants Stud. 2013 1(3):41–47

Balunas MJ, Kinghorn AD (2005) Drug discovery from medicinal plants. Life Sci 78(5):431–441

Birla H, Rai SN, Singh SS, Zahra W, Rawat A, Tiwari N, Singh RK, Pathak A, Singh SP (2019) *Tinospora cordifolia* suppresses neuroinflammation in Parkinsonian Mouse Model. NeuroMolecular Med 14:1–2

Bukar BB, Dayom DW, Uguru MO (2016) The growing economic importance of medicinal plants and the need for developing countries to harness from it: A mini review. IOSR J Pharm 6(5):42–42

Cameron A, Ewen M, Ross-Degnan D, Ball D, Laing R (2009) Medicine prices, availability, and affordability in 36 developing and middle-income countries: a secondary analysis. Lancet 373(9659):240–249

Chapman K, Chomchalow N (2003) Production of medicinal plants in Asia. In: III WOCMAP congress on medicinal and aromatic plants-volume 5: quality, efficacy, safety, processing and trade in medicinal, 679, pp 45–59

Cohen Y (2014) *Cannabis* plant named 'avidekel'. U.S. Patent Application 14/193,252

Cohen Y (2015) *Cannabis* plant named 'Erez'. U.S. Patent Application 14/757,040

Cunningham AB (1998) Medicinal plants for forest conservation and health care Author: Food and Agriculture Organization of the United Nations. ISBN: 925104063X

Dahanukar SA, Kulkarni RA, Rege NN (2000) Pharmacology of medicinal plants and natural products. Indian J Pharmacol 32(4):S81–S118

Dahlberg AC, Trygger SB (2009) Indigenous medicine and primary health care: the importance of lay knowledge and use of medicinal plants in rural South Africa. Hum Ecol 37(1):79–94

Das A, Chaudhuri D, Sarkar R, Ghate NB, Panja S, Mandal N (2017) Plants of Indian traditional medicine with antioxidant activity. In: Nutritional antioxidant therapies: treatments and perspectives. Springer, Cham, pp 27–64

Dastagir G, Rizvi MA (2016) Glycyrrhiza glabra L. (Liquorice). Pak J Pharm Sci 29(5):1727

Dhawan OP, Shahabuddin S, Trivedi M, Sattar A, Alam M, SamadA, Zaim M, Dwivedi S, Singh SP, Singh HP, Khanuja SPS (2003). Method of producing a poppy plant. U.S. Patent 6,534,696

Dubey NK, Kumar R, Tripathi P (2004) Global promotion of herbal medicine: India's opportunity. Curr Sci 86(1):37–41

Duke JA (1990) Promising phytomedicinals. In: Advances in new crops. Proceedings of the first national symposium 'new crops: research, development, economics', Indianapolis, Indiana, USA, 23–26 October 1988, Timber Press, pp 491–498

Dwivedi S, Singh M, Singh AP, Singh V, Khanuja SPS, Naqvi AA, Kumar S, Council of Scientific & Industrial Research (2003) Peppermint plant named 'Pranjal'. U.S. Patent PP14,090

Egger JN, Terra Nova Nurseries, Inc. (2014) *Heuchera* plant named 'Ginger Snap'. U.S. Patent PP24,545

Farnsworth NR, Soejarto DD (1991) Global importance of medicinal plants. Conserv Med Plants 26:25–51

Farnsworth NR, Akerele O, Bingel AS, Soejarto DD, Guo Z (1985) Medicinal plants in therapy. Bull World Health Organ 63(6):965

Firenzuoli F, Gori L (2007) Herbal medicine today: clinical and research issues. Evid Based Complement Alternat Med 4(S1):37–40

Gawade BV, Fegade SA (2012) Rauwolfia (reserpine) as a potential antihypertensive agent: A review. Int J Pharm Phytopharmacol Res 2(1):46–49

Ghimire SK, McKey D, Aumeeruddy-Thomas Y (2004) Heterogeneity in ethnoecological knowledge and management of medicinal plants in the Himalayas of Nepal: implications for conservation. Ecol Soc 9(3)

Ghorbani A (2005) Studies on pharmaceutical ethnobotany in the region of Turkmen Sahra, north of Iran:(Part 1): general results. J Ethnopharmacol 102(1):58–68

Ghosh JS, Patil PJ (2010) Antimicrobial activity of *Catharanthus roseus*–a detailed study. Br J Pharmacol Toxicol 1(1):40–44

Gupta AK, Khanuja SPS, Gupta MM, Shasany AK, Jain N, VermaRK, Darokar MP, Bagchi GD, Kumar S, Council Of Scientific and Industrial Research (2008) High herb, phyllanthin and hypophyllanthin yielding cultivar of *Phyllanthus amarus* 'CIM-Jeevan'. U.S. Patent 7,446,243

Gurib-Fakim A (2006) Medicinal plants: traditions of yesterday and drugs of tomorrow. Mol Asp Med 27(1):1–93

Hamilton AC (2004) Medicinal plants, conservation and livelihoods. Biodivers Conserv 13(8):1477–1517

Hoareau L, DaSilva EJ (1999) Medicinal plants: a re-emerging health aid. Electron J Biotechnol 2(2):3–4

Kala CP (2000) Status and conservation of rare and endangered medicinal plants in the Indian trans-Himalaya. Biol Conserv 93(3):371–379

Kala CP (2005a) Ethnomedicinal botany of the Apatani in the Eastern Himalayan region of India. J Ethnobiol Ethnomed 1:11

Kala CP (2005b) Indigenous uses, population density, and conservation of threatened medicinal plants in protected areas of the Indian Himalayas. Conserv Biol 19(2):368–378

Kaplan W, Mathers C (2011) The world medicines situation 2011. Global health trends: global burden of disease and pharmaceutical needs. World Health Organization, Geneva

Karki M, Tiwari BK, Badoni A, Bhattarai N (2003) Creating livelihoods and enhancing biodiversity-rich production systems based on medicinal and aromatic plants: preliminary lessons from South Asia. In: III WOCMAP congress on medicinal and aromatic plants-volume 4: targeted screening of medicinal and aromatic plants, economics, 678, pp 37–43

Keswani C, Bisen K, Singh SP, Singh HB (2017) Traditional knowledge and medicinal plants of India in intellectual property landscape. Med Plants-Int J Phytomeds Relat Ind 9(1):1–1

Khan BA, Abdukadir A, Qureshi R, Mustafa GH (2011) Medicinal uses of plants by the inhabitants of Khunjerab National Park, Gilgit, Pakistan. Pak J Bot 43(5):2301–2310

Khanuja SPS, Kumar S, Shasany AK, Dhawan S, Darokar MP,Naqvi AA, Dhawan OP, Singh AK, Patra NK, Bahl JR Bansal RP (2002) Mint plant named 'Saksham'. U.S. Patent PP13,279

Khanuja SPS, Shasany AK, Dhawan S, Darokar MP, Satapathy S,Kumar TRS, Saikia D, Patra NK, Bahl JR, Tripathy AK, Kumar S (2004) Mint plant named 'Sambhav'. U.S. Patent PP14,538

Khanuja SPS, Patra NK, Shasany AK, Kumar B, Gupta S, Upadhyay RK, Priya TP, Singh AK, Darokar MP, Tomar VKS Bahal JR (2006a) Mint plant named 'Cim Indus'. U.S. Patent PP16,474

Khanuja SPS, Shasany AK, Yadav U, Dhawan S, Darokar MP, Bahl JR, Gupta S, Pandey S, Singh AK, Bansal RP, Lal RK(2006b) Mint plant 'Kushal' for late transplanting. U.S. Patent PP16,566

Khanuja SPS, Lal RK, Agnihotri AK, Shasany AK, Naqvi AA,Dwivedi S, Misra HO, Dhawan, OP, Kalra A, Singh A, Bahl JR (2008a) Distinct type cultivar of *Ocimum basilicum* "CIM-SAUMYA". U.S. Patent 7,435,877

Khanuja SPS, Paul S, Shasany AK, Gupta AK, Darokar MP, Gupta MM, Verma RK, Ram G, Kumar A, Lal RK, Bansal RP(2008b) High artemisinin yielding Artemisia plant named 'CIM-Arogya'. U.S. Patent 7,375,260

Khare CP (2008) Indian medicinal plants: an illustrated dictionary. Springer Science & Business Media, Berlin

Khushbu C, Roshni S, Anar P, Carol M, Mayuree P (2011) Phytochemical and therapeutic potential of *Piper longum Linn* a review. Int J Res Ayurveda Pharm 2(1):157–161

Kuipers SE (1997) Trade in medicinal plants. Medicinal plants for forest conservation and health care (11)

Kulkarni RNR, Baskaran K, Chandrashekara RSR, Khanuja SPS, Darokar MP, Shasany AK, Uniyal GC, Gupta MM, Kumar S, Council of Scientific and Industrial Research, (2003) 'Dhawal', a high alkaloid producing periwinkle plant. U.S. Patent 6,548,746

Kumar S, Patra NK, Khanuja SPS, Shasany AK, Kalra A, Singh HB, Singh HP, Singh VR, Mengi N, Tanveer H, Naqvi AA(2001) Hybrid mint plant named 'Neerkalka'. U.S. Patent PP12,03

Kumar S, Patra NK, Singh HP, Kalra A, Singh HB, Ram P, Singh VR, Mengi N, Singh VP, Ram M, Shukla RS (2002) Mintplant named 'Kosi'. U.S. Patent PP12,426

Kuniyal CP, Bisht VK, Negi JS, Bhatt VP, Bisht DS, Butola JS, Sundriyal RC, Singh SK (2015) Progress and prospect in the integrated development of medicinal and aromatic plants (MAPs) sector in Uttarakhand, Western Himalaya. Environ Dev Sustain 17(5):1141–1162

Lal RK, Misra HO, Sharma JR, Singh N, Shasany AK, Naqvi AA, Bahl JR, Prasad A, Khanuja SPP, Council of Scientific & Industrial Research (2006) Citral rich high yielding Lemongrass plant 'Nima'of *Cymbopogon flexuosus*. U.S. Patent PP16,712

Lal R, Khanuja S, Agnihotri A, Misra H, Shasany A, Naqvi A, Dhawan O, Kalra A, Bahl J, Darokar M, Council of Scientific (2007a) High essential oil and eugenol yielding cultivar of ocimum sanctum'CIM-AYU'. U.S. Patent Application 11/825,452

Lal RK, Singh N, Misra HO, Sharma JR, Bahl JR, Shasany AK, Khanuja SPS, Council of Scientific Industrial Research (2007b) *Plantago ovata* plant named 'Mayuri'. U.S. Patent PP17,505

Lambert J, Srivastava JP, Vietmeyer N (1997) Medicinal plants: rescuing a global heritage. The World Bank, Washington, DC

Lavania UC, Rai SK, Lavania S, Basu S, Dubey BK, Ujagir R, Council of Scientific and Industrial Research(2016) Autotetraploid *Vetiveria zizanioides* plant useful for carbon sequestration and soil conservation named 'CIMAPKH 40'. U.S. Patent PP26,474

Lev E, Amar Z (2002) Ethnopharmacological survey of traditional drugs sold in the Kingdom of Jordan. J Ethnopharmacol 82(2–3):131–145

Lim TK (2016) *Dioscorea alata*. In: Edible medicinal and non medicinal plants. Springer, Dordrecht, pp 218–234

Liu J (1995) Pharmacology of oleanolic acid and ursolic acid. J Ethnopharmacol 49(2):57–68

Lubbe A, Verpoorte R (2011) Cultivation of medicinal and aromatic plants for specialty industrial materials. Ind Crop Prod 34(1):785–801

Ma JK, Chikwamba R, Sparrow P, Fischer R, Mahoney R, Twyman RM (2005) Plant-derived pharmaceuticals–the road forward. Trends Plant Sci 10(12):580–585

Maikhuri RK, Nautiyal S, Rao KS, Saxena KG (1998) Medicinal plant cultivation and biosphere reserve management: a case study from the Nanda Devi Biosphere Reserve, Himalaya. Curr Sci 25:157–163

Mati E, de Boer H (2011) Ethnobotany and trade of medicinal plants in the Qaysari Market, Kurdish Autonomous Region, Iraq. J Ethnopharmacol 133(2):490–510

Meena AK, Bansal P, Kumar S (2009) Plants-herbal wealth as a potential source of ayurvedic drugs. Asian J Tradit Med 4(4):152–170

Messiaen CM, Rouamba A (2004) Allium cepa L. Plant Resour Trop Afr 2:44–52

Nirmal SA, Pal SC, Otimenyin SO, Thanda Aye, Mostafa Elachouri, Sukalyan Kumar Kundu et al. (2013) Contribution of herbal products in global market. The Pharma Rev [cited 2019 March 7]. Available from: https://www.researchgate.net/publication/320357308

Okoli RI, Aigbe O, Ohaju-Obodo JO, Mensah JK (2007) Medicinal herbs used for managing some common ailments among Esan people of Edo State, Nigeria. Pak J Nutr 6(5):490–496

Pandey MM, Rastogi S, Rawat AK (2013) Indian traditional ayurvedic system of medicine and nutritional supplementation. Evid-Based Complement Alternat Med 2013:1–12

Parrotta JA, Agnoletti M (2007) Traditional forest knowledge: challenges and opportunities. For Ecol Manag 249:1–4

Patra NK, Kumar S, Khanuja SPS, Shasney AK, Kalra A, Singh HB, Singh HP, Singh VR, Tanveer H, Mengi N, Rajput DK (2002a) High yielding and stable plant of *Cymbopogon flexuosus* called 'Chirharit'. U.S. Patent PP13,336

Patra NK, Kumar S, Khanuja SPS, Shasney AK, Kalra A, Singh HB, Singh HP, Singh VR, Tanveer H, Mengi N, Rajput DK (2002b) Novel, high yielding stable *Mentha arvensis* plant named 'Damroo'. U.S. Patent PP12,791

Patra NK, Kumar S, Kalra A, Singh HB, Singh HP, Singh VR,Tanveer H, Mengi N, Dhawan OP, Negi MS, Ram P(2004) Vaishnavi, a high yielding self-pollinated *Cymbopogon martinii*. U.S. Patent 6,831,214

Ragot P, Pons E, Mompon B, Rousseau C, Schweitzer-MauduitInternational, Inc. and SWM Luxembourg sarl (2014) Product comprising a plant for medicinal, cosmetic, coloring or dermatologic use. U.S. Patent Application 14/462,213

Rai SN, Birla H, Singh SS, Zahra W, Patil RR, Jadhav JP, Gedda MR, Singh SP (2017) *Mucuna pruriens* protects against MPTP intoxicated neuroinflammation in Parkinson's disease through NF-κB/pAKT signaling pathways. Front Aging Neurosci 9:421

Ramakrishnan PS (1992) Shifting agriculture and sustainable development: an interdisciplinary study from north-eastern India. Unesco

Rao MR, Palada MC, Becker BN (2004) Medicinal and aromatic plants in agroforestry systems. In: New vistas in agroforestry 2004. Springer, Dordrecht, pp 107–122

Reynolds T, Sofowora A (2007) Medicinal plants and traditional medicine in Africa. Kew Bull. https://doi.org/10.2307/4108615. ISSN: 00755974

Schippmann U, Leaman DJ, Cunningham AB, Walter S (2003) Impact of cultivation and collection on the conservation of medicinal plants: global trends and issues. In: III WOCMAP congress on medicinal and aromatic plants-volume 2: conservation, cultivation and sustainable use of medicinal and 676, pp 31–44

Schippmann UW, Leaman D, Cunningham AB (2006) A comparison of cultivation and wild collection of medicinal and aromatic plants under sustainability aspects. Frontis 1:75–95

Sher H, Hussain F (2009) Ethnobotanical evaluation of some plant resources in Northern part of Pakistan. Afr J Biotechnol 8(17):4066

Sher H, Aldosari A, Ali A, de Boer HJ (2014) Economic benefits of high value medicinal plants to Pakistani communities: an analysis of current practice and potential. J Ethnobiol Ethnomed 10(1):71

Shukla S, Singh SP, Singh HB, Pushpangadan P, Council Of Scientific & Industrial Research (2008) High yielding multiple disease resistant/tolerant stable variety 'Madakini' of opium poppy. U.S. Patent 7,442,854

Singh HB, Jha A, Keswani C (2016a) Intellectual property issues in biotechnology. CABI, Wallingford, 304 pages. ISBN:13: 9781780646534

Singh HB, Jha A, Keswani C (2016b) Biotechnology in agriculture, medicine and industry: an overview. In: Singh HB, Jha A, Keswani C (eds) Intellectual property issues in biotechnology. CABI, Wallingford, pp 1–4

Singh SS, Rai SN, Birla H, Zahra W, Kumar G, Gedda MR, Tiwari N, Patnaik R, Singh RK, Singh SP (2018) Effect of chlorogenic acid supplementation in MPTP-intoxicated mouse. Front Pharmacol 9:757

Singh HB, Keswani C, Singh SP (2019a) Intellectual property issues in microbiology. Springer-Nature, Singapore, 425 pages. ISBN:9789811374654

Singh HB, Keswani C, Reddy MS, Royano ES, García-Estrada C (2019b) Secondary metabolites of plant growth promoting rhizomicroorganisms: discovery and applications. Springer-Nature, Singapore, 392 pages. ISBN:978-981-13-5861-6

Sofowora A, Ogunbodede E, Onayade A (2013) The role and place of medicinal plants in the strategies for disease prevention. Afr J Tradit Complement Altern Med 10(5):210–229

Umadevi M, Rajeswari R, Rahale CS, Selvavenkadesh S, Pushpa R, Kumar KS, Bhowmik D (2012) Traditional and medicinal uses of *Withania somnifera*. Pharma Innov 1(9, Part A):102

Vaidya AD, Devasagayam TP (2007) Recent advances in Indian herbal drug research guest editor: thomas paul asir devasagayam current status of herbal drugs in India: an overview. J Clin Biochem Nutr 41(1):1–1

Verma S, Singh SP (2008) Current and future status of herbal medicines. Vet World 1(11):347

Wakdikar S (2004) Global health care challenge: Indian experiences and new prescriptions. Electron J Biotechnol 7(3):02–03

World Health Organization (2007) WHO monographs on selected medicinal plants, vol 3. ISSN: 11667699

Wurtele E, Chappell J, Jones A, Celiz M, Ransom N, Hur M, Rizshsky L, Crispin M, Dixon P, Liu J, P Widrlechner M (2012) Medicinal plants: a public resource for metabolomics and hypothesis development. Meta 2(4):1031–1059

Chemotherapeutic Drugs and Gallbladder Cancer: Market Potential in India

20

Ruhi Dixit, Manoj Pandey, Punam Pandey, and V. K. Shukla

Abstract

Gallbladder cancer (GBC) is the most fatal cancer of the biliary tract with poor prognosis. Silent in its infancy, this malignancy remains asymptomatic until aggressive disease has progressed to an advanced and non-curative stage. The overall mean survival rate for patients with GBC is less than 6 months, with a 5-year survival rate of approximately 5% with a high relapse rate. Treatment depends upon the stage of disease, patient's age, nutritional status, performance status, and cardiopulmonary, hepatic, and renal functions. Complete surgical resection is considered the most curative modality for GBC. Chemotherapy has recently shown its effect on gallbladder cancer. Therapeutic agents, targeting cellular and molecular pathways, can effectively impede tumor growth. Newer drugs are being developed that work which target specific parts of cancer cells or their surrounding environments like tumor blood vessels. This chapter discusses market of chemotherapeutic drugs in India and market drives and government initiatives for the promotion of pharmaceutical sector in India with special emphasis on GBC.

Keywords

Gallbladder cancer · Therapeutics · Market drivers · Family health · Bioeconomy

R. Dixit (✉) · P. Pandey · V. K. Shukla
Department of General Surgery, Institute of Medical Sciences, Banaras Hindu University, Varanasi, Uttar Pradesh, India

M. Pandey
Department of Surgical Oncology, Institute of Medical Sciences, Banaras Hindu University, Varanasi, Uttar Pradesh, India

C. Keswani (ed.), *Bioeconomy for Sustainable Development*,
https://doi.org/10.1007/978-981-13-9431-7_20

20.1 Introduction

Gallbladder cancer (GBC) was first described in 1777 by Maximilian de Stoll (DeStoll 1788). GBC is the most common malignancy of the biliary tract, representing 80–95% of biliary tract cancers worldwide (Shukla et al. 1985; Rakic et al. 2014). GBC has marked geographical variation both globally and within the Indian subcontinent (Shukla et al. 1985; Dixit et al. 2012; Dixit and Shukla 2014; National Cancer Registry Programme 2013; Mhatre et al. 2016).

GBC has an abysmal prognosis. Silent in its infancy, this malignancy remains asymptomatic until aggressive disease has progressed to an advanced and non-curative stage. (Hundal and Shaffer 2014). In fact, early gallbladder cancer is discovered often when the gallbladder is removed as a treatment for gallstones. Otherwise, gallbladder cancer is often at its advanced stage at the time when it is diagnosed. The overall mean survival rate for patients with GBC is less than 6 months, with a 5-year survival rate of approximately 5% with a high relapse rate.

20.2 Treatment

Treatment depends upon the stage of disease (T status), patient's age, nutritional status, performance status, and cardiopulmonary, hepatic, and renal functions. Complete surgical resection is considered the most curative modality for GBC. But, most of the cases are diagnosed in advanced or metastatic stage. Only 10–15% of GBCs are amenable to surgery at initial presentation (Dixit 2015; Zhao and Lim 2017). Chemotherapy has recently shown its effect on gallbladder cancer (Qin et al. 2008; Dixit et al. 2012; Dwivedi et al. 2015). Only 10–30% of patients can be considered for surgery on presentation (Misra et al. 2003; Dixit et al. 2016).

GBC patients typically present late in the course of their disease and often are not candidates for curative surgical resection. Patients not fit for such major resection or found unresectable on imaging or exploration are usually offered palliative treatment. This may be in the form of surgical palliation (e.g., palliative bypass for gastric outlet, bowel, or biliary tract obstruction), endoscopic biliary stenting (for obstructive jaundice), or palliative chemotherapy (Misra et al. 2002; Todoroki 2000). Chemotherapy has recently shown its effect on gallbladder cancer. Chemotherapy protocols for GBC can be divided broadly into (1) 5-fluorouracil (5-FU)-based, (2) gemcitabine-based, (3) capecitabine-based, and (4) oxaliplatin, single-agent, or combination regimens. Either in combination or as a single agent, 5-FU has been used in the management of GBC for almost 30 years (Misra et al. 2003). Response rates range from 20% to 40%, with a median overall survival of 8–14 months (Xiu et al. 2010). 5-Fluorouracil, when used as a single agent, has been shown to have only a 20% response rate; gemcitabine had a limited response rate of 36% (Dutta 2012). Combination chemotherapy using gemcitabine and cisplatin offers a significant survival advantage for patients with advanced disease. The median progression-free survival in one trial was longer at 8 months for patients treated with cisplatin and gemcitabine compared with 5 months for those receiving only gemcitabine (Valle et al. 2010).

Reports of radiotherapy for GBC are few, with a small number of patients and conflicting results. A meta-analysis of reports concerning the role of radiotherapy in treatment of GBC from 1974 to 2000 reported a slight improvement in survival after adjuvant or palliative radiation (Houry et al. 2001; Dixit et al. 2017; Pandey et al. 2017). Adjuvant chemoradiotherapy with 5-FU has shown some survival benefit in completely resected GBC (Kresl et al. 2002; Czito et al. 2005); however, randomized trials currently are lacking.

Therapeutic agents, targeting cellular and molecular pathways, can effectively impede tumor growth (Aaronson 1991; Pandey et al. 2018; Kumar et al. 2012). Newer drugs are being developed that work which target specific parts of cancer cells or their surrounding environments like tumor blood vessels. The drug sorafenib (Nexavar), which is already used for some liver cancers, works in part by hindering new blood vessel growth (angiogenesis), and it is now being studied in case of GBC. Bevacizumab (Avastin), another drug that targets blood vessel growth, is also being studied against gallbladder cancer.

Other new drugs have different targets. As epidermal growth factor receptor (EGFR) is expressed in 38.5% of GBCs, targeted therapy against EGFR provides some optimism for a changing treatment paradigm in the future (Pignochino et al. 2010). Drugs that target EGFR are erlotinib (Tarceva), cetuximab (Erbitux), and lapatinib (Tykerb) and are now being studied for use in people with gallbladder cancer, usually in combination with chemotherapy or other targeted drugs (American Cancer Society 2014, 2017).

20.3 Anticancer Drugs Market: Segmentation

Cancer is due to abnormal growth of the cells. Anticancer drugs are also known as antineoplastic drugs. These drugs prevent and inhibit the growth of cancer cells. Based on drug types, global anticancer drugs market is segmented into cytotoxic drugs, targeted drugs, and hormonal drugs. The targeted drugs segment is further segmented into monoclonal antibodies, tyrosine kinase inhibitors, and others. The segment is anticipated to hold a significant share of the global market. The targeted drugs have a significant demand and are expected to continue dominating the global market throughout the forecast period. Based on therapy type, global anticancer drugs market is segmented into chemotherapy, targeted therapy, immunotherapy, and others. Targeted therapy held the highest share of the market and is anticipated to dominate the market during the forecast period. Side effects of chemotherapy and specificity of the targeted drug are the factors driving the growth of global market (Global Industry Analysis 2018). Based on cancer types, the global anticancer drugs market is segmented into lung cancer, breast cancer, leukemia, colorectal cancer, gallbladder cancer, and others.

Traditionally, surgery and radiotherapy have been the primary treatment, with anticancer drugs largely being used in metastatic cancers. Although chemotherapy has been successfully used for inhibiting cell growth over the last couple of decades, side effects of chemotherapy have forced researchers to look for some alternative medications for all types of cancer. With medications and therapies, the challenge with drug developers has been to effectively administer the drugs at the disease site in human body. To overcome this challenge, earlier while a majority of drugs were administered through injectable medium, the focus of drug manufacturers has currently shifted toward developing effective needle-free delivery systems, as driven by their patient-centric healthcare approach (Frost and Sullivan 2017).

Strong emphasis of drug manufacturers on continuous improvements in cancer treatment has resulted in the development of novel drug delivery approaches, enabling targeted administration of drug compounds. This transformation has pushed the growth of oncology drug delivery market—which is expected to exhibit a compound annual growth of ~11% during 2016–2020. While this growth will benefit all types of drug delivery mediums, oral and inhalers are expected to be among the leading gainers (Frost and Sullivan 2017).

Despite a strong growth in number of oral drug prescriptions, several injectable companies have invested to develop advanced product pipelines. Over the next 5 years, although intravenous infusion drug delivery is expected to dominate the market, subcutaneous (SC), intradermal, and intramuscular (IM) modes of drug delivery are expected to gain a strong market hold, as 30–35 new products are likely to be launched from various companies (Frost and Sullivan 2017).

In addition to traditional oral and inhaler methods, research on some innovative delivery methods including nanogel and transdermal patches is also underway, and scientists are evaluating the bioavailability of drug compounds on target sites, when drug is delivered using these mediums. Clinical effectiveness of these mediums will largely decide the market success of these new methods. A range of breakthrough innovations have been going around like wearable injectors, nanoparticle-based oral drugs, arrayed microinjections, implantable devices, intranasal drug delivery, transdermal patches, etc. (Frost and Sullivan 2017).

20.4 Market of Chemotherapeutic Drugs in India

The Indian pharmaceutical industry is one of the developing world's largest and most developed, ranking 4th in the world in terms of production volume and 13th in domestic consumption value (National Pharmaceutical Policy 2006). The Indian pharmaceutical market, along with the markets of China, Brazil, and Russia, will spearhead growth within these markets.

The Indian pharmaceutical market has characteristics that make it unique. First, branded generics dominate, making up for 70–80% of the retail market. Second, local players have enjoyed a dominant position driven by formulation development capabilities and early investments. Third, price levels are low, driven by intense competition. While India ranks tenth globally in terms of value, it is ranked third in volumes. These characteristics present their own opportunities and challenges (McKinsey and Company Report 2009).

The Indian pharmaceutical industry has become an important part of global pharmaceutical R&D. Indian firms have already entered regulated markets moving recently from nonregulated to semi-regulated markets mainly in European and US markets (Chaturvedi et al. 2007). Various industry reports suggest that the industry has been growing at 13–14% over the last 5 years. The treatment of chronic diseases has gone up (McKinsey and Company Report 2009). The report also suggested specialty therapies to increase their share and reach a scale comparable to that of the mass therapies by 2020. Super-specialty therapies comprised of therapeutic areas such as oncology are considerably niche and derive less than 20% of their prescriptions from generalists. Though these therapies comprise a small part of the market, they have grown at nearly double the market growth rate over the last few years (McKinsey and Company Report 2009).

India's pharmaceutical industry market for drugs to treat cancer has outstripped that of all other leading countries in recent year and has evolved from almost non-existent to a world's leader in the production of high-quality, low-cost, non-branded, or generic drugs, accounting for nearly 20% of the world's production. India currently produces almost all its own drug needs, and domestic companies control over 80% of the Indian market. There are tens of thousands of companies producing pharmaceuticals in India because of low barriers to entry and low capital requirements. The vast majority of them are small by Western standards with revenues of less than USD 5million (Greene 2007). India is the biggest provider of generic medications internationally using all the Indian generics accounting for 20% of global exports concerning volume.

Current oncology market in India is found to be growing at 20% every year and will be increasing for the coming 3–5 years. The cancer market in 2012 is valued at USD 172 million. There are many treatments available for cancer like chemotherapy, biologics, targeted therapy, hormonal therapy, and supportive therapy, but among them in the year 2012, chemotherapy occupied the highest market value of USD 104.97 million. In India oncology market is forecasted to grow to a 3831 Cr with a compounded annual growth rate (CAGR) of 15.46% by 2017 (Frost and Sullivan 2017).

Generic substitution for frequently used chemotherapy drugs in the treatment of common cancers has an enormous potential to generate significant cost savings and increase access to cancer treatments in India and other low- and middle-income countries.

Based on moving annual turnover, gastrointestinals (11.5%) had one of the biggest market shares in the Indian pharma market in 2018 (Indian Pharmaceuticals Industry Report 2019). McKinsey and Company report analysis observed that the Indian pharmaceutical market will grow to USD 55 billion by 2020 driven by a steady increase in affordability and a step jump in market access. India will be at the top, a close second only to the US market in terms of volumes. This combination of value and volume provides interesting opportunities for upgrading therapy and treatment levels.

20.5 Market Drivers

India is a populous country with increasing cancer incidence which serves as the key driver for growth of cancer market in India. The major market drivers are increasing number of cancer cases in India, development of alternative cancer therapies, many available options for treatment of the diseases, and increase in foreign direct investment (FDI) for oncology treatments. Various advanced-stage GBC cases and huge potential for clinical research data and trained researchers with quality data offer a unique opportunity for conducting research and clinical trials in different parts of the country (Shukla et al. 2012).

Rise in awareness in people who take treatment not considering the disease as "terminal illness" but as "treatable illness" makes the patients to opt for treatment rather than ignoring it. Rise in technology and advanced treatments with better outcomes is a key market trend. The government has been taking many initiatives to reduce the prices of the drugs, and negotiations are underway with pharma companies which make the treatment affordable and accessible to the patients. Many cancer centers are set up to treat patients with many facilities under one roof which makes the treatment feasible. Rise in particular number of gallbladder cancer cases provides the chance for pharma companies to invest in their R&D.

20.6 Market Challenges

In India, restriction on the use of new therapies in oncology and pricing structure of expensive cancer drugs are the most important challenges facing the market. Other restraints of the cancer market are unclear patent law, restrictive pricing and reimbursement policies, low patient affordability, ailing infrastructure, and insufficient government aid. Further growth of cancer market can occur if the reimbursement and insurance policies are solved. Health insurance is another factor which hampers the growth of cancer market questioning their affordability to patients. Only 15% of patients are covered under health insurance, and majority are underinsured. Average bill of cancer patient is Rs 10lakh, while the average amount of policy of health plan sold is 1.9 lakhs.

20.7 Critical Success Factors

The critical success factors of pharmaceutical market in India are as follows:

1. Establishment of National Cancer for Research and Development in Bulk Drug (NCRDBD) at National Institute of Pharmaceutical Education and Research (NIPER), Hyderabad.

2. Indian government has implemented financial assistance programs like the Credit Linked Capital Subsidy Scheme (CLCSS) for upgrading the technology of SME pharma companies to enable them to comply with GMP standards with the revised Schedule M norms under the Drugs and Cosmetics Act.
3. Big pharma companies are investing in India as the cost of R&D, and production is comparatively low in India. Qualified population base is providing technical manpower to the industry.
4. Around 300,000 postgraduates and 1500 PhDs are qualifying in biosciences in India every year.
5. Better regulatory awareness and technical skills have enabled Indian companies to penetrate the high value regulated market such as the EU and the USA, over the years.

20.8 Government Initiatives

Some of the initiatives taken by the government to promote the pharmaceutical sector in India are as follows:

- The allocation to the Ministry of Health and Family Welfare has increased by 13.1% to Rs. 61,398 crore (~USD 8.98 billion) in Union Budget 2019–2020.
- The Uttar Pradesh government announced in October 2018 that it will set up six pharma parks in the state and has received investment commitments of more than Rs. 5000–6000 crore (~USD 712–855 million) for the same.
- The National Health Protection Scheme is the largest government-funded healthcare program in the world, which is expected to benefit 100 million poor families in the country by providing a cover of up to Rs. 5 lakh (~USD 7723.2) per family per year for secondary and tertiary care hospitalization. The program was announced in Union Budget 2018–2019.
- In March 2018, the Drug Controller General of India (DCGI) announced its plans to start a single-window facility to provide consents, approvals, and other information. The move is aimed at giving a push to the Make in India initiative.
- The government of India is planning to set up an electronic platform to regulate online pharmacies under a new policy, in order to stop any misuse due to easy availability.
- The government of India unveiled "Pharma Vision 2020" aimed at making India a global leader in end-to-end drug manufacture. Approval time for new facilities has been reduced to boost investments.
- The government introduced mechanisms such as the Drug Price Control Order and the National Pharmaceutical Pricing Authority to deal with the issue of affordability and availability of medicines.

20.9 Future Vision

Cancer market captures a lion's share in therapeutic segment of India. The market is predicted to rise in the coming years, but due to certain forces like low affordability and unorganized healthcare services, the growth can't be at the rate predicted. The growing number of cancer patient in India is mainly due to poor strategy where physicians concentrate on treatment rather than prevention and screening. A large group of patients remain untapped and do not receive medical treatment, due to insufficient number of oncologist and drugs. Public health activities are given less importance. There is a lack of proper health laws and national health policy which mainly hampers the cancer market in India.

Due to the rising incidence of various cancer complications and booming of various advanced biological and drug targeted therapies, patent expiry of branded drugs is paralleled with commercialization of biosimilars. Commercialization of immunotherapies and targeted therapies may reduce the restraint factors and accelerate the growth. As oncology is a hospital-based business, thereby if a pharma company builds a strong sales network, it can tap into new opportunities. Huge gap in demand and supply of treatment makes affordability and accessibility a problem. This huge gap presents an opportunity for hospital administrators and entrepreneurs to dig into new opportunities to accommodate more number of cancer patients.

With pharma players such as Dr. Reddy's, Sun Pharma, and Zydus, among others, the Indian pharma industry is expected to show a robust growth in the next few years. India has the potential to emerge as the most preferred destination for highly potent API (HPAPI) manufacturing. HPAPIs are a relatively new segment in the pharma contract manufacturing sector with high profit margins and demand. Indian players have the complex chemistry manufacturing skills to differentiate themselves from competitors in this segment. Increase in growth of the biologic segment resulted in an increase in the application of HPAPI in segments such as hormonal therapies, glaucoma, and targeted release therapies for cancer. There is high demand for bio-pharma formulations with smaller, highly potent dosages with specialized release characteristics. Affordability and availability of high potency anti-cancer drugs in India, China and pharmerging markets is expected to increase with the anticipated strengthening of these markets through various governmental initiatives in the form of healthcare reforms (Frost and Sullivan 2017).

India is the largest provider of generic drugs globally. Indian pharmaceutical sector industry supplies over 50% of global demand for various vaccines, 40% of generic demand in the USA, and 25% of all medicine in the UK. The Indian government has taken many steps to reduce costs and bring down healthcare expenses. Speedy introduction of generic drugs into the market has remained in focus and is expected to benefit the Indian pharmaceutical companies. In addition, the thrust on rural health programs, lifesaving drugs, and preventive vaccines also augurs well for the pharmaceutical companies (Indian Pharmaceuticals Industry Report 2019).

Acknowledgment RD is grateful to the Department of Health Research, New Delhi, for providing financial support through the Women Scientists Scheme.

References

Aaronson S (1991) Growth factors and Cancer. Science 254(5035):1146–1153

American Cancer Society (2014) Cancer facts & figures 2014. American Cancer Society, Atlanta

American Cancer Society (2017) Cancer facts & figures 2017. American Cancer Society, Atlanta

Anticancer Drugs Market (Drug Type – Cytotoxic Drugs (Alkylating Agents, and Antimetabolites), Targeted Drugs (Monoclonal Antibodies and Tyrosine Kinase Inhibitors), and Hormonal Drugs; Therapy Type – Chemotherapy, Targeted Therapy, and Immunotherapy; Cancer Type – Lung Cancer, Breast Cancer, Leukemia, and Colorectal Cancer) – Global Industry Analysis, Size, Share, Growth, Trends, and Forecast 2017–2025 (2018)

Chaturvedi K, Chataway J, Wield D (2007) Policy, markets and knowledge: strategic synergies in Indian pharmaceutical firms. Tech Anal Strat Manag 19(5):565–588

Czito BG, Hurwitz HI, Clough RW, Tyler DS, Morse MA, Clary BM, Pappas TN, Fernando NH, Willett CG (2005) Adjuvant external-beam radiotherapy with concurrent chemotherapy after resection of primary gallbladder carcinoma: a 23-year experience. Int J Radiat Oncol Biol Phys 62(4):1030–1034

DeStoll M (1788) Rationismendendi, in nosocomiopracticovendobonensi. part 1 lugdunibatavarum, Haak et Socios et A et J Honkoop

Dixit R, Pandey M, Tripathi SK, Dwivedi AND, Shukla VK (2017) Comparative analysis of mutational profile of sonic hedgehog gene in gallbladder cancer. Dig Dis Sci 62(3):708–714

Dixit R, Shukla VK, Pandey M (2012) Molecular alterations in gallbladder cancer. World J Pathol 1:7

Dixit R, Shukla VK (2014) Why is gallbladder cancer common in the Gangetic Belt? In: Perspectives in cancer prevention-translational cancer research. Springer, New Delhi, pp 145–151

Dixit R, Singh G, Pandey M, Basu S, Bhartiya SK, Singh KK, Shukla VK (2016) Association of methylenetetrahydrafolate reductase gene polymorphism (MTHFR) in patients with gallbladder cancer. J Gastrointest Cancer 47(1):55–60

Dixit R (2015) Gene profiling of gallbladder cancer. Ph.D. thesis, Banaras Hindu University, Varanasi, India

Dutta U (2012) Gallbladder cancer. Can newer insights improve the outcome? J Gastroenterol Hepatol 27(4):642–653

Dwivedi AND, Jain S, Dixit R (2015) Gall bladder carcinoma: aggressive malignancy with protean loco-regional and distant spread. World J Clin Cases 3(3):231–244

Frost & Sullivan (2017) Preference to targeted therapies and patient centric approaches drive transformations in oncology drug delivery market

Greene W (2007) The emergence of India's pharmaceutical industry and implications for the U.S. Generic Drug Market. U.S. International Trade Commission

Houry S, Barrier A, Huguier M (2001) Irradiation therapy for gallbladder carcinoma: recent advances. J Hepato-Biliary-Pancreat Surg 8(6):518–524

Hundal R, Shaffer EA (2014) Gallbladder cancer: epidemiology and outcome. Clin Epidemiol 6:99–109

Indian Pharmaceuticals Industry Report (2019) India brand equity foundation

Kresl JJ, Schild SE, Henning GT, Gunderson LL, Donohue J, Pitot H, Haddock MG, Nagorney D (2002) Adjuvant external beam radiation therapy with concurrent chemotherapy in the management of gallbladder carcinoma. Int J Radiat Oncol Biol Phys 52(1):167–175

Kumar A, Kumar M, Dixit R, Jaiswal R, Srivastava V, Pandey M (2012) Presence of human papilloma virus and EGFR expression does not predict response to neoadjuvant chemotherapy in oral cancer. World J Surg Med Radiat Oncol 1:103–110

McKinsey & Company report (2009) India Pharma 2020: propelling access and acceptance, realising true potential

Mhatre SS, Nagrani RT, Budukh A, Chiplunkar S, Badwe R, Patil P, Laversanne M, Rajaraman P, Bray F, Dikshit R (2016) Place of birth and risk of gallbladder cancer in India. Indian J Cancer 53:304–308

Misra A, Misra S, Chaturvedi A, Srivastava PK (2002) Case report. Orbital metastasis from gall bladder carcinoma. Br J Radiol 75(889):72–73

Misra S, Chaturvedi A, Misra NC, Sharma ID (2003) Carcinoma of the gallbladder. Lancet Oncol 4:167–176

National Cancer Registry Programme (2013) Three years report of population based cancer registries; 2009–2011. Indian Council of Medical Research, New Delhi

National Pharmaceutical Policy (2006) Department of Chemicals 2 and Petrochemicals, Government of India, Dec. 28, 2005

Pandey M, Kannepali KK, Dixit R, Kumar M (2018) Effect of neoadjuvant chemotherapy and its correlation with HPV status, EGFR, Her-2-neu, and GADD45 expression in oral squamous cell carcinoma. World J Surg Oncol 16:20

Pandey P, Pandey M, Singh KK, Dixit R, Shukla VK (2017) Health related quality of life in patients of the gallbladder cancer with treatment. Int J Biol Med Res 8(2):5948–5953

Pignochino Y, Sarotto I, Peraldo-Neia C, Penachioni JY, Cavalloni G, Migliardi G, Casorzo L, Chiorino G, Risio M, Bardelli A, Aglietta M, Leone F (2010) Targeting EGFR/HER2 pathways enhances the antiproliferative effect of gemcitabine in biliary tract and gallbladder carcinomas. BMC Cancer 10:631

Qin TJ, Zhao XH, Yun J, Zhang LX, Ruan ZP, Pan BR (2008) Efficacy and safety of gemcitabine-oxaliplatin combined with huachansu in patients with advanced gallbladder carcinoma. World J Gastroenterol 14(33):5210–5216

Rakić M, Patrlj L, Kopljar M, Kliček R, Kolovrat M, Loncar B, Busic Z (2014) Gallbladder cancer. Hepatobiliary Surg Nutr 3(5):221–226

Shukla VK, Das PC, Dixit R, Bhartiya SK, Basu S, Raman MJ (2012) Study of AP endonuclease (APEX1/REF1), a DNA repair enzyme, in gallbladder carcinoma. Anticancer Res 32(4):1489–1492

Shukla VK, Khandelwal C, Roy SK, Vaidya MP (1985) Primary carcinoma of the gall bladder: a review of a 16-year period at the University Hospital. J Surg Oncol 28(1):32–35

Todoroki T (2000) Chemotherapy for gallbladder carcinoma – a surgeon's perspective. Hepato-Gastroenterology 47:948–955

Valle J, Wasan H, Palmer DH, Cunningham D, Anthoney A, Maraveyas A, Madhusudan S, Iveson T, Hughes S, Pereira SP, Roughton M, Bridgewater J, ABC-02 Trial Investigators (2010) Cisplatin plus gemcitabine versus gemcitabine for biliary tract cancer. N Engl J Med 362(14):1273–1281

Xiu AX, Hong TS, Hezel AF, Kooby DA (2010) Current management of gallbladder carcinoma. Oncologist 15:168–181

Zhao DY, Lim KH (2017) Current biologics for treatment of biliary tract cancers. J Gastrointest Oncol 8(3):430–440

MIX
Papier aus verantwortungsvollen Quellen
Paper from responsible sources
FSC® C105338

If you have any concerns about our products,
you can contact us on
ProductSafety@springernature.com

In case Publisher is established outside the EU,
the EU authorized representative is:
Springer Nature Customer Service Center GmbH
Europaplatz 3, 69115 Heidelberg, Germany

Printed by Libri Plureos GmbH
in Hamburg, Germany